Methods in Molecular Biology™

AF768378

Series Editor
John M. Walker
School of Life Sciences
University of Hertfordshire
Hatfield, Hertfordshire, AL10 9AB, UK

For other titles published in this series, go to
www.springer.com/series/7651

Bioinformatics for Comparative Proteomics

Edited by

Cathy H. Wu

Department of Computer and Information Sciences,
Center for Bioinformatics and Computational Biology,
University of Delaware, Newark, DE, USA

Chuming Chen

Department of Computer and Information Sciences,
Center for Bioinformatics and Computational Biology,
University of Delaware, Newark, DE, USA

Humana Press

Editors
Cathy H. Wu, Ph.D.
Center for Bioinformatics
and Computational Biology
University of Delaware
15 Innovation Way, Suite 205
Newark, DE 19711
USA
wuc@dbi.udel.edu

Chuming Chen, Ph.D.
Center for Bioinformatics
and Computational Biology
University of Delaware
15 Innovation Way, Suite 205
Newark, DE 19711
USA
chenc@dbi.udel.edu

ISSN 1064-3745 e-ISSN 1940-6029
ISBN 978-1-4939-5790-3 ISBN 978-1-60761-977-2 (eBook)
DOI 10.1007/978-1-60761-977-2
Springer New York Dordrecht Heidelberg London

Printed on acid-free paper

Humana Press is part of Springer Science+Business Media (www.springer.com)

Preface

With the rapid development of proteomic technologies in life sciences and in clinical applications, many bioinformatics methodologies, databases, and software tools have been developed to support comparative proteomics study. This volume aims to highlight the current status, challenges, open problems, and future trends in developing bioinformatics tools and resources for comparative proteomics research and to serve as a definitive source of reference providing both the breadth and depth needed on the subject of *Bioinformatics for Comparative Proteomics.*

The volume is structured to introduce three major areas of research methods: (1) basic bioinformatics frameworks related to comparative proteomics, (2) bioinformatics databases and tools for proteomics data analysis, and (3) integrated bioinformatics systems and approaches for studying comparative proteomics in the systems biology context.

Part I (Bioinformatics Framework for Comparative Proteomics) consists of seven chapters:

Chapter 1 presents a comprehensive review (with categorization and description) of major protein bioinformatics databases and resources that are relevant to comparative proteomics research.

Chapter 2 provides a practical guide to the comparative proteomics community for exploiting the knowledge captured from and the services provided in UniProt databases.

Chapter 3 introduces the InterPro protein classification system for automatic protein annotation and reviews the signature methods used in the InterPro database.

Chapter 4 introduces the Reactome Knowledgebase that provides an integrated view of the molecular details of human biological processes.

Chapter 5 introduces eFIP (extraction of Functional Impact of Phosphorylation), a Web-based text mining system that can aid scientists in quickly finding abstracts from literature related to the phosphorylation (including site and kinase), interactions, and functional aspects of a given protein.

Chapter 6 presents a tutorial for the Protein Ontology (PRO) Web resources to help researchers in their proteomic studies by providing key information about protein diversity in terms of evolutionary-related protein classes based on full-length sequence conservation and the various protein forms that arise from a gene along with the specific functional annotation.

Chapter 7 describes a method for the annotation of functional residues within experimentally uncharacterized proteins using position-specific site annotation rules derived from structural and experimental information.

Part II (Proteomic Bioinformatics) consists of ten chapters:

Chapter 8 describes how the detailed understanding of information value of mass spectrometry-based proteomics data can be elucidated by performing simulations using synthetic data.

Chapter 9 describes the concepts, prerequisites, and methods required to analyze a shotgun proteomics data set using a tandem mass spectrometry search engine.

Chapter 10 presents computational methods for quantification and comparison of peptides by label-free LC–MS analysis, including data preprocessing, multivariate statistical methods, and detection of differential protein expression.

Chapter 11 proposes an alternative to MS/MS spectrum identification by combining the uninterpreted MS/MS spectra from overlapping peptides and then determining the consensus identifications for sets of aligned MS/MS spectra.

Chapter 12 describes the Trans-Proteomic Pipeline, a freely available open-source software suite that provides uniform analysis of LC–MS/MS data from raw data to quantified sample proteins.

Chapter 13 provides an overview of a set of open-source software tools and steps involved in ELISA microarray data analysis.

Chapter 14 presents the state of the art on the Proteomics Databases and Repositories.

Chapter 15 is a brief guide to preparing both large- and small-scale protein interaction data for publication.

Chapter 16 demonstrates a new graphical user interface tool called PRIDE Converter, which greatly simplifies the submission of MS data to PRIDE database for submitted proteomics manuscripts.

Chapter 17 presents a method for describing a protein's posttranslational modifications by integrating the top–down and bottom–up MS data using the Protein Inference Engine.

Chapter 18 describes an integrated top–down and bottom–up approach facilitated by concurrent liquid chromatography–mass spectrometry analysis and fraction collection for comprehensive high-throughput intact protein profiling.

Part III (Comparative Proteomics in Systems Biology) consists of four chapters:

Chapter 19 gives an overview of the content and usage of the PhosphoPep database, which supports systems biology signaling research by providing interactive interrogation of MS-derived phosphorylation data from four different organisms.

Chapter 20 describes "omics" data integration to map a list of identified proteins to a common representation of the protein and uses the related structural, functional, genetic, and disease information for functional categorization and pathway mapping.

Chapter 21 describes a knowledge-based approach relying on existing metabolic pathway information and a direct data-driven approach for a metabolic pathway-centric integration of proteomics and metabolomics data.

Chapter 22 provides a detailed description of a method used to study temporal changes in the endoplasmic reticulum (ER) proteome of fibroblast cells exposed to ER stress agents (tunicamycin and thapsigargin).

This volume targets the readers who wish to learn about state-of-the-art bioinformatics databases and tools, novel computational methods and future trends in proteomics data analysis, and comparative proteomics in systems biology. The audience may range from graduate students embarking upon a research project, to practicing biologists working on proteomics and systems biology research, and to bioinformaticians developing advanced databases, analysis tools, and integrative systems. With its interdisciplinary nature, this volume is expected to find a broad audience in biotechnology and pharmaceutical companies and in various academic departments in biological and medical sciences (such as biochemistry, molecular biology, protein chemistry, and genomics) and computational sciences and engineering (such as bioinformatics and computational biology, computer science, and biomedical engineering).

We thank all the authors and coauthors who had contributed to this volume. We thank our series editor, Dr. John M. Walker, for reviewing all the chapter manuscripts and providing constructive comments. We also thank Dr. Winona C. Barker from Georgetown University for reviewing the manuscripts. We thank Dr. Qinghua Wu for proof reading the book draft. Finally, we would like to extend our thanks to David C. Casey and Anne Meagher of Springer US, Jeya Ruby and Ravi Amina of SPi for their help in the compilation of this book.

Newark, DE, USA *Cathy H. Wu and Chuming Chen*

Contents

Contributors

RUEDI AEBERSOLD • *Institute of Molecular Systems Biology, ETH Zurich, Zurich, Switzerland*
ROLF APWEILER • *The European Bioinformatics Institute, Cambridge, UK*
CECILIA N. ARIGHI • *Department of Computer and Information Sciences, University of Delaware, Newark, DE, USA*
NUNO BANDEIRA • *Center for Computational Mass Spectrometry, University of California, San Diego, La Jolla, CA, USA*
HARALD BARSNES • *Department of Informatics, University of Bergen, Bergen, Norway*
BERND BODENMILLER • *Institute of Molecular Systems Biology, ETH Zurich, Zurich, Switzerland*
KRISTY J. BROWN • *Center for Genetic Medicine Research, Children's National Medical Center, Washington, DC, USA*
CHUMING CHEN • *Department of Computer and Information Sciences, University of Delaware, Newark, DE, USA*
NILS CHRISTIAN • *Max-Planck-Institute for Molecular Plant Physiology, Potsdam-Golm, Germany*
DON S. DALY • *Pacific Northwest National Laboratory, Richland, WA, USA*
PETER D'EUSTACHIO • *Department of Biochemistry, New York University School of Medicine, New York, NY, USA*
OLIVER EBENHÖH • *Max-Planck-Institute for Molecular Plant Physiology, Potsdam-Golm, Germany*
NATHAN J. EDWARDS • *Department of Biochemistry and Molecular & Cellular Biology, Georgetown University Medical Center, Washington, DC, USA*
INGVAR EIDHAMMER • *Department of Informatics, University of Bergen, Bergen, Norway*
JAN ERIKSSON • *Swedish University of Agricultural Sciences, Uppsala, Sweden*
DAVID FENYÖ • *The Rockefeller University, New York, NY, USA*
CATHERINE A. FORMOLO • *Center for Genetic Medicine Research, Children's National Medical Center, Washington, DC, USA*
MORGAN C. GIDDINGS • *Departments of Microbiology & Immunology and Biomedical Engineering, The University of North Carolina at Chapel Hill, Chapel Hill, NC, USA*
BRIAN HALLIGAN • *Bioinformatics, Human and Molecular Genetics Center, Medical College of Wisconsin, Milwaukee, WI, USA*
YETRIB HATHOUT • *Center for Genetic Medicine Research, Children's National Medical Center, Washington, DC, USA*
HENNING HERMJAKOB • *EMBL Outstation, European Bioinformatics Institute (EBI), Cambridge, UK*
HONGZHAN HUANG • *Department of Computer and Information Sciences, University of Delaware, Newark, DE, USA*

Sarah Hunter • *EMBL Outstation, European Bioinformatics Institute (EBI), Cambridge, UK*
Stuart R. Jefferys • *Department of Bioinformatics & Computational Biology, The University of North Carolina at Chapel Hill, Chapel Hill, NC, USA*
Robel Y. Kahsay • *DuPont Central Research & Development, Wilmington, DE, USA*
Andrew Keller • *Rosetta Biosoftware, Seattle, WA, USA*
Lennart Martens • *EMBL Outstation, European Bioinformatics Institute (EBI), Cambridge, UK*
Patrick May • *Max-Planck-Institute for Molecular Plant Physiology, Potsdam-Golm, Germany*
Jennifer McDowall • *EMBL Outstation, European Bioinformatics Institute (EBI), Cambridge, UK*
Peter B. McGarvey • *Department of Biochemistry and Molecular & Cellular Biology, Georgetown University Medical Center, Washington, DC, USA*
Michelle Mintz • *Center for Genetic Medicine Research, Children's National Medical Center, Washington, DC, USA*
Darren A. Natale • *Department of Biochemistry and Molecular & Cellular Biology, Georgetown University Medical Center, Washington, DC, USA*
Jules A. Nchoutmboube • *Department of Computer and Information Sciences, University of Delaware, Newark, DE, USA*
Claire O'Donovan • *The European Bioinformatics Institute, Cambridge, UK*
Sandra Orchard • *EMBL Outstation, European Bioinformatics Institute (EBI), Cambridge, UK*
Ljiljana Paša-Tolić • *Pacific Northwest National Laboratory, Richland, WA, USA*
Florian Reisinger • *EMBL Outstation, European Bioinformatics Institute (EBI), Cambridge, UK*
Habtom W. Ressom • *Department of Oncology, Georgetown University Medical Center, Washington, DC, USA*
Errol W. Robinson • *Pacific Northwest National Laboratory, Richland, WA, USA*
Vijay K. Shanker • *Department of Computer and Information Sciences, University of Delaware, Newark, DE, USA*
David Shteynberg • *Institute for Systems Biology, Seattle, WA, USA*
Amy Y. Siu • *Department of Computer and Information Sciences, University of Delaware, Newark, DE, USA*
Asako Takanohashi • *Center for Genetic Medicine Research, Children's National Medical Center, Washington, DC, USA*
Zhixin Tian • *Pacific Northwest National Laboratory, Richland, WA, USA*
Nikola Tolić • *Pacific Northwest National Laboratory, Richland, WA, USA*
Catalina O. Tudor • *Department of Computer and Information Sciences, University of Delaware, Newark, DE, USA*
Adeline Vanderver • *Center for Genetic Medicine Research, Children's National Medical Center, Washington, DC, USA*
Rency S. Varghese • *Department of Oncology, Georgetown University Medical Center, Washington, DC, USA*

Sona Vasudevan • *Department of Biochemistry and Molecular & Cellular Biology, Georgetown University Medical Center, Washington, DC, USA*
C.R. Vinayaka • *Department of Biochemistry and Molecular & Cellular Biology, Georgetown University Medical Center, Washington, DC, USA*
Juan Antonio Vizcaíno • *EMBL Outstation, European Bioinformatics Institute (EBI), Cambridge, UK*
Dirk Walther • *Max-Planck-Institute for Molecular Plant Physiology, Potsdam-Golm, Germany*
Wolfram Weckwerth • *Molecular Systems Biology, University of Vienna, Vienna, Austria*
Amanda M. White • *Pacific Northwest National Laboratory, Richland, WA, USA*
Cathy H. Wu • *Department of Computer and Information Sciences, University of Delaware, Newark, DE, USA*
Si Wu • *Pacific Northwest National Laboratory, Richland, WA, USA*
Richard C. Zangar • *Pacific Northwest National Laboratory, Richland, WA, USA*
Jian Zhang • *Department of Biochemistry and Molecular & Cellular Biology, Georgetown University Medical Center, Washington, DC, USA*

Part I

Bioinformatics Framework for Comparative Proteomics

Chapter 1

Protein Bioinformatics Databases and Resources

Chuming Chen, Hongzhan Huang, and Cathy H. Wu

Abstract

In the past decades, a variety of publicly available data repositories and resources have been developed to support protein related information management, data-driven hypothesis generation and biological knowledge discovery. However, there is also an increasing confusion for the researchers who are trying to quickly find the appropriate resources to help them solve their problems. In this chapter, we present a comprehensive review (with categorization and description) of major protein bioinformatics databases and resources that are relevant to comparative proteomics research. We conclude the chapter by discussing the challenges and opportunities for developing new protein bioinformatics databases.

Key words: Bioinformatics, Database, Protein sequence, Protein family, Protein structure, Protein function, Proteomics, Data integration, Comparative analysis

1. Introduction

Advances of high-throughput technologies in the study of molecular biology systems in the past decades have marked the beginning of a new era of research, in which biological researchers systematically study organisms on the levels of genomes (complete genetic sequences) (1), transcriptomes (gene expressions) (2) and proteomes (protein expressions) (3). Because proteins occupy a middle ground molecularly between gene and transcript information and higher levels of molecular and cellular structure and organization, and most physiological and pathological processes are manifested at the protein level, biological scientists are growingly interested in applying proteomics techniques to foster a better understanding of basic molecular biology, disease processes and discovery of new diagnostic, prognostic and therapeutic targets for numerous diseases (4, 5).

Cathy H. Wu and Chuming Chen (eds.), *Bioinformatics for Comparative Proteomics*, Methods in Molecular Biology, vol. 694,
DOI 10.1007/978-1-60761-977-2_1, © Springer Science+Business Media, LLC 2011

Recently, proteomics data analysis has moved toward information integration of multiple studies including cross-species analyses (6–9). The richness of proteomics data allows researchers to ask complex biological questions and gain new scientific insights. To support comparative proteomics, data-driven hypothesis generation, and biological knowledge discovery, many protein-related bioinformatics databases, query facilities, and data analysis software tools have been developed. These organize and provide biological annotations for individual proteins to support sequence, structural, functional and evolutionary analyses in the context of pathway, network and systems biology. However, it is not always easy for researchers to quickly find the pieces of related information. In this chapter, we present a comprehensive review (with categorization and description) of major protein bioinformatics databases and resources that are relevant to comparative proteomics research. We highlight some of these databases, and focus on the types of data stored and related data access and data analysis supports. We also discuss the challenges and opportunities for developing new protein bioinformatics databases in terms of supporting data integration and comparative analysis, maintaining data provenance and managing biological knowledge.

2. Overview

Our coverage of protein bioinformatics databases in this chapter is by no means exhaustive. We refer the readers to ref. 10 for a more complete list. Our intention is to cover those that are recent, high quality, publicly available, and are expected to be of interest to more users in the comparative proteomics community. Based on the topics and data stored, protein bioinformatics databases can be primarily classified as sequence databases, family databases, structure databases, function databases and proteomics databases as shown in Table 1. It is worth noting that certain databases can be classified into more than one category. Please visit http://www.proteininformationresource.org/staff/chenc/MiMB/dbSummary.html to access the databases reviewed in this chapter through their corresponding web addresses (URLs).

3. Databases and Resources Highlights

3.1. Protein Sequence Databases

Protein sequence databases serve as the archival repositories for collections of protein sequences as well as their associated annotations. These databases are also the primary sources for developing other

Table 1
Overview of protein bioinformatics databases

Primary category	Secondary category	Database name	Database content	URL	References
Sequence	NCBI	Reference Sequence (RefSeq)	Biologically non-redundant collection of DNA, RNA, and protein sequences	http://www.ncbi.nlm.nih.gov/RefSeq/	(11)
		Entrez Protein Database	Collection of protein sequences from a variety of sources, and translations from annotated coding regions in GenBank and RefSeq	http://www.ncbi.nlm.nih.gov/sites/entrez?db,protein	(20)
	UniProt	UniProt Knowledgebase (UniProtKB)	Collection of functional information on proteins with accurate, consistent and rich annotation	http://www.uniprot.org/help/uniprotkb	(13)
		UniProt Archive (UniParc)	Comprehensive and non-redundant database that contains most of the publicly available protein sequences in the world	http://www.uniprot.org/help/uniparc	(14)
		UniProt Reference Clusters (UniRef)	Clustered sets of sequences from UniProt Knowledgebase (including splice variants and isoforms) and selected UniParc records	http://www.uniprot.org/help/uniref	(15)
Family	Whole protein	PIRSF	Comprehensive and non-overlapping clustering of UniProtKB sequences into a hierarchical order to reflect their evolutionary relationships based on whole protein sequences	http://www.pir.georgetown.edu/pirwww/dbinfo/pirsf.shtml	(18)
		Clusters of Orthologous Groups of proteins (COGs)	Phylogenetic classification of proteins encoded in complete genomes	http://www.ncbi.nlm.nih.gov/COG/	(64)

(continued)

Table 1 (continued)

Primary category	Secondary category	Database name	Database content	URL	References
		Protein ANalysis THrough Evolutionary Relationships Classification System (PANTHER)	Proteins are classified by expert biologists into families and subfamilies of shared function and further categorized by GO terms	http://www.pantherdb.org/	(29)
		ProtoNet	Automatic hierarchical classification of protein sequences	http://www.protonet.cs.huji.ac.il/index.php	(65)
	Protein domain	Pfam	Protein families of domains each represented by multiple sequence alignments and Hidden Markov Models (HMMs)	http://www.pfam.sanger.ac.uk/	(19)
		ProDom	Comprehensive set of protein domain families automatically generated from the UniProtKB	http://www.prodom.prabi.fr/prodom/current/html/home.php	(21)
		Conserved Domains Database (CDD)	Collections of multiple sequence alignments representing conserved domains	http://www.ncbi.nlm.nih.gov/sites/entrez?db=cdd	(66)
		Simple Modular Architecture Research Tool (SMART)	Resource for identification and annotation of protein domains and the analysis of domain architectures	http://www.smart.embl.de/	(31)
	Protein motif	PRINTS	Group of conserved motifs used to characterize a protein family	http://www.bioinf.manchester.ac.uk/dbbrowser/PRINTS/index.php	(30)
		PROSITE	Protein domains, families and functional sites as well as associated patterns and profiles to identify them	http://www.ca.expasy.org/prosite/	(24)

	Integrative	InterPro	Integrated resource of protein families, domains and functional sites from Pfam, PRINTS, PROSITE, ProDom, SMART, PIRSF etc.	http://www.ebi.ac.uk/interpro/	(27)
Structure	3D structure	Worldwide Protein Data Bank (wwPDB)	Repository for the 3D coordinates and related information on more than 38,000 macromolecular structures, including proteins, nucleic acids and large macromolecular complexes that have been determined using X-ray crystallography, NMR and electron microscopy techniques	http://www.wwpdb.org/	(23)
		Molecular Modeling Database (MMDB)	3D macromolecular structures, including proteins and polynucleotides.	http://www.ncbi.nlm.nih.gov/sites/entrez?db=structure	(67)
		ModBase	3D protein models calculated by comparative modeling	http://www.modbase.compbio.ucsf.edu/modbase-cgi/index.cgi	(68)
		SWISS-MODEL Repository	Annotated protein 3D models	http://www.swissmodel.expasy.org/repository/	(69)
	Structural classification	CATH	Hierarchical classification of protein domain structures in the Protein Data Bank	http://www.cathdb.info/	(37)
		Structural Classification Of Proteins (SCOP)	Description of the evolutionary and structural relationships of the proteins of known structures	http://www.scop.mrc-lmb.cam.ac.uk/scop/	(22)
		SUPERFAMILY	Structural and functional annotation for all proteins and genomes based on a collection of Hidden Markov Models, which represents structural protein domains at the SCOP superfamily level	http://www.supfam.org/SUPERFAMILY/	(32)

(continued)

Table 1 (continued)

Primary category	Secondary category	Database name	Database content	URL	References
	Protein folding	Protein Folding Database (PFD)	Repository of available experimental protein folding data	http://www.pfd.med.monash.edu.au/public_html/index.php	(38)
		KineticDB	Experimental data on protein folding kinetics	http://www.kineticdb.protres.ru/db/index.pl	(70)
	Protein modification	RESID	Collection of annotations and structures for protein pre-, co- and post-translational modifications	http://www.ebi.ac.uk/RESID/	(71)
		Phospho3D	3D structures of phosphorylation sites that stores information retrieved from the phospho.ELM database	http://www.cbm.bio.uniroma2.it/phospho3d/	(40)
Function	Inter-molecular interactions	IntAct	Protein interaction data from literature and user submission	http://www.ebi.ac.uk/intact/main.xhtml	(42)
		Database of Interacting Proteins (DIP)	Experimentally determined protein–protein interactions	http://www.dip.doe-mbi.ucla.edu/dip/Main.cgi	(72)
		Reactome	A curated knowledgebase of biological pathways	http://www.reactome.org/	(47)
		Biological General Repository for Interaction Datasets (BioGRID)	Collections of protein and genetic interactions from major model organism species	http://www.thebiogrid.org	(73)
	Metabolic pathways	Kyoto Encyclopedia of Genes and Genomes (KEGG)	Pathway maps on the molecular interaction and reaction networks for metabolism	http://www.genome.jp/kegg/pathway.html	(74)

		BioCyc	Pathway/Genome Databases (PGDBs) on the pathways and genomes of different organisms	http://www.biocyc.org/	(51)
		MetaCyc	Nonredundant, experimentally elucidated metabolic pathways	http://www.metacyc.org/	(51)
	Integrative	Michigan molecular interactions (MiMI)	Merged view of several popular interaction databases including: BIND, HPRD, IntAct, GRID, and others	http://www.mimitest.ncibi.org/MimiWeb/main-page.jsp	(75)
Proteomics	Gel electrophoresis	WORLD-2DPAGE Constellation	List of World-2DPAGE database servers, World-2DPAGE Portal that queries simultaneously world-wide proteomics databases, and World-2DPAGE Repository	http://www.world-2dpage.expasy.org/	(52)
	Mass spectrometry	Global Proteome Machine Database (GPMDB)	Mass spectral library for data from a variety of organisms, the identified peptides are matched to the Ensembl genome database	http://www.thegpm.org/GPMDB/index.html	(76)
		PRoteomics IDEntifications database (PRIDE)	Protein and peptide identifications that have been described in the scientific literature together with the evidence supporting these identifications	http://www.ebi.ac.uk/pride/	(54)
		PeptideAtlas	Peptides identified in a large set of LC–MS/MS proteomics experiments	http://www.peptideatlas.org/	(77)
		Peptidome	Tandem mass spectrometry peptide and protein identification data generated by the scientific community	http://www.ncbi.nlm.nih.gov/peptidome/	(78)

resources such as protein family databases, and the foundation for medical and functional studies.

3.1.1. RefSeq

The National Center for Biotechnology Information Reference Sequence (NCBI RefSeq) database provides curated non-redundant sequences for genomic regions, transcripts and proteins (11). RefSeq collection is derived from the sequence data available in the redundant archival database GenBank (12). RefSeq sequences include coding regions, conserved domains, variations, references, names, and database cross-references. The sequences are annotated using a combined approach of collaboration, automated prediction, and manual curation (11). The RefSeq release 37 of September 11, 2009 includes 8,835,796 proteins and 9,005 organisms. The RefSeq data can be accessed from NCBI web sites by Entrez query, BLAST, FTP download etc.

3.1.2. UniProt

The UniProt Consortium consists of groups from the European Bioinformatics Institute (EBI), the Swiss Institute of Bioinformatics (SIB) and the Protein Information Resource (PIR). The UniProt Consortium provides a central resource for protein sequences and functional annotations with four database components to support protein bioinformatics research:

- The UniProt Knowledgebase (UniProtKB) is the predominant data store for functional information on proteins (13). The UniProtKB consists of two sections: UniProtKB/Swiss-Prot, which contains manually annotated records with information extracted from literature and curator-evaluated computational analysis, and UniProtKB/TrEMBL, which contains computationally analyzed records with rule-based automatic annotation. Comparative analysis and query across databases are supported by the UniProtKB extensive cross-references, functional and feature annotations, classification, and literature-based evidence attribution. The UniProtKB release 15.9 of October 13, 2009 includes 510,076 UniProtKB/Swiss-Prot sequence entries, comprising 179,409,349 amino acids abstracted from 183,725 references, and 9,501,907 UniProtKB/TrEMBL sequence entries comprising 3,068,281,486 amino acids.
- The UniProt archive (UniParc) (14) is an archival protein sequence database from all major publicly accessible resources. UniParc contains protein sequences and database cross-references to the provenance of the sequences. Text- and sequence-based searches are available from UniParc database web site.
- The UniProt Reference Clusters (UniRef) (15) merge sequences and sub-sequences that are 100% (UniRef100), ≥90%

(UniRef90), or ≥50% (UniRef50) identical, regardless of source organism to speed up searches.

- The UniProt Metagenomic and Environmental Sequences (UniMES) database is a repository specifically developed for Metagenomic and environmental data. UniMES currently contains data from the Global Ocean Sampling Expedition (GOS) (16), which predicts nearly six million proteins, primarily from oceanic microbes (13).

The UniProt web site (http://www.uniprot.org) is the primary access point to the data and documentation. The site also provides batch retrieval using UniProt identifiers, BLAST-based sequence similarity search, ClustalW-based sequence alignment, and Database identifier mapping. The UniProt FTP download site provides batch download of protein sequence data in various formats, including flat file text, XML, RDF, FASTA, and GFF. Programmatic access to the data and search results is supported via simple HTTP RESTful web services or UniProtJAPI (17) for Java-based applications.

3.2. Protein Family Databases

The primary protein sequence databases can be used to develop new resources with value-added information by either classifying protein sequences into families or assigning certain properties to the sequences by detecting specific sequence features such as domains, motifs, and functional sites.

3.2.1. PIRSF

The PIRSF classification system provides comprehensive and non-overlapping clustering of UniProtKB (13) sequences into a hierarchical order to reflect their evolutionary relationships based on whole proteins rather than on the component domains. The PIRSF system classifies the protein sequences into families, whose members are both homologous (evolved from a common ancestor) and homeomorphic (sharing full-length sequence similarity and a common domain architecture) (18). The PIRSF family classification results are expert-curated based on literature review and integrative sequence and functional analysis. The classification report shows the information on PIRSF members and general statistics, family and function/structure relationships, database cross-references and graphical display of domain and motif architecture of seed members or all members. The web-based PIRSF system has been shown as a useful tool for studying the function and evolution of protein families (18). It provides batch retrieval of entries from the PIRSF database. The PIRSF scan allows searching a query sequence against the set of fully curated PIRSF families with benchmarked Hidden Markov Models. The PIRSF membership hierarchy data is also available for FTP download.

3.2.2. Pfam

Pfam is a database of protein domains and families represented as multiple sequence alignments and Hidden Markov Models (HMMs) (19). Pfam is built based on the protein sequence data from UniProtKB (13), NCBI GenPept (20) and selected Metagenomics projects. The Pfam database contains two components: Pfam-A and Pfam-B. Pfam-A entries are manually curated high-quality representative seed alignments, profile HMMs built from the seed alignments, and an automatically generated full alignment for all detectable family member protein sequences. Pfam-B entries are automatically generated from the ProDom database (21). The Pfam release 24.0 of October 2009 contains 11,912 families. The Pfam database is further organized into higher-level hierarchical groupings of related families called clan (19), which are collections of related Pfam-A entries built manually based on the similarity of their sequences, known structures, profile-HMMs, and other databases such as SCOP (22). The Pfam database web site provides a set of query and browsing interfaces for analyzing protein sequences for Pfam matches, for viewing Pfam family annotations, alignments, groups of related families, and the domains of a protein sequence, as well as for finding the domains on a PDB (23) structure. The Pfam data can be downloaded from its FTP site or programmatically accessed through RESTful and SOAP based web services.

3.2.3. PROSITE

PROSITE (24) is a database of annotated motif descriptors (patterns or profiles), which can be used for the identification of protein domains and families. The motif descriptors are derived from multiple alignments of homologous sequences and have the advantage of identifying distant relationships among sequences (25). A set of ProRules providing additional information about the functionally and/or structurally critical amino acids are used to increase the discriminatory power of the motif descriptors (24). The PROSITE web site provides keywords-based search and allows browsing of motif entries, ProRule description, taxonomic scope, and number of positive hits. The ScanProsite (26) tool allows one either to scan protein sequences for the occurrence of PROSITE motifs by entering UniProtKB AC and/or ID, PDB identifier(s) or protein sequence(s), or to scan the UniProtKB or PDB databases for the occurrence of a pattern by entering the PROSITE AC and/or ID or user's own pattern(s). The ScanProsite (26) tool can also be accessed programmatically through a simple HTTP web service. The PROSITE documentation entries and related tools can be downloaded from its FTP site.

3.2.4. InterPro

InterPro (27) is an integrated resource of predictive models or "signatures" representing protein domains, families, regions, repeats and sites from major protein signature databases including Gene3D (28), PANTHER (29), Pfam (19), PIRSF (18),

PRINTS (30), ProDom (21), PROSITE (24), SMART (31), SUPERFAMILY (32) and TIGRFAMs (33). Each entry in the InterPro database is annotated with a descriptive abstract name and cross-references to the original data sources, as well as to specialized functional databases. The InterPro release 23.0 of September 23, 2009 includes 19,150 entries containing 434 new signatures. The database is available via a web interface and anonymous FTP download. The software tool InterProScan (34) is provided as a protein sequence classification and comparison package that can be used via a web interface and SOAP-based Web Services or can be installed locally for bulk operations. The InterPro BioMart (35) allows users to retrieve InterPro data from a query-optimized data warehouse that is synchronized with the main InterPro database, and to build simple or complex queries and control the query results through a unified interface.

3.3. Protein Structure Databases

Many bioinformatics studies are based on the premise that proteins of similar sequences carry out similar functions whereas those with different sequences carry out different functions. More and more experimental data support the notion that structure of a protein reflects the nature of the role it is playing, therefore, determining its function in the biological process. The protein structure databases organize and annotate various experimentally determined protein structures, providing the biological community access to the experimental data in a useful way.

3.3.1. worldwide PDB

The worldwide PDB (wwPDB) was established in 2003 as an international collaboration to maintain a single and publicly available Protein Data Bank Archive (PDB Archive) of macro-molecular structural data (23). The wwPDB member includes RCSB PDB (USA), the Macromolecular Structure Database at the European Bioinformatics Institute (MSD-EBI) (UK), the Protein Data Bank Japan (PDBj) at Osaka University (Japan) and the BioMagResBank (BMRB) at the University of Wisconsin – Madison (USA). The "PDB Archive" is a collection of flat files in three different formats: the legacy PDB file format; the PDB exchange format that follows the mmCIF syntax (http://www.deposit.pdb.org/mmcif/); and the PDBML/XML format (36). Each member site serves as a deposition, data processing and distribution site for the PDB Archive and each provides its own view of the primary data and a variety of tools and resources. As of October 27, 2009, there are 61,086 structures in the wwPDB database.

3.3.2. CATH

CATH (Class, Architecture, Topology, Homology) is a database of protein domain structures in the Protein Data Bank, where domains are hierarchically classified by the curators guided by prediction algorithms (such as structure comparison). CATH clusters proteins at four major levels (37):

- *Class (C)*: secondary structure composition and packing within the structure.
- *Architecture (A)*: orientations of the secondary structures ignoring the connectivity among the secondary structures.
- *Topology (T)*: whether they share the same topology in the core of the domain.
- *Homologous superfamily (H)*: sequence and structural similarities.

The CATH release 3.2.0 of July 14, 2008 contains 114,215 assigned domains. CATH provides the SSAP server, which allows users to compare the structures of two proteins and view the subsequent structural alignment.

3.3.3. SCOP

The SCOP (Structural Classification of Proteins) database provides a comprehensive and detailed description of the evolutionary and structural relationships of the proteins of known structures. The SCOP classification hierarchy is constructed based on a domain in the experimentally determined protein structure and includes the following levels (22):

- *Species*: distinct protein sequence and its naturally occurring or artificially created variants.
- *Protein*: similar sequences of essentially the same functions.
- *Family*: proteins with related sequences but typically distinct functions.
- *Superfamily*: protein families with common evolutionary ancestor.
- *Fold*: superfamilies with structural similarity (same major secondary structures in the same arrangement and with the same topological connections, not necessarily with common evolutionary origin).
- *Class*: based on the secondary structure content and organization of folds.

The SCOP release 1.75 of June 2009 includes 38,221 PDB entries, 1,195 folds, 1,962 superfamilies and 3,902 families.

3.3.4. PFD

The Protein Folding Database (PFD) is a publicly searchable repository that collects experimental thermodynamic and kinetic data for the folding of proteins. Experimenters deposit data including Constructor, Mutations, Equilibrium Method, Kinetic Method, Equilibrium Data, Kinetic Data, and Publications (38). The PFD database uses the International Foldeomics Consortium standards (39) for data deposition, analysis and reporting to facilitate the comparison of folding rates, energies and structure across diverse sets of proteins (38). The PFD release 2.2 of June 8, 2009 contains 296 entries, 70 proteins, 53 families, 30 species and 230 (five proteins) φ values. The web site provides advanced text searches of protein names, literature references, and experimental details with search results displayed in a tabular view. The graphical visualization tools have been built for raw equilibrium data, chevron data, contact order and folding rates with the hyperlinks on the graph directly link to the data in the text format.

3.3.5. Phospho3D

Phospho3D (40) is a database of 3D structures of phosphorylation sites. Phospho3D is constructed by using the data collected from the phospho.ELM (41) database of experimentally verified phosphorylation sites in eukaryotic proteins, and is enriched with structural information and annotations at the residue level. The basic information unit in the Phospho3D database consists of the instance, its flanking sequence (ten residues) and its "zone," a 3D neighborhood including any residue whose distance does not exceed 12 Å (40). For each zone, structural similarity and biochemical similarity are used to collect the results of a large-scale local structural comparison versus a representative dataset of PDB (23) protein chains, which provide the clues for the identification of new putative phosphorylation sites. Users can browse the data in Phospho3D database or search the database using kinase name, PDB identification code or keywords.

3.4. Protein Function Databases

The unique feature of proteins that allows their diverse functions is the ability to bind to other molecules specifically. For example, proteins can be enzymes to catalyze the chemical reactions in the cell or to manipulate the replication and transcription of DNA. Many proteins are also involved in the process of cell signaling and signal transduction. Protein function databases maintain information about metabolic pathways, enzymes, compounds, and the inter-molecular interactions and regulatory pathways mechanisms underlying many biological processes.

3.4.1. IntAct

IntAct is an open source database and toolkit for the storage, presentation and analysis of protein interaction data (42). IntAct provides all relevant experimental details of protein interactions described in the originating publication. All the entries in the database are fully compliant with the IMEx (43) guidelines and MIMIx (44) standard. The technical details of the experiment, binding sites, protein tags and mutations are annotated with the Molecular Interaction ontology of the Proteomics Standard Initiative (PSI-MI) (45). The latest database contains 202,419 binary interactions, 60,310 proteins, 11,119 experiments and 1,509 controlled vocabulary terms. The IntAct web site provides both textual and graphical views of protein interactions, and allows exploring interaction networks in the context of the Gene Ontology (46) controlled vocabulary and InterPro (27) domains of the interacting proteins. IntAct data and source code are available for downloading from its web site. In addition, a set of tools have been developed by the IntAct project:

- *ProViz*: visualization of protein–protein interaction graphs.
- *MiNe*: compute the minimal connecting networks for a given set of proteins.
- *PSI-MI Semantic Validator*: validate files in PSI-MI XML 2.5 and PSI-PAR format.

3.4.2. Reactome

Reactome is an open source, expert-curated and peer-reviewed database of biological reactions and pathways with cross-references to major molecular databases (47). The basic information in the Reactome database is provided by either publications or sequence similarity-based inference. The Reactome release 30 of September 30, 2009 contains 3,916 proteins, 2,955 complexes, 3,541 reactions, and 1,045 pathways for *Homo sapiens*. Reactome data can be exported in SBML (48), Protégé (49), Cytoscape (50) and BioPax (http://www.biopax.org) formats. Software tools like PathFinder, SkyPainter and Reactome BioMart (35) have been developed to support data mining and analysis of large-scale data sets.

3.4.3. MetaCyc and BioCyc

MetaCyc is a database of non-redundant, experimentally elucidated metabolic pathways and enzymes curated from the scientific literature (51). MetaCyc stores pathways involved in Primary and Secondary metabolism. It also stores compounds, proteins, protein complexes and genes associated with these pathways with extensive links to other biological databases of protein sequences, nucleic acid sequences, protein structures and literature. BioCyc is a collection of Pathway/Genome Databases (PGDBs) (51). Each BioCyc PGDB contains the metabolic network of one organism predicted by the Pathway tool software using MetaCyc as a reference database. Web-based query, browsing, visualization and comparative analysis tools are also provided on the MetaCyc and BioCyc web sites. A collection of data files is also available for downloading.

3.5. Proteomics Databases

The advent of high-throughput 2D-gel and mass spectrometry based analytical techniques and the available protein sequence databases have created massive amount of proteomics data. To facilitate the sharing and further computational analysis of published proteomics data, several repositories have been created.

3.5.1. World-2DPAGE

The World-2DPAGE Constellation (52) is an effort of the Swiss Institute of Bioinformatics (SIB) to promote and publish two-dimensional gel electrophoresis proteomics data online through the ExPASy proteomics server. The World-2DPAGE Constellation consists of three components:

- *WORLD-2DPAGE List* (http://www.world-2dpage.expasy.org/list/) contains references to known federated 2D PAGE databases, as well as to 2D PAGE-related servers and services.

- *World-2DPAGE Portal* (http://www.world-2dpage.expasy.org/portal/) is a dynamic portal that serves as a single interface to query simultaneously world-wide gel-based proteomics databases that are built using the Make2D-DB package (53).
- *World-2DPAGE Repository* (http://www.world-2dpage.expasy.org/repository/) is a public repository for gel-based proteomics data with protein identifications published in the literature. Mass-spectrometry based proteomics data from related studies can also be submitted to the PRIDE database (54) so that interested readers can explore the data in the views of 2D-gel and/or MS.

3.5.2. PRIDE

The PRoteomics IDEntifications database (PRIDE) is a repository for mass-spectrometry based proteomics data including identifications of proteins, peptides and post-translational modifications that have been described in the scientific literature, together with supporting mass spectra (54). The PRIDE team has built an infrastructure and a set of software tools to facilitate the data submissions in PRIDE XML or mzData XML format from labs using different MS-based proteomics technologies. The PRIDE database can be queried by experiment accession number, protein accession number, literature reference, and sample parameters including species, tissue, sub-cellular location and disease state. The query results can be retrieved as PRIDE XML, mzData XML, or HTML. The PRIDE database includes a BioMart (35) interface that provides access to public PRIDE data from a query-optimized data warehouse as well as programmatic web service access. The PRIDE project also provides the Protein Identifier Cross-Reference Service (PICR) (55), which maps protein sequence identifiers from over 60 different databases via the UniParc (14) database. The Database on Demand (DoD, http://www.ebi.ac.uk/pride/dod) service provides custom FASTA formatted sequence databases according to a set of user-selectable criteria to optimize the search engine results. By November 19, 2009, the PRIDE database contains 10,329 experiments, 2,827,384 identified proteins, 12,542,472 identified peptides, 1,891,670 unique peptides and 56,703,344 Spectra.

4. Discussion

Although a variety of protein bioinformatics databases and resources have been developed to catalog and store different information about proteins, there are still opportunities to develop new solutions to facilitate comparative analysis, data-driven hypothesis generation, and biological knowledge discovery.

4.1. Data Integration and Comparative Analysis

As the volume and diversity of data and the desire to share those data increase, we inevitably encounter the problem of combining heterogeneous data generated from many different but related sources and providing the users with a unified view of this combined data set. This problem emerges in the biological and biomedical research community, where research data from different bioinformatics data repositories and laboratories need to be combined and analyzed. There are urgent needs for developing computational methods to integrate data from multiple studies and to answer more complex biological questions than traditional methods can tackle. Comparing experimental results across multiple laboratories and data types can also help forming new hypotheses for further experimentation (56–58). Different laboratories use different experimental protocols, instruments and analysis techniques, which make direct comparisons of their experimental results difficult. However, having related data in one place can make queries and comparisons of combined protein and gene data sets and further analysis possible.

In general, there are two types of data integration approaches. The data warehouse approach puts data sources into a centralized location with a global data schema and an indexing system for fast data retrieval. An example of this approach is the NIAID (National Institute for Allergy and Infectious Diseases) Biodefense Resource Center (http://www.proteomicsresource.org), which uses a protein-centric data warehouse (Master Protein Directory) to integrate and support mining and comparative analysis of large and heterogeneous "omics" data across different experiments and organisms (59). Another approach to data integration involves the federation of data across multiple sources. An example of this approach is the BioMart (35), an open source database management system that uses integrated query interfaces to query different BioMarts and allows users to group and refine their query results. The BioMart can also be accessed programmatically through web services or software libraries written in programming languages Java or Perl.

4.2. Data Provenance and Biological Knowledge

In many cases, the most difficult tasks in protein bioinformatics data management and analysis are not mapping biological entities from different sources or managing and processing large set of experimental data, such as gel images and mass spectra. Rather, it is in recording the detailed provenance of data, i.e., what was done, why it was done, where it was done, which instrument was used, what settings were used, how it was done. The provenance of experimental data is an important aspect of scientific best practice and is central to scientific discovery (60).

In proteomics studies, although great efforts have been made to develop and maintain data format standards, such as mzXML (61) and HUPO PSI (HUPO Proteomics Standards Initiative) (62), and minimal information standards for describing such data, for example, MIAPE (Minimum Information About a Proteomics Experiment) (63), the ontologies and related tools that provide formal representation of a set of concepts and their relationships within the domain of "omics" experiments still lag behind the current development of experimental protocols and methods. The standardization of data provenance remains a somewhat manual process, which depends on the efforts of database maintainers and data submitters.

The general biological and biomedical scientists are more interested in finding and viewing the "knowledge" contained in an already analyzed data set. However, much of the protein data generated in high-throughput research is insignificant in the conclusions of an analysis. Unfortunately, this information seldom comes with the standard data files and formats and is usually not easily found in omics repositories unless a reanalysis is performed or the data is annotated by a curator. For example, tables of proteins present in a given proteomics experiment are routinely found as supplemental data in scientific publications, but are not available in a searchable or easily computable format. This is unfortunate as this supplemental information is the result of considerable analysis by the original authors of a study to minimize false positive and false negative results, thus often representing the "knowledge" that underlies additional analysis and conclusions reached in a publication.

The NIAID Biodefense Resource Center developed a simple set of defined fields called "structured assertion" that could be used across proteomics, microarray and possibly other data types (59). A "structured assertion" can represent the results in a simple form like "protein V (presented) in experimental condition W," where V represents any valid identifier and W represents a value in a simple experimental ontology. A simple two-field assertion for the analyzed results of proteomics and microarray data and an "experimental condition" field containing simple keywords was implemented to describe the key experimental variables (growth conditions, sample fractionation, time, temperature, infection status and others) and "Expression Status," which has three values: increase, decrease or present. Although seemingly simple, the approach provides unique analytical power in the form of enabling simple queries across results from different data types and laboratories.

Acknowledgment

We would like to thank Dr. Winona C. Barker for reviewing the manuscript and providing constructive comments.

References

1. Ridley, M. (2006) *Genome*. Harper Perennial, New York.
2. Velculescu, V. E., Zhang, L., Zhou, W., Vogelstein, J., Basrai, M. A., Bassett, D. E. Jr, Hieter, P., Vogelstein, B., Kinzler, K. W. (1997) Characterization of the yeast transcriptome. *Cell* **2,** 243–251.
3. Anderson, N. L., Anderson, N. G. (1998) Proteome and proteomics: new technologies, new concepts, and new words. *Electrophoresis* **11,** 1853–1861.
4. Hye, A., Lynham, S., Thambisetty, M., Causevic, M., Campbell, J., Byers, H. L., Hooper, C., Rijsdijk, F., Tabrizi, S. J., Banner, S., Shaw, C. E., Foy, C., Poppe, M., Archer, N., Hamilton, G., Powell, J., Brown, R. G., Sham, P., Ward, M., Lovestone, S. (2006) Proteome-based plasma biomarkers for Alzheimer's disease. *Brain* **11,** 3042–3050.
5. Decramer, S., Wittke, S., Mischak, H., Zürbig, P., Walden, M., Bouissou, F., Bascands, J. L., Schanstra, J. P. (2006) Predicting the clinical outcome of congenital unilateral ureteropelvic junction obstruction in newborn by urinary proteome analysis. *Nat. Med.* **4,** 398–400.
6. Savidor, A., Donahoo, R. S., Hurtado-Gonzales, O., Land, M. L., Shah, M. B., Lamour, K. H., McDonald, W. H. (2008) Cross-species global proteomics reveals conserved and unique processes in *Phytophthora sojae* and *Phytophthora ramorum*. *Mol. Cell Proteomics* **8,** 1501–1516.
7. Huang, M., Chen, T., Chan, Z. (2006) An evaluation for cross-species proteomics research by publicly available expressed sequence tag database search using tandem mass spectral data. *Rapid Commun. Mass Spectrom.* **18,** 2635–2640.
8. Ishii, A., Dutta, R., Wark, G. M., Hwang, S. I., Han, D. K., Trapp, B. D., Pfeiffer, S. E., Bansal, R. (2009) Human myelin proteome and comparative analysis with mouse myelin. *Proc. Natl. Acad. Sci. U. S. A.* **34,** 14605–14610.
9. Irmler, M., Hartl, D., Schmidt, T., Schuchhardt, J., Lach, C., Meyer, H. E., Hrabé, de Angelis M., Klose, J., Beckers, J. (2008) An approach to handling and interpretation of ambiguous data in transcriptome and proteome comparisons. *Proteomics* **6,** 1165–1169.
10. Galperin, M. Y., Cochrane, G. R. (2009) Nucleic acids research annual database issue and the NAR online molecular biology database collection in 2009. *Nucleic Acids Res.* **37,** D1–D4.
11. Pruitt, K. D., Tatusova, T., Maglott, D. R. (2007) NCBI reference sequences (RefSeq): a curated non-redundant sequence database of genomes, transcripts and proteins. *Nucleic Acids Res.* **35,** D61–D65.
12. Benson, D. A., Karsch-Mizrachi, I., Lipman, D. J., Ostell, J., Wheeler, D. L. (2008) GeneBank. *Nucleic Acids Res.* **36,** D25–D30.
13. The UniProt Consortium. (2010) The universal protein resource (UniProt) in 2010. *Nucleic Acids Res.* **38,** D142–D148.
14. Leinonen, R., Diez, F. G., Binns, D., Fleischmann, W., Lopez, R., Apweiler, R. (2004) UniProt archive. *Bioinformatics* **20,** 3236–3237.
15. Suzek, B. E., Huang, H., McGarvey, P., Mazumder, R., Wu, C. H. (2007) UniRef: comprehensive and non-redundant UniProt reference clusters. *Bioinformatics* **23,** 1282–1288.
16. Yooseph, S., Sutton, G., Rusch, D. B., Halpern, A. L., Williamson, S. J., Remington, K., Eisen, J. A., Heidelberg, K. B., Manning, G., Li, W., Jaroszewski, L., Cieplak, P., Miller, C. S., Li, H., Mashiyama, S. T., Joachimiak, M. P., van Belle, C., Chandonia, J. M., Soergel, D. A., Zhai, Y., Natarajan, K., Lee, S., Raphael, B. J., Bafna, V., Friedman, R., Brenner, S. E., Godzik, A., Eisenberg, D., Dixon, J. E., Taylor, S. S., Strausberg, R. L., Frazier, M., Venter, J. C. (2007) The Sorcerer II global ocean sampling expedition: expanding the universe of protein families. *PLoS Biol.* **5,** e16.
17. Patient, S., Wieser, D., Kleen, M., Kretschmann, E., Martin, M. J., Apweiler, R. (2008) UniProtJAPI: a remote API for accessing UniProt data. *Bioinformatics* **24,** 1321–1322.
18. Nikolskaya, A. N., Arighi, C. N., Huang, H., Barker, W. C., Wu, C. H. (2006) PIRSF family classification system for protein functional and evolutionary analysis. *Evol. Bioinform. Online* **2,** 197–209.
19. Finn, R. D., Tate, J., Mistry, J., Coggill, P. C., Sammut, S. J., Hotz, H. R., Ceric, G., Forslund, K., Eddy, S. R., Sonnhammer, E. L., Bateman, A. (2008) The Pfam protein families database. *Nucleic Acids Res.* **36,** D281–D288.
20. Wheeler, D. L., Barrett, T., Benson, D. A., Bryant, S. H., Canese, K., Chetvernin, V., Church, D. M., DiCuccio, M., Edgar, R., Federhen, S., Geer, L. Y., Kapustin, Y., Khovayko, O., Landsman, D., Lipman, D. J., Madden, T. L., Maglott, D. R., Ostell, J., Miller, V., Pruitt, K. D., Schuler, G. D.,

Sequeira, E., Sherry, S. T., Sirotkin, K., Souvorov, A., Starchenko, G., Tatusov, R. L., Tatusova, T. A., Wagner, L., Yaschenko, E. (2007) Database resources of the National Center for Biotechnology Information. *Nucleic Acids Res.* **35,** D5–D12.

21. Bru, C., Courcelle, E., Carrère, S., Beausse, Y., Dalmar, S., Kahn, D. (2005) The ProDom database of protein domain families: more emphasis on 3D. *Nucleic Acids Res.* **33,** D212–D215.

22. Andreeva, A., Howorth, D., Chandonia, J. M., Brenner, S. E., Hubbard, T. J., Chothia, C., Murzin, A. G. (2008) Data growth and its impact on the SCOP database: new developments. *Nucleic Acids Res.* **36,** D419–D425.

23. Berman, H., Henrick, K., Nakamura, H., Markley, J. L. (2007) The worldwide Protein Data Bank (wwPDB): ensuring a single, uniform archive of PDB data. *Nucleic Acids Res.* **35,** D301–D303.

24. Hulo, N., Bairoch, A., Bulliard, V., Cerutti, L., Cuche, B. A., de Castro, E, Lachaize, C., Langendijk-Genevaux, P. S., Sigrist, C. J. (2008) The 20 years of PROSITE. *Nucleic Acids Res.* **36,** D245–D249.

25. Sigrist, C. J. A., Cerutti, L., Hulo, N., Gattiker, A., Falquet, L., Pagni, M., Bairoch, A., Bucher, P. (2002) PROSITE: a documented database using patterns and profiles as motif descriptors. *Brief. Bioinform.* **3,** 265–274.

26. De Castro, E., Sigrist, C. J. A., Gattiker, A., Bulliard, V., Langendijk-Genevaux, P. S., Gasteiger, E., Bairoch, A., Hulo, N. (2006) ScanProsite: detection of PROSITE signature matches and ProRule-associated functional and structural residues in proteins. *Nucleic Acids Res.* **34,** W362–W365.

27. Hunter, S., Apweiler, R., Attwood, T. K., Bairoch, A., Bateman, A., Binns, D., Bork, P., Das, U., Daugherty, L., Duquenne, L., Finn, R. D., Gough, J., Haft, D., Hulo, N., Kahn, D., Kelly, E., Laugraud, A., Letunic, I., Lonsdale, D., Lopez, R., Madera, M., Maslen, J., McAnulla, C., McDowall, J., Mistry, J., Mitchell, A., Mulder, N., Natale, D., Orengo, C., Quinn, A. F., Selengut, J. D., Sigrist, C. J., Thimma, M., Thomas, P. D., Valentin, F., Wilson, D., Wu, C. H., Yeats, C. (2009) InterPro: the integrative protein signature database. *Nucleic Acids Res.* **37,** D211–D215.

28. Yeats, C., Lees, J., Reid, A., Kellam, P., Martin, N., Liu, X., Orengo, C. (2008) Gene3D: comprehensive structural and functional annotation of genomes. *Nucleic Acids Res.* **36,** D414–D418.

29. Mi, H., Guo, N., Kejariwal, A., Thomas, P. D. (2007) PANTHER version 6: protein sequence and function evolution data with expanded representation of biological pathways. *Nucleic Acids Res.* **35,** D247–D252.

30. Attwood, T. K. (2002) The PRINTS database: a resource for identification of protein families. *Brief. Bioinform.* **3,** 252–263.

31. Letunic, I., Doerks, T., Bork, P. (2009) SMART 6: recent updates and new developments. *Nucleic Acids Res.* **37,** D229–D232.

32. Wilson, D., Pethica, R., Zhou, Y., Talbot, C., Vogel, C., Madera, M., Chothia, C., Gough, J. (2009) SUPERFAMILY – sophisticated comparative genomics, data mining, visualization and phylogeny. *Nucleic Acids Res.* **37,** D380–D386.

33. Haft, D. H., Selengut, J. D., White, O. (2003) The TIGRFAMs database of protein families. *Nucleic Acids Res.* **31,** D371–D373.

34. Mulder, N., Apweiler, R. (2007) InterPro and InterProScan: tools for protein sequence classification and comparison. *Methods Mol Biol.* **396,** 59–70.

35. Smedley, D., Haider, S., Ballester, B., Holland, R., London, D., Thorisson, G., Kasprzyk, A. (2009) BioMart – biological queries made easy. *BMC Genomics* **10,** 22.

36. Westbrook, J., Ito, N., Nakamura, H., Henrick, K., Berman, H. M. (2005) PDBML: the representation of archival macromolecular structure data in XML. *Bioinformatics* **21,** 988–992.

37. Cuff, A. L., Sillitoe, I., Lewis, T., Redfern, O. C., Garratt, R., Thornton, J., Orengo, C. A. (2009) The CATH classification revisited – architectures reviewed and new ways to characterize structural divergence in superfamilies. *Nucleic Acids Res.* **37,** D310–D314.

38. Fulton, K. F., Bate, M. A., Faux, N. G., Mahmood, K., Betts, C., Buckle, A. M. (2007) Protein Folding Database (PFD 2.0): an online environment for the International Foldeomics Consortium. *Nucleic Acids Res.* **35,** D304–D307.

39. Maxwell, K. L., Wildes, D., Zarrine-Afsar, A., De Los Rios, M. A., Brown, A. G., Friel, C. T., Hedberg, L., Horng, J. C., Bona, D., Miller, E. J., Vallée-Bélisle, A., Main, E. R., Bemporad, F., Qiu, L., Teilum, K., Vu, N. D., Edwards, A. M., Ruczinski, I., Poulsen, F. M., Kragelund, B. B., Michnick, S. W., Chiti, F., Bai, Y., Hagen, S. J., Serrano, L., Oliveberg, M., Raleigh, D. P., Wittung-Stafshede, P., Radford, S. E., Jackson, S. E., Sosnick, T. R., Marqusee, S., Davidson, A. R., Plaxco, K. W. (2005) Protein folding: defining a "standard" set of experimental conditions and a preliminary kinetic data set of two-state proteins. *Protein Sci.* **14,** 602–616.

40. Zanzoni, A., Ausiello, G., Via, A., Gherardini, P. F., Helmer-Citterich, M. (2007) Phospho3D: a database of three-dimensional structures of protein phosphorylation sites. *Nucleic Acids Res.* **35,** D229–D231.
41. Diella, F., Cameron, S., Gemünd, C., Linding, R., Via, A., Kuster, B., Sicheritz-Pontén, T., Blom, N., Gibson, T. J. (2004) Phospho. ELM: a database of experimentally verified phosphorylation sites in eukaryotic proteins. *BMC Bioinformatics* **5,** 79.
42. Aranda, B., Achuthan, P., Alam-Faruque, Y., Armean, I., Bridge, A., Derow, C., Feuermann, M., Ghanbarian, A. T., Kerrien, S., Khadake, J., Kerssemakers, J., Leroy, C., Menden, M., Michaut, M., Montecchi-Palazzi, L., Neuhauser, S. N., Orchard, S., Perreau, V., Roechert, B., van Eijk, K., Hermjakob, H. (2010) The IntAct molecular interaction database in 2010. *Nucleic Acids Res.* **38,** D525–D531.
43. Orchard, S., Kerrien, S., Jones, P., Ceol, A., Chatr-Aryamontri, A., Salwinski, L., Nerothin, J., Hermjakob, H. (2007) Submit your interaction data the IMEx way: a step by step guide to trouble-free deposition. *Proteomics* 7 **Suppl 1,** 28–34.
44. Orchard, S., Salwinski, L., Kerrien, S., Montecchi-Palazzi, L., Oesterheld, M., Stümpflen, V., Ceol, A., Chatr-aryamontri, A., Armstrong, J., Woollard, P., Salama, J. J., Moore, S., Wojcik, J., Bader, G. D., Vidal, M., Cusick, M. E., Gerstein, M., Gavin, A. C., Superti-Furga, G., Greenblatt, J., Bader, J., Uetz, P., Tyers, M., Legrain, P., Fields, S,, Mulder, N., Gilson, M., Niepmann, M., Burgoon, L., De Las Rivas, J., Prieto, C., Perreau, V. M., Hogue, C., Mewes, H. W., Apweiler, R., Xenarios, I., Eisenberg, D., Cesareni, G., Hermjakob, H. (2007) The minimum information required for reporting a molecular interaction experiment (MIMIx). *Nat. Biotechnol.* **25,** 894–898.
45. Kerrien, S., Orchard, S., Montecchi-Palazzi, L., Aranda, B., Quinn, A. F., Vinod, N., Bader, G. D., Xenarios, I., Wojcik, J., Sherman, D., Tyers, M., Salama, J. J., Moore, S., Ceol, A., Chatr-Aryamontri, A., Oesterheld, M., Stümpflen, V., Salwinski, L., Nerothin, J., Cerami, E., Cusick, M. E., Vidal, M., Gilson, M., Armstrong, J., Woollard, P., Hogue, C., Eisenberg, D., Cesareni, G., Apweiler, R., Hermjakob, H. (2007) Broadening the horizon – level 2.5 of the HUPO-PSI format for molecular interactions. *BMC Biol.* **5,** 44.
46. Ashburner M, Ball CA, Blake JA, Botstein D, Butler H, Cherry JM, Davis AP, Dolinski K, Dwight SS, Eppig JT, Harris MA, Hill DP, Issel-Tarver L, Kasarskis A, Lewis S, Matese JC, Richardson JE, Ringwald M, Rubin GM, Sherlock G. (2000) Gene ontology: tool for the unification of biology. The Gene Ontology Consortium. *Nat. Genet.* **25,** 25–29.
47. Matthews, L., Gopinath, G., Gillespie, M., Caudy, M., Croft, D., de Bono, B., Garapati, P., Hemish, J., Hermjakob, H., Jassal, B., Kanapin, A., Lewis, S., Mahajan, S., May, B., Schmidt, E., Vastrik, I., Wu, G., Birney, E., Stein, L., D'Eustachio, P. (2009) Reactome knowledgebase of human biological pathways and processes. *Nucleic Acids Res.* **37,** D619–D622.
48. Hucka, M., Finney, A., Sauro, H. M., Bolouri, H., Doyle, J. C., Kitano, H., Arkin, A. P., Bornstein, B. J., Bray, D., Cornish-Bowden, A., Cuellar, A. A., Dronov, S., Gilles, E. D., Ginkel, M., Gor, V., Goryanin, II., Hedley, W. J., Hodgman, T. C., Hofmeyr, J. H., Hunter, P. J., Juty, N. S., Kasberger, J. L., Kremling, A., Kummer, U., Le Novère, N., Loew, L. M., Lucio, D., Mendes, P., Minch, E., Mjolsness, E. D., Nakayama, Y., Nelson, M. R., Nielsen, P. F., Sakurada, T., Schaff, J. C., Shapiro, B. E., Shimizu, T. S., Spence, H. D., Stelling, J., Takahashi, K., Tomita, M., Wagner, J., Wang, J., SBML Forum. (2003) The systems biology markup language (SBML): a medium for representation and exchange of biochemical network models. *Bioinformatics* **19,** 524–531.
49. Noy, N. F., Crubezy, M., Fergerson, R. W., Knublauch, H., Tu, S. W., Vendetti, J., Musen, M. A. (2003) Protégé-2000: an open-source ontology-development and knowledge-acquisition environment. *AMIA. Annu Symp Proc.* 953.
50. Cline, M. S., Smoot, M., Cerami, E., Kuchinsky, A., Landys, N., Workman, C., Christmas, R., Avila-Campilo, I., Creech, M., Gross, B., Hanspers, K., Isserlin, R., Kelley, R., Killcoyne, S., Lotia, S., Maere, S., Morris, J., Ono, K., Pavlovic, V., Pico, A. R., Vailaya, A., Wang, P. L., Adler, A., Conklin, B. R., Hood, L., Kuiper, M., Sander, C., Schmulevich, I., Schwikowski, B., Warner, G. J., Ideker, T., Bader, G. D. (2007) Integration of biological networks and gene expression data using Cytoscape. *Nat. Protoc.* **2,** 2366–2382.
51. Caspi, R., Foerster, H., Fulcher, C. A., Kaipa, P., Krummenacker, M., Latendresse, M., Paley, S., Rhee, S. Y., Shearer, A., Tissier, C., Walk, T. C., Zhang, P. and Karp, P. D. (2008) The MetaCyc Database of metabolic pathways and enzymes and the BioCyc collection of Pathway/Genome Databases. *Nucleic Acids Res.* **36,** D623–D631.
52. Hoogland, C., Mostaguir, K., Appel, R. D., Lisacek, F. (2008) The World-2DPAGE Constellation to promote and publish gel-based

proteomics data through the ExPASy server. *J. Proteomics* **71**, 245–248.
53. Mostaguir, K., Hoogland, C., Binz, P. A., Appel, R. D. (2003) The Make 2D-DB II package: conversion of federated two-dimensional gel electrophoresis databases into a relational format and interconnection of distributed databases. *Proteomics* **3**, 1441–1444.
54. Vizcaíno, J. A., Côté, R., Reisinger, F., Barsnes, H., Foster, J. M., Rameseder, J., Hermjakob, H., Martens, L. (2009) The proteomics identifications database: 2010 update. *Nucleic Acids Res.* **38**, D736–D742.
55. Côté, R. G., Jones, P., Martens, L., Kerrien, S., Reisinger, F., Lin, Q., Leinonen, R., Apweiler, R., Hermjakob, H. (2007) The protein identifier cross-referencing (PICR) service: reconciling protein identifiers across multiple source databases. *BMC Bioinformatics* **8**, 401–414.
56. Burgun, A., Bodenreider, O. (2008) Accessing and integrating data and knowledge for biomedical research. *Yearb Med Inform.* 91–101.
57. Hwang, D., Rust, A. G., Ramsey, S., Smith, J. J., Leslie, D. M., Weston, A. D., de Atauri, P., Aitchison, J. D., Hood, L., Siegel, A. F., Bolouri, H. (2005) A data integration methodology for systems biology. *Proc. Natl Acad. Sci. U. S. A.* **102**, 17296–17301.
58. Mathew, J. P., Taylor, B. S., Bader, G. D., Pyarajan, S., Antoniotti, M., Chinnaiyan, A. M., Sander, C., Burakoff, S. J., Mishra, B. (2007) From bytes to bedside: data integration and computational biology for translational cancer research. *PLoS Comput. Biol.* **3**, e12.
59. McGarvey, P. B., Huang, H., Mazumder, R., Zhang, J., Chen, Y., Zhang, C., Cammer, S., Will, R., Odle, M., Sobral, B., Moore, M., Wu, C. H. (2009) Systems integration of biodefense omics data for analysis of pathogen–host interactions and identification of potential targets. *PLoS One* **4**, e7162.
60. Stevens, R., Zhao, J., Goble, C. (2007) Using provenance to manage knowledge of in silico experiments. *Brief. Bioinform.* **8**, 183–194.
61. Pedrioli, P. G., Eng, J. K., Hubley, R., Vogelzang, M., Deutsch, E. W., Raught, B., Pratt, B., Nilsson, E., Angeletti, R. H., Apweiler, R., Cheung, K., Costello, C. E., Hermjakob, H., Huang, S., Julian, R. K., Kapp, E., McComb, M. E., Oliver, S. G., Omenn, G., Paton, N. W., Simpson, R., Smith, R., Taylor, C. F., Zhu, W., Aebersold, R. (2004) A common open representation of mass spectrometry data and its application to proteomics research. *Nat. Biotechnol.* **22**, 1459–1466.
62. Orchard, S., Montechi-Palazzi, L., Deutsch, E. W., Binz, P. A., Jones, A. R., Paton, N., Pizarro, A., Creasy, D. M., Wojcik, J., Hermjakob, H. (2007) Five years of progress in the standardization of proteomics data 4(th) annual spring workshop of the HUPO-proteomics standards initiative April 23–25, 2007 Ecole Nationale Supérieure (ENS), Lyon, France. *Proteomics* **7**, 3436–3440.
63. Taylor, C. F., Paton, N. W., Lilley, K. S., Binz, P. A., Julian, R. K. Jr, Jones, A. R., Zhu, W., Apweiler, R., Aebersold, R., Deutsch, E. W., Dunn, M. J., Heck, A. J., Leitner, A., Macht, M., Mann, M., Martens, L., Neubert, T. A., Patterson, S. D., Ping, P., Seymour, S. L., Souda, P., Tsugita, A., Vandekerckhove, J., Vondriska, T. M., Whitelegge, J. P., Wilkins, M. R., Xenarios, I., Yates, J. R. 3rd, Hermjakob, H. (2007) The minimum information about a proteomics experiment (MIAPE). *Nat. Biotechnol.* **25**, 887–893.
64. Tatusov, R. L., Fedorova, N. D., Jackson, J. D., Jacobs, A. R., Kiryutin, B., Koonin, E. V., Krylov, D. M., Mazumder, R., Mekhedov, S. L., Nikolskaya, A. N., Rao, B. S., Smirnov, S., Sverdlov, A. V., Vasudevan, S., Wolf, Y. I., Yin, J. J., Natale, D. A. (2003) The COG database: an updated version includes eukaryotes. *BMC Bioinformatics* **4**, 41–54.
65. Kaplan, N., Sasson, O., Inbar, U., Friedlich, M., Fromer, M., Fleischer, H., Portugaly, E., Linial, N., Linial, M. (2005) ProtoNet 4.0: a hierarchical classification of one million protein sequences. *Nucleic Acids Res.* **33**, D216–D218.
66. Marchler-Bauer, A., Anderson, J. B., Chitsaz, F., Derbyshire, M. K., DeWeese-Scott, C., Fong, J. H., Geer, L. Y., Geer, R. C., Gonzales, N. R., Gwadz, M., He, S., Hurwitz, D. I., Jackson, J. D., Ke, Z., Lanczycki, C. J., Liebert, C. A., Liu, C., Lu, F., Lu, S., Marchler, G. H., Mullokandov, M., Song, J. S., Tasneem, A., Thanki, N., Yamashita, R. A., Zhang, D., Zhang, N., Bryant, S. H. (2009) CDD: specific functional annotation with the conserved domain database. *Nucleic Acids Res.* **37**, D205–D210.
67. Wang, Y., Addess, K. J., Chen, J., Geer, L. Y., He, J., He, S., Lu, S., Madej, T., Marchler-Bauer, A., Thiessen, P. A., Zhang, N., Bryant, S. H. (2007) MMDB: annotating protein sequences with Entrez's 3D-structure database. *Nucleic Acids Res.* **35**, D298–D300.
68. Pieper, U., Eswar, N., Webb, B. M., Eramian, D., Kelly, L., Barkan, D. T., Carter, H., Mankoo, P., Karchin, R., Marti-Renom, M. A., Davis, F. P., Sali, A. (2009) MODBASE, a

database of annotated comparative protein structure models and associated resources. *Nucleic Acids Res.* **37,** D347–D354.

69. Kiefer, F., Arnold, K., Künzli, M., Bordoli, L., Schwede, T. (2009) The SWISS-MODEL repository and associated resources. *Nucleic Acids Res.* **37,** D387–D392.

70. Bogatyreva, N. S., Osypov, A. A., Ivankov, D. N. (2009) KineticDB: a database of protein folding kinetics. *Nucleic Acids Res.* **37,** D342–D346.

71. Garavelli, J. S. (2004) The RESID database of protein modifications as a resource and annotation tool. *Proteomics* **4,** 1527–1533.

72. Salwinski, L., Miller, C. S., Smith, A. J., Pettit, F. K., Bowie, J. U., Eisenberg, D. (2004) The database of interacting proteins: 2004 update. *Nucleic Acids Res.* **32,** D449–D451.

73. Breitkreutz, B. J., Stark, C., Reguly, T., Boucher, L., Breitkreutz, A., Livstone, M., Oughtred, R., Lackner, D. H., Bähler, J., Wood, V., Dolinski, K., Tyers, M. (2008) The BioGRID interaction database: 2008 update. *Nucleic Acids Res.* **36,** D637–D640.

74. Kanehisa, M., Goto, S. (2000) KEGG: Kyoto encyclopedia of genes and genomes. *Nucleic Acids Res.* **28,** 27–30.

75. Tarcea, V. G., Weymouth, T., Ade, A., Bookvich, A., Gao, J., Mahavisno, V., Wright, Z., Chapman, A., Jayapandian, M., Ozgür, A., Tian, Y., Cavalcoli, J., Mirel, B., Patel, J., Radev, D., Athey, B., States, D., Jagadish, H. V. (2009) Michigan molecular interactions r2: from interacting proteins to pathways. *Nucleic Acids Res.* **37,** D642–D646.

76. Craig, R., Cortens, J. C., Fenyo, D., Beavis, R. C. (2006) Using annotated peptide mass spectrum libraries for protein identification. *J. Proteome Res.* **5,** 1843–1849.

77. Deutsch, E. W., Lam, H., Aebersold, R. (2008) PeptideAtlas: a resource for target selection for emerging targeted proteomics workflows. *EMBO Rep.* **9,** 429–434.

78. Slotta, D. J., Barrett, T., Edgar, R. (2009) NCBI peptidome: a new public repository for mass spectrometry peptide identifications. *Nat. Biotechnol.* **27,** 600–601.

Chapter 2

A Guide to UniProt for Protein Scientists

Claire O'Donovan and Rolf Apweiler

Abstract

One of the essential requirements of the proteomics community is a high quality annotated nonredundant protein sequence database with stable identifiers and an archival service to enable protein identification and characterization. The scope of this chapter is to illustrate how Universal Protein Resource (UniProt) (The UniProt Consortium, Nucleic Acids Res. 38:D142–D148, 2010) can be best utilized for proteomics purposes with a particular focus on exploiting the knowledge captured in the UniProt databases, the services provided and the availability of complete proteomes.

Key words: Protein sequence database, Annotation, Stable identifiers, Complete proteome, Archive, Nonredundant

1. Introduction

The Proteomics community has evolved intensively over the last decade but one constant is the need to identify the resulting proteins and their potential functions. This requires the availability of a nonredundant protein sequence database, with maximal coverage including splice isoforms, disease variant(s) and posttranslational modifications. Sequence archiving is an essential feature in order to be able to interpret and maintain the proteomic set results. Stable identifiers, consistent nomenclature and controlled vocabularies are highly beneficial for protein identification. The last but by no means least requirement is the provision of detailed information on protein function, biological processes, and molecular interactions and pathways cross-referenced to appropriate external sources. In this chapter, we will show how the Universal Protein Resource fulfils these criteria.

Cathy H. Wu and Chuming Chen (eds.), *Bioinformatics for Comparative Proteomics*, Methods in Molecular Biology, vol. 694, DOI 10.1007/978-1-60761-977-2_2,

2. Materials

The mission of the Universal Protein Resource (UniProt) is to provide the scientific community with a comprehensive, high-quality and freely accessible resource of protein sequence and functional information, which is essential for modern biological research. UniProt is produced by the UniProt Consortium, which consists of groups from the European Bioinformatics Institute (EBI), the Protein Information Resource (PIR), and the Swiss Institute of Bioinformatics (SIB). Its activities are mainly supported by the National Institutes of Health (NIH) with additional funding from the European Commission and the Swiss Federal Government.

It has five components optimized for different uses. The UniProt Knowledgebase (UniProtKB) (1) is an expertly curated database, a central access point for integrated protein information with cross-references to multiple sources. The UniProt Archive (UniParc) (2) is a comprehensive sequence repository, reflecting the history of all protein sequences. UniProt Reference Clusters (UniRef) (3) merge closely related sequences based on sequence identity to speed up searches whereas the UniProt Metagenomic and Environmental Sequences database (UniMES) was created to respond to the expanding area of metagenomic data. UniProtKB Sequence/Annotation Version Archive (UniSave) is the UniProtKB protein entry archive, which contains all versions of each protein entry (Fig. 1).

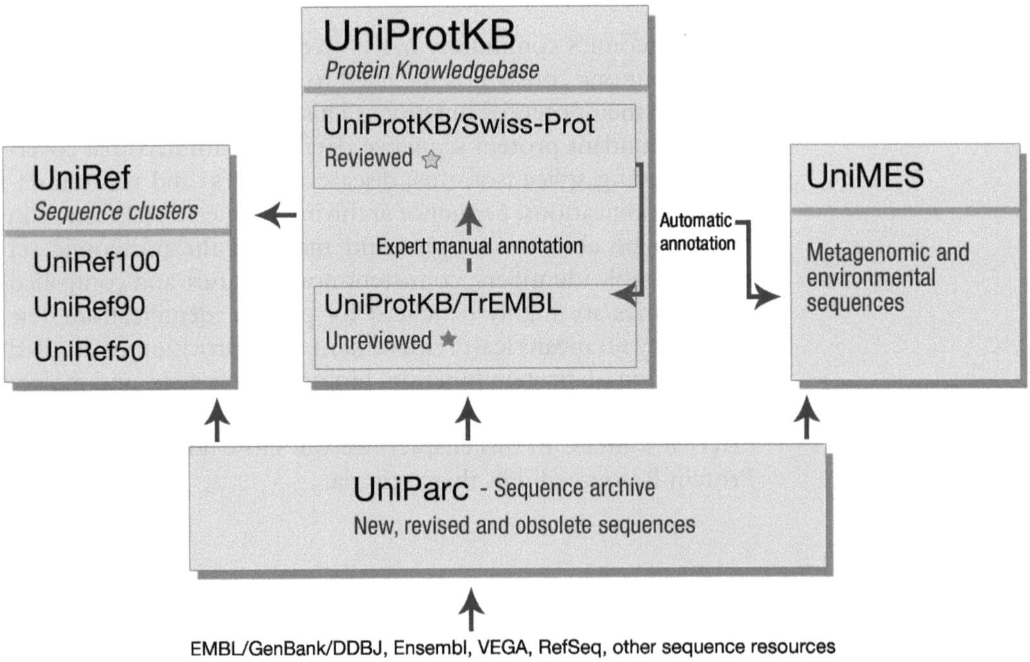

Fig. 1. UniProt databases.

2.1. The UniProt Archive

UniParc is the main sequence storehouse and is a comprehensive repository that reflects the history of all protein sequences. UniParc contains all new and revised protein sequences from all publicly available sources (http://www.uniprot.org/help/uniparc) to ensure that complete coverage is available at a single site. To avoid redundancy, all sequences 100% identical over the entire length are merged, regardless of source organism. New and updated sequences are loaded on a daily basis, cross-referenced to the source database accession number, and provided with a sequence version that increments on changes to the underlying sequence. The basic information stored within each UniParc entry is the identifier, the sequence, cyclic redundancy check number, source database(s) with accession and version numbers, and a time stamp. If a UniParc entry lacks a cross-reference to a UniProtKB entry, the reason for its exclusion from UniProtKB is provided (e.g., pseudogene). In addition, each source database accession number is tagged with its status in that database, indicating if the sequence still exists or has been deleted in the source database and cross-references to NCBI GI and TaxId if appropriate.

2.2. The UniProt Knowledgebase

UniProtKB consists of two sections, UniProtKB/Swiss-Prot and UniProtKB/TrEMBL. The former contains manually annotated records with information extracted from literature and curator-evaluated computational analysis. Annotation is done by biologists with specific expertise to achieve accuracy. In UniProtKB/Swiss-Prot, annotation consists of the description of the following: function(s), enzyme-specific information, biologically relevant domains and sites, post-translational modifications, subcellular location(s), tissue specificity, developmental specific expression, structure, interactions, splice isoform(s), associated diseases or deficiencies, or abnormalities etc. The UniProt Knowledgebase aims to describe, in a single record, all protein products derived from a certain gene from a certain species. After an inspection of the sequences, the curator selects the reference sequence, does the corresponding merging, and lists the splice and genetic variants along with disease information when available. This results in not only the whole record having an accession number but also unique identifiers for each protein form derived by alternative splicing, proteolytic cleavage, and posttranslational modification. The freely available tool VARSPLIC (4) enables the recreation of all annotated splice variants from the feature table of a UniProt Knowledgebase entry, or for the complete database. A FASTA-formatted file containing all splice variants annotated in the UniProt Knowledgebase can be downloaded for use with similarity search programs.

UniProtKB/TrEMBL contains high quality computationally analyzed records enriched with automatic annotation and classification. The computer-assisted annotation is created using both automatically generated rules as well as manually curated rules

(UniRule) based on protein families (5–8). UniProtKB/TrEMBL contains the translations of all coding sequences (CDS) present in the EMBL/GenBank/DDBJ Nucleotide Sequence Databases and, with some defined exclusions, *Arabidopsis thaliana* sequences from The Arabidopsis Information Resource (TAIR) (9), yeast sequences from the Saccharomyces Genome Database (SGD) (10) and *Homo sapiens* sequences from the Ensembl database (11). It will soon be extended to include other Ensembl organism sets and RefSeq records. Records are selected for full manual annotation and integration into UniProtKB/Swiss-Prot according to defined annotation priorities.

Integration between the three types of sequence-related databases (nucleic acid sequences, protein sequences, and protein tertiary structures) as well as with specialized data collections is important for the UniProt users. UniProtKB is currently cross-referenced with more than ten million links to 114 different databases with regular update cycles. This extensive network of cross-references allows UniProt to act as a focal point of biomolecular database interconnectivity. All cross-referenced databases are documented at http://www.uniprot.org/docs/dbxref and if appropriate are included in the UniProt ID mapping tool at http://www.uniprot.org/help/mapping with the file for download at ftp://ftp.uniprot.org/pub/databases/uniprot/current_release/knowledgebase/idmapping.

2.3. The UniProt Reference Clusters

UniRef provides clustered sets of all sequences from the UniProt Knowledgebase (including splice forms as separate entries) and selected records from the UniProt Archive to achieve complete coverage of sequence space at identity levels of 100, 90, and 50% while hiding redundant sequences (3). The UniRef clusters are generated in a hierarchical manner; the UniRef100 database combines identical sequences and sub-fragments into a single UniRef entry, UniRef90 is built from UniRef100 clusters and UniRef50 is built from UniRef90 clusters. Each individual member sequence can exist in only one UniRef cluster at each identity level and have only one parent or child cluster at another identity level. UniRef100, UniRef90, and UniRef50 yield a database size reduction of ~10, 40, and 70%, respectively. Each cluster record contains source database, protein name, and taxonomy information on each member sequence but is represented by a single selected representative protein sequence and name; the number of members and lowest common taxonomy node for the membership is also included. The representative protein sequence or cluster representative is automatically selected using an algorithm that accounts for (1) Quality of entry annotation: order of preference is a member from UniProtKB/Swiss-Prot, UniProtKB/TrEMBL, then UniParc; (2) Meaningful name: members with protein names that do not contain words such as "hypothetical" or "probable"

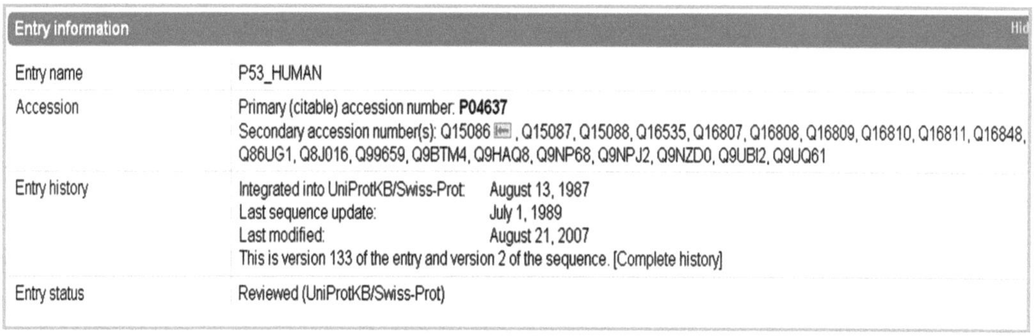

Entry information	
Entry name	P53_HUMAN
Accession	Primary (citable) accession number: **P04637** Secondary accession number(s): Q15086, Q15087, Q15088, Q16535, Q16807, Q16808, Q16809, Q16810, Q16811, Q16848, Q86UG1, Q8J016, Q99659, Q9BTM4, Q9HAQ8, Q9NP68, Q9NPJ2, Q9NZD0, Q9UBI2, Q9UQ61
Entry history	Integrated into UniProtKB/Swiss-Prot: August 13, 1987 Last sequence update: July 1, 1989 Last modified: August 21, 2007 This is version 133 of the entry and version 2 of the sequence. [Complete history]
Entry status	Reviewed (UniProtKB/Swiss-Prot)

Fig. 2. UniSave link.

are preferred; (3) Organism: members from model organisms are preferred; (4) Sequence length: longest sequence is preferred. UniRef100 is one of the most comprehensive and nonredundant protein sequence dataset available. The reduced size of the UniRef90 and UniRef50 datasets provide faster sequence similarity searches and reduce the research bias in similarity searches by providing a more even sampling of sequence space.

2.4. The UniProt Metagenomic and Environmental Sequences

The UniProt Knowledgebase contains entries with a known taxonomic source. However, the expanding area of metagenomic data has necessitated the creation of a separate database, the UniProt Metagenomic and Environmental Sequences database (UniMES). UniMES currently contains data from the Global Ocean Sampling Expedition (GOS), which predicts nearly six million proteins, primarily from oceanic microbes. By combining the predicted protein sequences with automatic classification by InterPro, the integrated resource for protein families, domains and functional sites, UniMES uniquely provides free access to the array of genomic information gathered.

2.5. The UniProtKB Sequence/Annotation Version Archive

UniSave is a repository of UniProtKB/Swiss-Prot and UniProtKB/TrEMBL entry versions and provides the backend to the UniProtKB entry history service (Fig. 2) and is also provided as a standalone service at http://www.ebi.ac.uk/uniprot/unisave.

These descriptions of our databases should illustrate that UniProt does provide a high quality annotated nonredundant database with maximal coverage and sequence archiving.

3. Methods

This section will describe particular features of the UniProt activities, which fulfill the proteomics community requirements of detailed information on protein function, biological processes, molecular

Names and origin	
Protein names	*Recommended name:* **Glutamate carboxypeptidase 2** EC=3.4.17.21 *Alternative name(s):* **Glutamate carboxypeptidase II** **Membrane glutamate carboxypeptidase** Short name=mGCP **N-acetylated-alpha-linked acidic dipeptidase I** Short name=NAALADase I **Pteroylpoly-gamma-glutamate carboxypeptidase** **Folylpoly-gamma-glutamate carboxypeptidase** Short name=FGCP **Folate hydrolase 1** **Prostate-specific membrane antigen** Short name=PSMA Short name=PSM
Gene names	Name: **FOLH1** Synonyms: FOLH, NAALAD1, PSM, PSMA

Fig. 3. UniProt nomenclature.

interactions and pathways cross-referenced to appropriate external sources and stable identifiers, consistent nomenclature and controlled vocabularies.

3.1. Protein Annotation

UniProtKB consists of two sections, Swiss-Prot and TrEMBL.

UniProtKB/Swiss-Prot contains manually annotated records with information extracted from literature and curator-evaluated computational analysis. Manual annotation consists of a critical review of experimentally proven or computer-predicted data about each protein. An essential aspect of the annotation protocol is the use of official nomenclatures and controlled vocabularies that facilitate consistent and accurate identification (Fig. 3).

Annotation consists of the description of the following: functions(s), enzyme-specific information, biologically relevant domains and sites, posttranslation modifications, subcellular location(s), tissue specificity, developmental specific expression, structure, interactions, splice isoforms(s), associated diseases or deficiencies, or abnormalities etc (Fig. 4).

Another important part of the annotation process involves merging of different reports for a single protein. After an inspection of the sequences the curator selects the reference sequence, does the corresponding merging and lists the splice and genetic variants along with disease information when available (Fig. 5). Data are continuously updated by an expert team of biologists.

General annotation (Comments)	Hide | To
Function	Photoreceptor required for image-forming vision at low light intensity. Required for photoreceptor cell viability after birth. Light-induced isomerization of 11-cis to all-trans retinal triggers a conformational change leading to G-protein activation and release of all-trans retinal.
Subcellular location	Membrane; Multi-pass membrane protein.
Tissue specificity	Rod shaped photoreceptor cells which mediates vision in dim light.
Post-translational modification	Phosphorylated on some or all of the serine and threonine residues present in the C-terminal region.
Involvement in disease	Defects in RHO are the cause of retinitis pigmentosa type 4 (RP4) [MIM:180380]. RP leads to degeneration of retinal photoreceptor cells. Patients typically have night vision blindness and loss of midperipheral visual field. As their condition progresses, they lose their far peripheral visual field and eventually central vision as well. RP4 inheritance is autosomal dominant. Ref.7 Ref.8 Ref.9 Ref.10 Ref.11 Ref.12 Ref.13 Ref.14 Ref.15 Ref.16 Ref.17 Ref.18 Ref.20 Ref.21 Ref.22 Ref.23 Ref.24 Ref.25 Ref.26 Ref.28 Ref.30 Ref.31 Defects in RHO are a cause of retinitis pigmentosa autosomal recessive (ARRP) [MIM:268000]. Ref.27 Defects in RHO are the cause of congenital stationary night blindness autosomal dominant type 1 (CSNBAD1) [MIM:610445]; also known as rhodopsin-related congenital stationary night blindness. Congenital stationary night blindness is a non-progressive retinal disorder characterized by impaired night vision. Ref.19 Ref.29 Ref.32
Sequence similarities	Belongs to the G-protein coupled receptor 1 family. Opsin subfamily.
biophysicochemical properties	Absorption: Abs(max)=495 nm

Fig. 4. Protein annotation.

3.2. The Gene Ontology Consortium and UniProt

To promote database interoperability and provide consistent annotation, the UniProt Consortium is a key member of the Gene Ontology Consortium (12) and benefits from the presence of the GO editorial office at the EBI. UniProt curators will continue to assign Gene Ontology (GO terms) to the gene products in UniProtKB during the UniProt manual curation process. UniProtKB also profits from GO annotation carried out by other GO Consortium members. Currently we include manual GO annotations from 19 GO Consortium annotation groups, and we further supplement this with high-quality annotations from other manual annotation sources (including the Human Protein Atlas and LIFEdb). In addition to this manually curated GO annotation, automatic GO annotation pipelines exist and will be further developed to ensure that the functional knowledge supplied by various UniProtKB ontologies, Ensembl orthology data, and InterPro matches are fully exploited to provide high-quality, comprehensive set of GO annotation predictions for all UniProtKB entries.

3.3. Cross-references to External Sources

One challenge in life sciences research is the ability to integrate and exchange data coming from multiple research groups. The UniProt Consortium is committed to fostering interaction and exchange with the scientific community, ensuring wide access to UniProt resources, and promoting interoperability between resources. An essential component of this interoperability is the provision of cross-references to these resources in UniProt entries (Fig. 6).

Alternative products Hide | Top

This entry describes 2 isoforms produced by **alternative splicing.** [Align] [Select]

Isoform 1 (identifier: **P04637-1**)

This isoform has been chosen as the 'canonical' sequence. All positional information in this entry refers to it. This is also the sequence that appears in the downloadable versions of the entry.

Isoform 2 (identifier: **P04637-2**)

Also known as: I9RET;

The sequence of this isoform differs from the canonical sequence as follows:
332-341: IRGRERFEMF → DGTSFQKENC
342-393: Missing.

Note: Seems to be non-functional. Expressed in quiescent lymphocytes.

Natural variations

■	Alternative sequence	332 – 341	10	IRGRERFEMF → DGTSFQKENC in isoform 2.		VSP_006535
□	Alternative sequence	342 – 393	52	Missing in isoform 2.		VSP_006536

□	DNA binding	102 – 292	191	
□	Region	1 – 83	83	Interaction with HRMT1L2
□	Region	1 – 44	44	Transcription activation (acidic)
□	Region	66 – 110	45	Interaction with WWOX
□	Region	100 – 370	271	Interaction with HIPK1 (By similarity)
□	Region	116 – 292	177	Interaction with AXIN1 (By similarity)
□	Region	241 – 248	8	Interacts with the 53BP2 SH3 domain
□	Region	300 – 393	94	Interaction with CARM1
□	Region	319 – 360	42	Interaction with HIPK2
□	Region	325 – 356	32	Oligomerization

Fig. 5. Sequence annotation.

3.4. Nonredundant Complete UniProt Proteome Sets

UniProt constructs complete nonredundant proteome sets. Each set and its analysis is made available shortly after the appearance of a new complete genome sequence in the nucleotide sequence databases. A standard procedure is used to create, from the UniProtKB, proteome sets for bacterial, archaeal and some eukaryotic genomes. Proteome sets for certain metazoan genomes are

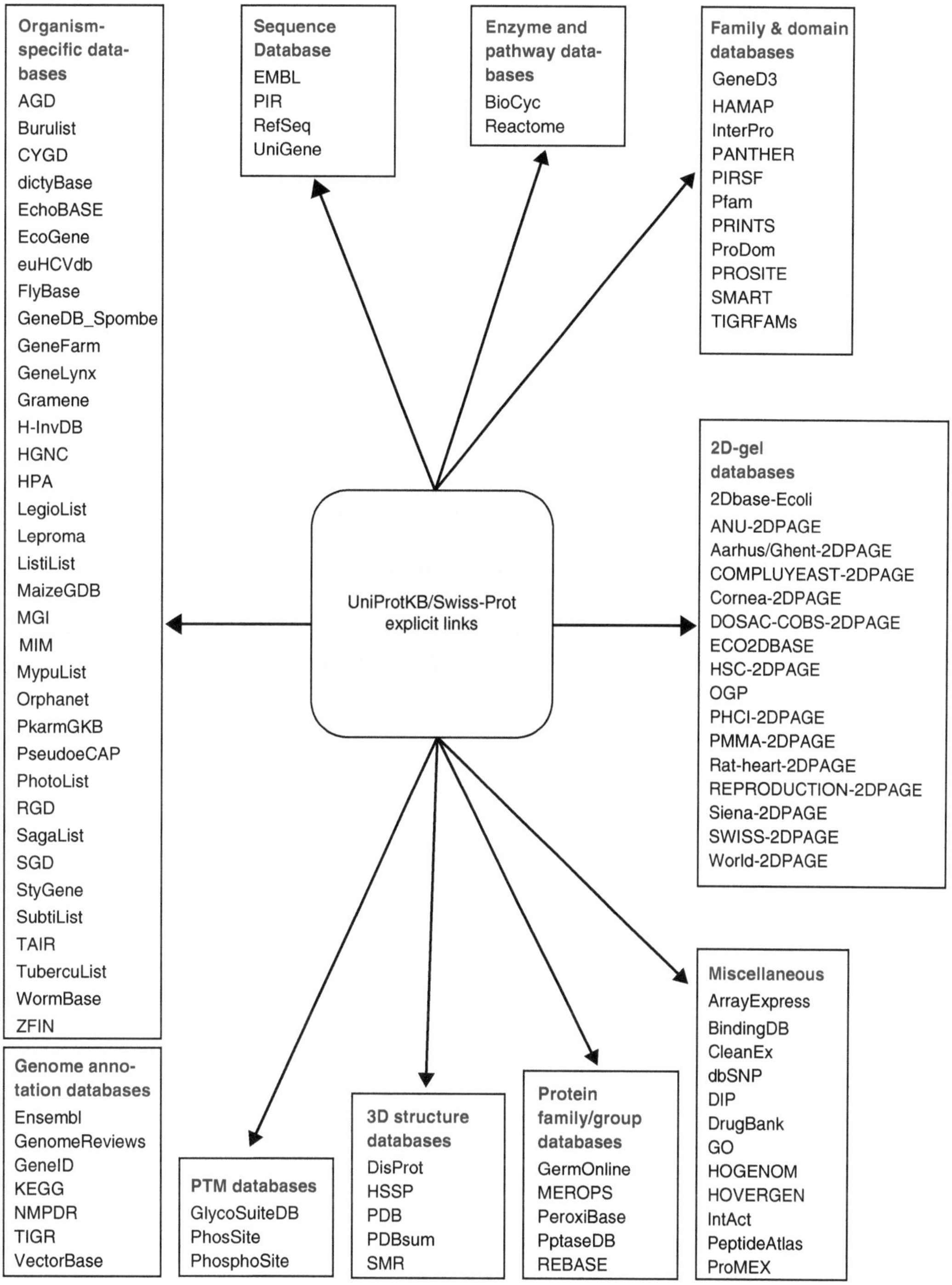

Fig. 6. UniProt cross-references.

produced by a separate procedure because a full and accurate set of coding sequence predictions are not yet available in the nucleotide sequence databases. Currently, the International Protein Index (IPI) (13) derived from the UniProt Knowledgebase, Ensembl, and the NCBI's RefSeq project (and cross-referenced to from UniProtKB) provides this data but development is currently underway to extend the UniProtKB production pipeline to replace IPI's functionality and for UniProt to provide these sets directly in collaboration with Ensembl and RefSeq. It is envisaged that this development should be complete by mid 2010. It is a core goal for UniProt to provide meaningful annotation for these complete proteomes with a combination of our manual and automatic annotation protocols.

3.5. Using the UniProt Website

The UniProt consortium released its new improved unified website in 2009: a new interface, a new search engine, and many new options to serve its user community better. User feedback and the analysis of the use of our previous sites have led us to put more emphasis on supporting the most frequently used functionalities: database searches with simple (and sometimes less simple) queries that often consist of only a few terms have been enhanced by a good scoring system and a suggestion mechanism. Searching with ontology terms is assisted by auto-completion, and we also provide the possibility of using ontologies to browse search results. The viewing of database entries is improved with configurable views, a simplified terminology and a better integration of documentation. Medium-to-large sized result sets can now be retrieved directly on the site, so people no longer need to be referred to commercial, third party services. Access to the following most common bioinformatics tools have been simplified: sequence similarity searches, multiple sequence alignments, batch retrieval, and a database identifier mapping tool can now be launched directly from any page, and the output of these tools can be combined, filtered, and browsed like normal database searches. Programatic access to all data and results is possible via simple HTTP (REST) requests (http://www.uniprot.org/help/technical). In addition to the existing formats that support the different data sets (e.g., plain text, FASTA, and XML for UniProtKB), now it also provides (configurable) tab-delimited, RSS and GFF downloads where possible, and all data is available in RDF (http://www.w3.org/RDF/), a W3C standard for publishing data on the Semantic Web. Extensive documentation on how to best use this resource is available at: http://www.uniprot.org/help/.

References

1. The UniProt Consortium. (2010) The Universal Protein Resource (UniProt) in 2010. *Nucleic Acids Res.* **38**, D142–D148.
2. Leinonen R, Diez FG, Binns D, Fleischmann W, Lopez R, Apweiler R. (2004) UniProt archive. *Bioinformatics* **20**, 3236–3237.
3. Suzek BE, Huang H, McGarvey P, Mazumder R, Wu CH. (2007) UniRef: comprehensive and non-redundant UniProt reference clusters. *Bioinformatics* **23**, 1282–1288.
4. Kersey P, Hermjakob H, Apweiler R. (2000) VARSPLIC: alternatively-spliced protein

sequences derived from Swiss-Prot and TrEMBL. *Bioinformatics* **16**, 1048–1049.

5. Gattiker A, Michoud K, Rivoire C, Auchincloss AH, Coudert E, Lima T, Kersey P, Pagni M, Sigrist CJ, Lachaize C, et al. (2003) Automated annotation of microbial proteomes in SWISS-PROT. *Comput. Biol. Chem.* **27**, 49–58.
6. Fleischmann W, Moller S, Gateau A, Apweiler R. (1999) A novel method for automatic functional annotation of proteins. *Bioinformatics* **15**, 228–233.
7. Wu CH, Nikolskaya A, Huang H, Yeh L-S, Natale DA, Vinayaka CR, Hu ZZ, Mazumder R, Kumar S, Kourtesis P, et al. (2004) PIRSF: family classification system at the Protein Information Resource. *Nucleic Acids Res.* **32**, D112–D114.
8. Natale DA, Vinayaka CR, Wu CH. (2004) Large-scale, classification-driven, rule-based functional annotation of proteins. In: *Encyclopedia of Genetics, Genomics, Proteomics and Bioinformatics* – Subramaniam S, ed. Bioinformatics John Wiley West Sussex, England.
9. Swarbreck D, Wilks C, Lamesch P, Berardini TZ, Garcia-Hernandez M, Foerster H, Li D, Meyer T, Muller R, Ploetz L, et al. (2008) The Arabidopsis Information Resource (TAIR): gene structure and function annotation. *Nucleic Acids Res.* **36**, D1009–D1014.
10. Hong EL, Balakrishnan R, Dong Q, Christie KR, Park J, Binkley G, Costanzo MC, Dwight SS, Engel SR, Fisk DG, et al. (2008) The Ontology annotations at SGD: new data sources and annotation methods. *Nucleic Acids Res.* **36**, D577–D581.
11. Flicek P, Aken BL, Beal K, Ballester B, Caccamo M, Chen Y, Clarke L, Coates G, Cunningham F, Cutts T, et al. (2008) Ensembl 2008. *Nucleic Acids Res.* **36**, D707–D714.
12. The Gene Ontology Consortium. (2000) Gene ontology: tool for the unification of biology. *Nat. Genet.* **25**, 25–29.
13. Kersey PJ, Duarte J, Williams A, Karavidopoulou Y, Birney E, Apweiler R. (2004) The International Protein Index: an integrated database for proteomics experiments. *Proteomics* **4**, 1985–1988.

Chapter 3

InterPro Protein Classification

Jennifer McDowall and Sarah Hunter

Abstract

Improvements in nucleotide sequencing technology have resulted in an ever increasing number of nucleotide and protein sequences being deposited in databases. Unfortunately, the ability to manually classify and annotate these sequences cannot keep pace with their rapid generation, resulting in an increased bias toward unannotated sequence. Automatic annotation tools can help redress the balance. There are a number of different groups working to produce protein signatures that describe protein families, functional domains or conserved sites within related groups of proteins. Protein signature databases include CATH-Gene3D, HAMAP, PANTHER, Pfam, PIRSF, PRINTS, ProDom, PROSITE, SMART, SUPERFAMILY, and TIGRFAMs. Their approaches range from characterising small conserved motifs that can identify members of a family or subfamily, to the use of hidden Markov models that describe the conservation of residues over entire domains or whole proteins. To increase their value as protein classification tools, protein signatures from these 11 databases have been combined into one, powerful annotation tool: the InterPro database (http://www.ebi.ac.uk/interpro/) (Hunter et al., Nucleic Acids Res 37:D211–D215, 2009). InterPro is an open-source protein resource used for the automatic annotation of proteins, and is scalable to the analysis of entire new genomes through the use of a downloadable version of InterProScan, which can be incorporated into an existing local pipeline. InterPro provides structural information from PDB (Kouranov et al., Nucleic Acids Res 34:D302–D305, 2006), its classification in CATH (Cuff et al., Nucleic Acids Res 37:D310–D314, 2009) and SCOP (Andreeva et al., Nucleic Acids Res 36:D419–D425, 2008), as well as homology models from ModBase (Pieper et al., Nucleic Acids Res 37:D347–D354, 2009) and SwissModel (Kiefer et al., Nucleic Acids Res 37:D387–D392, 2009), allowing a direct comparison of the protein signatures with the available structural information. This chapter reviews the signature methods found in the InterPro database, and provides an overview of the InterPro resource itself.

Key words: Protein family, Domain, Signature, Functional classification, Homology, Hidden Markov model, Profile, Clustering, Regular expression

1. Introduction

With the increasing number of unannotated protein and nucleotide sequences populating databases, there is a pressing need to elucidate functional information that goes beyond the capabilities of

Cathy H. Wu and Chuming Chen (eds.), *Bioinformatics for Comparative Proteomics*, Methods in Molecular Biology, vol. 694, DOI 10.1007/978-1-60761-977-2_3, © Springer Science+Business Media, LLC 2011

experimental work alone. The first step in characterizing unannotated proteins is to identify other proteins with which they share high sequence identity. In this way, well-characterized proteins are associated with uncharacterized ones, permitting the transfer of annotation. Even if there are no characterized members in a group, the identification of clusters of proteins bearing strong sequence similarity provides good targets for functional or structural studies, as the information can potentially be transferred among the group.

Sequence similarity searches, such as BLAST (7) or FASTA (8), have traditionally been used for the automatic classification of sequences, but their ability to detect homologues depends on the search algorithm used, as well as on the database searched. Furthermore, because they treat each position in the query sequence with equal importance, they have a limited ability to detect divergent homologues. By contrast, protein signatures use multiple sequence alignments as part of the model-building process, which enable them to take into account the level of conservation at different positions. Specific residues in a family of proteins will be highly conserved if they are important for structure or function, whereas less important regions may have fewer constraints. Through their ability to match proteins that retain conservation in important regions, even when the overall percent similarity is low, protein signatures can detect more divergent homologues than simple sequence searches can. As such, protein signatures provide a "description" of a protein family or domain that defines its characteristics.

InterPro integrates protein signatures from 11 major signature databases (CATH-Gene3D, HAMAP, PANTHER, Pfam, PIRSF, PRINTS, ProDom, PROSITE, SMART, SUPERFAMILY, and TIGRFAMs) into a single resource, taking advantage of the different areas of specialization of each to produce a resource that provides protein classification on multiple levels: protein families, structural superfamilies and functionally close subfamilies, as well as functional domains, repeats and important sites. By linking related signatures together, InterPro places them in a hierarchical classification scheme, reflecting their underlying evolutionary relationships. Consequently, one can address issues such as the co-evolution of domains, or the functional divergence of proteins based on domain composition. InterPro is used for automatic annotation of the UniProtKB/TrEMBL (9) database using rules and automatic annotation algorithms, thereby permitting new sequences to be automatically assigned to a protein family and receive predicted functional annotation.

2. Protein Classification Tools

Protein signatures provide a means of protein classification, and are useful for comparing the evolutionary relatedness of groups of proteins. Several protein signature databases have emerged. These databases use different methods for creating signatures, including sequence clustering, regular expression, profiles and hidden Markov models (HMMs). A brief description of each follows.

2.1. Sequence Clustering

Sequence clustering is an automated method that clusters together proteins sharing regions of high sequence similarity.

PRODOM (10) is a signature database of domain families constructed automatically by clustering homologous regions. ProDom starts with the smallest sequence (representing a single domain) in UniProt Knowledgebase (UniProtKB) and, after removing fragments, searches for significant matches using PSI-Blast (11). These matches are removed from the database and used to create a domain family. The process is then repeated using the next smallest sequence. The ProDom database has a high coverage of protein space and is good at identifying new domain families in otherwise uncharacterised sequence.

2.2. Regular Expressions

A regular expression is a computer-readable formula for a pattern, which can be used to search for matching strings in a text. Regular expression can be used to describe short, conserved motifs of amino acids found within a protein sequence. These motifs can describe active sites, binding sites or other conserved sites.

PROSITE (12) is a signature database that includes regular expressions, or patterns. These signatures are built from multiple sequence alignments of known families, which are searched for conserved motifs important for the biological function of the family. The pattern of conservation within the motif is modelled as a regular expression. For example, the following hypothetical pattern, C-{P}-x(2)-[LI], can be translated as Cys-{any residue except Pro}-any residue-any residue-[Leu or Ile]. By focusing on small conserved regions, PROSITE patterns can detect divergent homologues, as long as their motifs are recognized by the regular expression. To reduce the number of false matches to these short motifs, PROSITE use mini-profiles to support pattern matches.

2.3. Profiles

A profile is a matrix of position-specific amino acid weights and gap costs, in which each position in the matrix provides a score of the likelihood of finding a particular amino acid at a specific position in the sequence (i.e., residue frequency distributions) (13). A similarity score is calculated between the profile and a matching sequence for a given alignment using the scores in the matrix.

PROSITE produces profiles in addition to its regular expressions. These profiles generally model domains and repeats, although there are a few full-length protein family profiles as well. They use multiple sequence alignments of UniProtKB/SwissProt sequences to build their profiles.

HAMAP (14) is a signature database of profiles describing well-characterized microbial protein families. Their models take the full length of the proteins into consideration when making a match, applying length and taxonomy filters. Their focus is the annotation of bacterial and archaeal genomes from sequencing projects, as well as plastid genomes.

PRINTS (15) is a database of protein fingerprints, which are sets of conserved motifs used to define protein families, subfamilies, domains, and repeats. The individual motifs are similar in length to PROSITE patterns, but they are modelled using profiles of small conserved regions within multiple sequence alignments, not using regular expressions. True hits should match all the motifs in a given PRINTS fingerprint. A match to part of a fingerprint represents a hit to a more divergent homologue. PRINTS use of different combinations of motifs to define family and sibling subfamily classifications that can be grouped into PRINTS hierarchies.

2.4. Hidden Markov Models

Hidden Markov models (HMMs) (16) are based on Bayesian statistical methods that make use of probabilities rather than the scores found in Profiles. However, like profiles, HMMs are able to model divergent as well as conserved regions within an alignment, and take into account both insertions and deletions. As a result, HMMs are good for detecting more divergent homologues, and can be used to model either discrete domains or full-length protein alignments. HMMs are based on multiple sequence alignments (seed alignments). The HMMER package was written by Sean Eddy (http://hmmer.janelia.org).

PFAM (17) is a signature database of domains, repeats, motifs and families based on HMMs. There are two components to Pfam: PfamA models are high quality and manually curated, whereas PfamB models are automatically generated using the ADDA database (18). Pfam also generates clans that group together PfamA models related by sequence or structure. The Pfam database has a very high coverage of sequence space and is used for the annotation of genomes.

SMART (Simple Modular Architecture Research Tool) (19) is a database of protein domains and repeats based on HMMs. SMART make use of manually curated multiple sequence alignments of well-characterized protein families, focusing on domains found in signaling, extracellular, and chromatin-associated proteins.

TIGRFAMs (20) is an HMM signature database of full-length proteins and domains at the superfamily, subfamily, and equivalog levels, where equivalogs are groups of homologous proteins conserved with respect to their predicted function. These models are grouped into TIGFAMs hierarchies.

PIRSF (21) is an HMM signature database based on protein families. PIRSF focus on modeling full-length proteins that share the same domain composition (homeomorphs) to annotate the biological functions of these families. Matching proteins must satisfy a length restriction to be included as a family member.

PANTHER (22) is an HMM signature database of protein families and subfamilies that together form PANTHER hierarchies. The subfamilies model the divergence of specific functions within protein families. Where possible, these subfamilies are associated with biological pathways.

SUPERFAMILY (23) is an HMM signature database of structural domains. SUPERFAMILY uses the SCOP (Structural Classification Of Proteins) database domain definitions at both the family and superfamily levels to provide structural annotation. SUPERFAMILY models each sequence in a family or superfamily independently, and then combine the results, which increases the sensitivity of homolog detection. SUPERFAMILY is useful for genome comparisons of the distribution of structural domains.

CATH-Gene3D (24) is an HMM signature database of structural domains. CATH-Gene3D uses the CATH database domain definitions at both the family and homologous superfamily levels, modeling each sequence independently and then combing the results.

3. InterPro Protein Classification Resource

The InterPro database integrates protein signatures from CATH-Gene3D, HAMAP, PANTHER, Pfam, PIRSF, PRINTS, ProDom, PROSITE, SMART, SUPERFAMILY, and TIGRFAMs into one integrated resource, thereby enabling researchers to use all the major protein signature databases at once in a single format, with the added benefit of manual quality checks (see Note 1). InterPro is a consortium produced by the members of the individual protein signature databases. By combining over 60,000 signatures from its member databases, InterPro achieves greater sequence and taxonomic coverage than any one database could achieve on its own. In addition, InterPro capitalizes on the specialization of each database to produce a comprehensive resource that can classify and annotate proteins on multiple levels, including superfamily/family/subfamily classifications, domain organization, and the identification of repeats and important functional and structural sites (see Note 2). InterPro adds in structural information, manual annotation and database links to provide a comprehensive classification database.

3.1. Searching InterPro

The InterPro database is freely available to the public at: http://www.ebi.ac.uk/interpro. InterPro can accommodate both single sequence queries and its incorporation into a local pipeline for multiple sequence annotation. There are three ways to query the InterPro database:

1. *Text Search.* Searches can be made using straight text, InterPro entry accessions, signature accessions, Gene Ontology (GO) terms, and UniProtKB accessions or names. Searches return a list of relevant InterPro entries or, in the case of UniProtKB accessions, a detailed graphical view of all signature matches to the protein, as well as structural information when available.
2. *InterProScan Sequence Search.* InterProScan (25) combines all the search engines from the member databases. There is a website version, as well as a standalone version that can be downloaded and installed locally (ftp://ftp.ebi.ac.uk/pub/databases/interpro/iprscan/). The website version only accepts a single protein sequence at a time, whereas the standalone version accepts multiple or bulk protein and nucleotide sequences, the later being translated into all six frames and ORF length filtered. The InterProScan results display signature matches but not structural information, as the latter relates to specific proteins, not to sequences.
3. *BioMart* (26). This search engine allows a user to construct complex queries to quickly retrieve custom InterPro datasets without having to download the complete InterPro database. To build a query, choose the *Dataset* to query, select your *Filters* (input data), and then select your *Attributes* (output). Results are obtained in a variety of formats, and can be linked to related data in both PRIDE (Proteomics identification database) (27) and Reactome (Knowledgebase of biological pathways) (28).

3.2. InterPro Entry Content

InterPro groups together all the protein signatures that cover the same sequence in the same set of proteins into a single entry, thereby removing any redundancy between the member databases. Each InterPro entry is manually curated and supplemented with additional information, including: entry type (family, domain, repeat or site), abstract and references, taxonomy of matching proteins, and GO term annotation (including biological process, molecular function and cellular component) (29) (see Note 3). Each entry provides a list of matching proteins, and users can view the complete signature matches for each of these proteins, which are listed under "Protein Matches" (see Note 4).

There are several links to external databases. Structural links are provided to the PDB (Protein Data Bank), and to CATH and

SCOP classification databases. Additional database links are provided to: IntAct database (30), IntEnz database (31), CAZy (Carbohydrate-Active enzymes) database (32), IUPHAR receptor database (33), COMe database (34), MEROPS database (35), PANDIT database (36), PDBeMotif database (37), CluSTr database (38), Pfam Clan database (17), and Genome Properties database (39).

3.3. InterPro Protein Matches

The protein signatures in InterPro are run against the entire UniProtKB (SwissProt and TrEMBL) database. Protein matches can be viewed in graphical or tabular formats. The graphical view of a given protein displays all matching signatures, with the area of the sequence match shown as a colored bar. By combining all signatures for a given protein in a single view, users can cross-compare annotation from all the different member databases. Links are provided to UniProtKB, Dasty (DAS display of protein annotated features) (40), SPICE (DAS display of PDB, UniProtKB sequence and Ensembl) (41), GO annotation, taxonomy, and BioMart. The link to BioMart provides a summary of the matches in tabular form, including the *E*-values for each signature, where appropriate (see Note 5).

The view also displays the structural features immediately below the signature matches to enable users to directly compare the signatures with known structural information. The structural features are displayed as striped bars, showing the region of sequence covered by structures in PDB, homology models in ModBase and SwissModel, as well as the division of the PDB structure into structural domains by CATH and SCOP. CATH and SCOP classify protein domains according to their structure, placing them within family and homologous superfamily hierarchies. Clicking on the Astex (42) icon in the protein graphical view will launch AstexViewer, which will display a 3-dimentional view of the PDB structure with the CATH or SCOP domain highlighted in yellow.

Splice variants are also displayed in the protein graphical view, which allows a user to easily visualize differences in domain organization. Splice variants are displayed below the master sequence and show variant accessions (e.g., O15151-2).

3.4. InterPro Relationships

InterPro links related entries together when their protein signatures overlap significantly in both sequence and protein matches. There are two types of relationships: Parent/Child and Contains/Found In.

Parent/Child relationships occur between InterPro entries that describe the same sequence region in hierarchical protein sets. They resemble the biological definitions of superfamily, family, and subfamily, but may contain many more variant subsets than can be described using these biological terms. Parent/Child relationships apply to individual sites, repeats, domains or full-length protein models. The "Child" entry is a subset of its "Parent" entry. These important hierarchical relationships arise from the different specializations of the member databases, designing their signatures according to structural, sequence or functional conservation, and are unique to InterPro.

Contains/Found In relationships describe the subdivision of a protein into domains, regions or sites. Contains/Found In relationships occur between entries in which one set of entry signatures completely contains the sequence covered by the other set of entry signatures. The signature matching the longer sequence "Contains" the signature(s) matching the shorter sequence(s).

4. Notes

1. Protein signatures have higher sensitivity (find more divergent homologues) and specificity (make fewer false matches) than sequence similarity tools such as BLAST or FASTA. However, there still needs to be a balance, as pushing for greater sensitivity will increase the number of false positives, whereas pushing for greater specificity will increase the number of false negatives. High specificity is an important criterion for incorporating a signature into InterPro, which provides manual quality checks on all signatures. However, it is inevitable that some false positives will be present in the databases.
2. The transfer of annotation must be done with caution. Homologous proteins share a similar biology, but not necessarily a common function, even when their % identity is high. Homologs found in different species (orthologs) may have the same function if all other components of the system are present. However, homologs found in the same organism (paralogs) are more likely to be functionally distinct because of genetic drift, even when % identity is very high.
3. Although annotation is correct at the time of integration, annotation can change over time as new experimental evidence is presented, and entries are only revisited periodically.
4. Protein signatures are predictive methods, therefore all functional annotation based on signature matches must be confirmed experimentally.
5. An effective criterion for assessing the significance of a sequence match is the expectation value (*E*-value). Although simple % identity scores are useful, they become inaccurate measures of homology at lower values. By contrast, an *E*-value is the number of hits expected to have a score equal or better

by chance alone. The lower the *E*-value, the more significant the score. However, *E*-values depend on the length and composition of the model and the size of the database being searched. Therefore, it is not advisable to compare *E*-values between member databases as an estimate of which model is better, because the members use different model building processes and different search procedures. *E*-values are only relevant within a specific member database.

References

1. Hunter S, Apweiler R, Attwood TK, Bairoch A, Bateman A, Binns D, Bork P, Das U, Daugherty L, Duquenne L, Finn RD, Gough J, Haft D, Hulo N, Kahn D, Kelly E, Laugraud A, Letunic I, Lonsdale D, Lopez R, Madera M, Maslen J, McAnulla C, McDowall J, Mistry J, Mitchell A, Mulder N, Natale D, Orengo C, Quinn AF, Selengut JD, Sigrist CJ, Thimma M, Thomas PD, Valentin F, Wilson D, Wu CH, Yeats C. (2009) InterPro: the integrative protein signature database. *Nucleic Acids Res.* **37**, D211–D215.
2. Kouranov A, Xie L, de la Cruz J, Chen L, Westbrook J, Bourne PE, Berman HM. (2006) The RCSB PDB information portal for structural genomics. *Nucleic Acids Res.* **34**, D302–D305.
3. Cuff AL, Sillitoe I, Lewis T, Redfern OC, Garratt R, Thornton J, Orengo CA. (2009) The CATH classification revisited--architectures reviewed and new ways to characterize structural divergence in superfamilies. *Nucleic Acids Res.* **37**, D310–D314.
4. Andreeva A, Howorth D, Chandonia JM, Brenner SE, Hubbard TJ, Chothia C, Murzin AG. (2008) Data growth and its impact on the SCOP database: new developments. *Nucleic Acids Res.* **36**, D419–D425.
5. Pieper U, Eswar N, Webb BM, Eramian D, Kelly L, Barkan DT, Carter H, Mankoo P, Karchin R, Marti-Renom MA, Davis FP, Sali A. (2009) MODBASE, a database of annotated comparative protein structure models and associated resources. *Nucleic Acids Res.* **37**, D347–D354.
6. Kiefer F, Arnold K, Künzli M, Bordoli L, Schwede T. (2009) The SWISS-MODEL Repository and associated resources. *Nucleic Acids Res.* **37**, D387–D392.
7. Altschul SF, Gish W, Miller W, Myers EW, Lipman DJ. (1990) Basic local alignment search tool. *J Mol Biol.* **215**, 403–410.
8. Pearson WR. (1990) Rapid and sensitive sequence comparison with FASTP and FASTA. *Methods Enzymol.* **183**, 63–98.
9. UniProt Consortium. (2009) The Universal Protein Resource (UniProt) 2009. *Nucleic Acids Res.* **37**, D169–D174.
10. Servant F, Bru C, Carrère S, Courcelle E, Gouzy J, Peyruc D, Kahn D. (2002) ProDom: automated clustering of homologous domains. *Brief Bioinform.* **3**(3), 246–251.
11. Altschul SF, Madden TL, Schäffer AA, Zhang J, Zhang Z, Miller W, Lipman DJ. (1997) Gapped BLAST and PSI-BLAST: a new generation of protein database search programs. *Nucleic Acids Res.* **25**(17), 3389–3402.
12. Sigrist CJA, Cerutti L, Hulo N, Gattiker A, Falquet L, Pagni M, Bairoch A, Bucher P. (2002) PROSITE: a documented database using patterns and profiles as motif descriptors. *Brief Bioinform.* **3**, 265–274.
13. Gribskov M, Lüthy R, Eisenberg D. (1990) Profile analysis. *Methods Enzymol.* **183**, 146–159.
14. Lima T, Auchincloss AH, Coudert E, Keller G, Michoud K, Rivoire C, Bulliard V, de Castro E, Lachaize C, Baratin D, Phan I, Bougueleret L, Bairoch A. (2009) HAMAP: a database of completely sequenced microbial proteome sets and manually curated microbial protein families in UniProtKB/Swiss-Prot. *Nucleic Acids Res.* **37**, D471–D478.
15. Attwood TK. (2002) The PRINTS database: a resource for identification of protein families. *Brief Bioinform.* **3**(3), 252–263.
16. Krogh A, Brown M, Mian IS, Sjölander K, Haussler D. (1994) Hidden Markov models in computational biology. Applications to protein modeling. *J Mol Biol.* **235**(5), 1501–1531.
17. Finn RD, Tate J, Mistry J, Coggill PC, Sammut SJ, Hotz HR, Ceric G, Forslund K, Eddy SR, Sonnhammer EL, Bateman A. (2008) The Pfam protein families database. *Nucleic Acids Res.* **36**, D281–D288.
18. Heger A, Wilton CA, Sivakumar A, Holm L. (2005) ADDA: a domain database with global coverage of the protein universe. *Nucleic Acids Res.* **33**, D188–D191.

19. Letunic I, Doerks T, Bork P. (2009) SMART 6: recent updates and new developments. *Nucleic Acids Res.* **37**, D229–D232.
20. Haft DH, Selengut JD, White O. (2003) The TIGRFAMs database of protein families. *Nucleic Acids Res.* **31**(1), 371–373.
21. Wu CH, Nikolskaya A, Huang H, Yeh LS, Natale DA, Vinayaka CR, Hu ZZ, Mazumder R, Kumar S, Kourtesis P, Ledley RS, Suzek BE, Arminski L, Chen Y, Zhang J, Cardenas JL, Chung S, Castro-Alvear J, Dinkov G, Barker WC. (2004) PIRSF: family classification system at the Protein Information Resource. *Nucleic Acids Res.* **32**, D112–D114.
22. Mi H, Lazareva-Ulitsky B, Loo R, Kejariwal A, Vandergriff J, Rabkin S, Guo N, Muruganujan A, Doremieux O, Campbell MJ, Kitano H, Thomas PD. (2005) The PANTHER database of protein families, subfamilies, functions and pathways. *Nucleic Acids Res.* **33**, D284–D288.
23. Wilson D, Pethica R, Zhou Y, Talbot C, Vogel C, Madera M, Chothia C, Gough J. (2009) SUPERFAMILY – sophisticated comparative genomics, data mining, visualization and phylogeny. *Nucleic Acids Res.* **37**, D380–D386.
24. Yeats C, Lees J, Reid A, Kellam P, Martin N, Liu X, Orengo C. (2008) Gene3D: comprehensive structural and functional annotation of genomes. *Nucleic Acids Res.* **36**, D414–D418.
25. Quevillon E, Silventoinen V, Pillai S, Harte N, Mulder N, Apweiler R, Lopez R. (2005) InterProScan: protein domains identifier. *Nucleic Acids Res.* **33**, W116–W120.
26. Haider S, Ballester B, Smedley D, Zhang J, Rice P, Kasprzyk A. (2009) BioMart Central Portal – unified access to biological data. *Nucleic Acids Res.* **37**, W23–W27.
27. Jones P, Côté RG, Cho SY, Klie S, Martens L, Quinn AF, Thorneycroft D, Hermjakob H. (2008) PRIDE: new developments and new datasets. *Nucleic Acids Res.* **36**, D878–D883.
28. Joshi-Tope G, Gillespie M, Vastrik I, D'Eustachio P, Schmidt E, de Bono B, Jassal B, Gopinath GR, Wu GR, Matthews L, Lewis S, Birney E, Stein L. (2005) Reactome: a knowledgebase of biological pathways. *Nucleic Acids Res.* **33**, D428–D432.
29. Reference Genome Group of the Gene Ontology Consortium. (2009) The Gene Ontology's Reference Genome Project: a unified framework for functional annotation across species. *PLoS Comput Biol.* **5**(7), e1000431.
30. Kerrien S, Alam-Faruque Y, Aranda B, Bancarz I, Bridge A, Derow C, Dimmer E, Feuermann M, Friedrichsen A, Huntley R, Kohler C, Khadake J, Leroy C, Liban A, Lieftink C, Montecchi-Palazzi L, Orchard S, Risse J, Robbe K, Roechert B, Thorneycroft D, Zhang Y, Apweiler R, Hermjakob H. (2007) IntAct – open source resource for molecular interaction data. *Nucleic Acids Res.* **35**, D561–D565.
31. Fleischmann A, Darsow M, Degtyarenko K, Fleischmann W, Boyce S, Axelsen KB, Bairoch A, Schomburg D, Tipton KF, Apweiler R. (2004) IntEnz, the integrated relational enzyme database. *Nucleic Acids Res.* **32**, D434–D437.
32. Cantarel BL, Coutinho PM, Rancurel C, Bernard T, Lombard V, Henrissat B. (2009) The Carbohydrate-Active EnZymes database (CAZy): an expert resource for Glycogenomics. *Nucleic Acids Res.* **37**, D233–D238.
33. Harmar AJ, Hills RA, Rosser EM, Jones M, Buneman OP, Dunbar DR, Greenhill SD, Hale VA, Sharman JL, Bonner TI, Catterall WA, Davenport AP, Delagrange P, Dollery CT, Foord SM, Gutman GA, Laudet V, Neubig RR, Ohlstein EH, Olsen RW, Peters J, Pin JP, Ruffolo RR, Searls DB, Wright MW, Spedding M. (2009) IUPHAR-DB: the IUPHAR database of G protein-coupled receptors and ion channels. *Nucleic Acids Res.* **37**, D680–D685.
34. Degtyarenko K, Contrino S. (2004) COMe: the ontology of bioinorganic proteins. *BMC Struct Biol.* **4**, 3.
35. Rawlings ND, Morton FR, Kok CY, Kong J, Barrett AJ. (2008) MEROPS: the peptidase database. *Nucleic Acids Res.* **36**, D320–D325.
36. Whelan S, de Bakker PI, Quevillon E, Rodriguez N, Goldman N. (2006) PANDIT: an evolution-centric database of protein and associated nucleotide domains with inferred trees. *Nucleic Acids Res.* **34**, D327–D331.
37. Golovin A, Henrick K. (2008) MSDmotif: exploring protein sites and motifs. *BMC Bioinformatics.* **9**, 312.
38. Petryszak R, Kretschmann E, Wieser D, Apweiler R. (2005) The predictive power of the CluSTr database. *Bioinformatics.* **21**(18), 3604–3609.
39. Haft DH, Selengut JD, Brinkac LM, Zafar N, White O. (2005) Genome Properties: a system for the investigation of prokaryotic

genetic content for microbiology, genome annotation and comparative genomics. *Bioinformatics.* **21**(3), 293–306.

40. Jimenez RC, Quinn AF, Garcia A, Labarga A, O'Neill K, Martinez F, Salazar GA, Hermjakob H. (2008) Dasty2, an Ajax protein DAS client. *Bioinformatics.* **21**(14), 3198–3199.

41. Prlić A, Down TA, Hubbard TJ. (2005) Adding some SPICE to DAS. *Bioinformatics.* **21**(Suppl 2), ii40–ii41.

42. Hartshorn MJ. (2002) AstexViewer: a visualisation aid for structure-based drug design. *J Comput Aided Mol Des.* **16**(12), 871–881.

Chapter 4

Reactome Knowledgebase of Human Biological Pathways and Processes

Peter D'Eustachio

Abstract

The Reactome Knowledgebase is an online, manually curated resource that provides an integrated view of the molecular details of human biological processes that range from metabolism to DNA replication and repair to signaling cascades. Its data model allows these diverse processes to be represented in a consistent way to facilitate usage as online text and as a resource for data mining, modeling, and analysis of large-scale expression data sets over the full range of human biological processes.

Key words: BioMart, Data aggregation, Gene ontology, High-throughput expression data, Pathway analysis

1. Introduction

In a living cell, molecules are synthesized, covalently modified, degraded, transported from one location to another and bound to one another to form complexes. The Reactome Knowledgebase aims to systematically describe the functions of human proteins in these terms with manually curated data from the published literature, to generate a consistently annotated knowledgebase of human biological processes useful as an online reference for individual processes and as a data mining and analysis resource for systems biology. The Reactome data model is reductionist: all of biology can be represented as events that convert input physical entities into output physical entities, located in subcellular compartments, mediated by the action of other physical entities acting as catalysts and positive or negative regulators (Fig. 1). Data are organized to facilitate the superposition of user-generated tissue- and state-specific expression data on the Reactome generic

Cathy H. Wu and Chuming Chen (eds.), *Bioinformatics for Comparative Proteomics*, Methods in Molecular Biology, vol. 694, DOI 10.1007/978-1-60761-977-2_4, © Springer Science+Business Media, LLC 2011

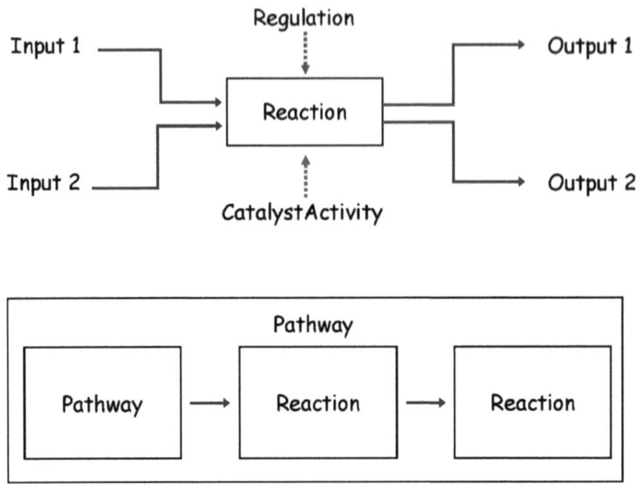

Fig. 1. Key elements of the Reactome data model. Physical entities (molecules and complexes) are transformed by single step reactions that can be ordered into multistep pathways.

parts list and to allow the export of the list in forms that enable model building and the integration of other data types. An introduction to our data model will provide a basis for describing methods for browsing and mining the Reactome Knowledgebase.

Reactome captures physical entities and their interactions in a frame-based data model. Classes (frames) describe concepts such as reaction, physical entity, subcellular location, and catalysis. Class attributes hold specific identifying information about the instances.

1.1. Events

Attributes of a reaction include its reactants (input), products (output), catalyst and subcellular location. Attributes of a catalyst instance are a physical entity and a Gene Ontology (GO) (1, 2) Molecular Function term that describes its activity. Instances of the regulation class link reactions to the factors that modulate them. The Reactome data model extends the concept of a biochemical reaction to include events such as the association of molecules to form a complex, the transport of a molecule between two cell compartments and the transduction of a signal.

A group of reactions can be organized into a goal-directed pathway. Attributes of a pathway instance are the names of the reactions and the smaller pathways it contains, as well as a GO Biological Process term. A pathway has no molecular attributes – these are inferred from the attributes of its included reactions. A single reaction may belong to one or more pathways.

1.2. Entities

Physical entities include proteins, nucleic acids, small molecules, and complexes of two or more molecules. Molecules are modified, moved from place to place, cleaved or take on different three-dimensional conformations. The Reactome data model captures this information in a computable format by treating each differently modified or located version of a molecule as a separate physical entity. The modification process itself is a reaction in which the input is the unmodified physical entity and the output is the modified one.

1.3. Subcellular Locations

The functions of biological molecules depend on their subcellular locations, so chemically identical entities located in different compartments are represented as distinct physical entities. Transport events are therefore ordinary reactions. Subcellular locations of molecules are annotated with terms from the GO Cellular Component Ontology.

1.4. Reference Entities

The annotation of alternative locations, posttranslational modifications and conformations of a molecule causes instances of a physical entity to proliferate. The basic chemical information that all forms share is stored in a separate class of reference physical entities, allowing information to be entered only once, reducing error, facilitating data maintenance and explicitly linking all the alternative forms of a single entity. The attributes of a reference entity include its name, reference chemical structure or sequence, and its accession numbers in reference databases: UniProt for proteins (3), ChEBI for small molecules (4) and EMBL for nucleic acids (5).

1.5. Complexes

Many biological reactions involve macromolecular complexes. The Reactome knowledgebase annotates these entities as instances of the complex class, the attributes of which are subcellular location and the identities of the complex's components (macromolecules, small molecules, and other complexes). Molecular assembly operations can then be described as reactions with complex components as input and the assembled complex as output. Complexes refer to all of the components they contain, so it is possible to fetch all complexes that involve a particular component or to dissect a complex to find its constituents.

1.6. Entity Sets

Often it is convenient to group physical entities based on common properties. For example, a transport protein at the plasma membrane may work equally well with any of an array of related small molecules. The Reactome data model allows the creation of entity sets, here one comprising the extracellular forms of the array of small molecules and another comprising their cytosolic forms. A single reaction is then annotated that converts the extracellular set into the cytosolic set. Sets are also used to describe protein paralogs that are functionally interchangeable.

Together, the entity, complex and set classes allow the detailed and flexible annotation and querying of physical entities and their interactions.

1.7. Evidence

Every reaction in the Reactome Knowledgebase is backed by direct or indirect evidence from the biomedical literature. Direct evidence for a human reaction comes from an assay on human cells described in a research publication whose PubMed identifier is stored as an attribute of the reaction. Much biomedical knowledge, however, derives from observations in experimentally tractable nonhuman systems that are thought to be good functional homologues of human ones. Such nonhuman data are used to document a human reaction in two steps. First, we annotate the reaction in the nonhuman species, using the physical entities of that organism – for example, the *Drosophila melanogaster* Notch protein –with appropriate literature reference attributes. Second, we annotate the human reaction, using human physical entities – for example, the four human Notch paralogs. The human reaction has no literature reference but instead has an attribute indicating its inference from the *Drosophila* reaction, and the complete chain of evidence is preserved from the primary experiment to the nonhuman reaction to the inferred human reaction.

1.8. Stable Identifiers

The Reactome data set is dynamic. New material is regularly added as new aspects of human biology are manually curated and existing material is revised and updated. To facilitate tracking of events (reactions, pathways and regulatory events) and physical entities (molecules and complexes) as they are revised and possibly merged or split, a stable identifier is assigned to each such instance so that it can be tracked between releases of the knowledgebase. Stable identifiers have the format REACT_XXX.YYY, where XXX is the identifier number assigned to the entity or event and YYY is the version number. Stable identifiers are assigned to entities and events when they are first released in the Reactome Database. When the annotations of a data instance are revised, the version number of the object's identifier is increased by one. The stable identifier of an event or entity is stored and displayed as one of its attributes, hyperlinked to a history page for the identifier.

2. Methods

2.1. Browsing the Reactome Knowledgebase

Access to the Reactome Knowledgebase is provided via its web site, http://www.reactome.org (Fig. 2). This web page is the starting point both for browsing the database for information on a specific topic and for launching data mining and pathway analysis operations.

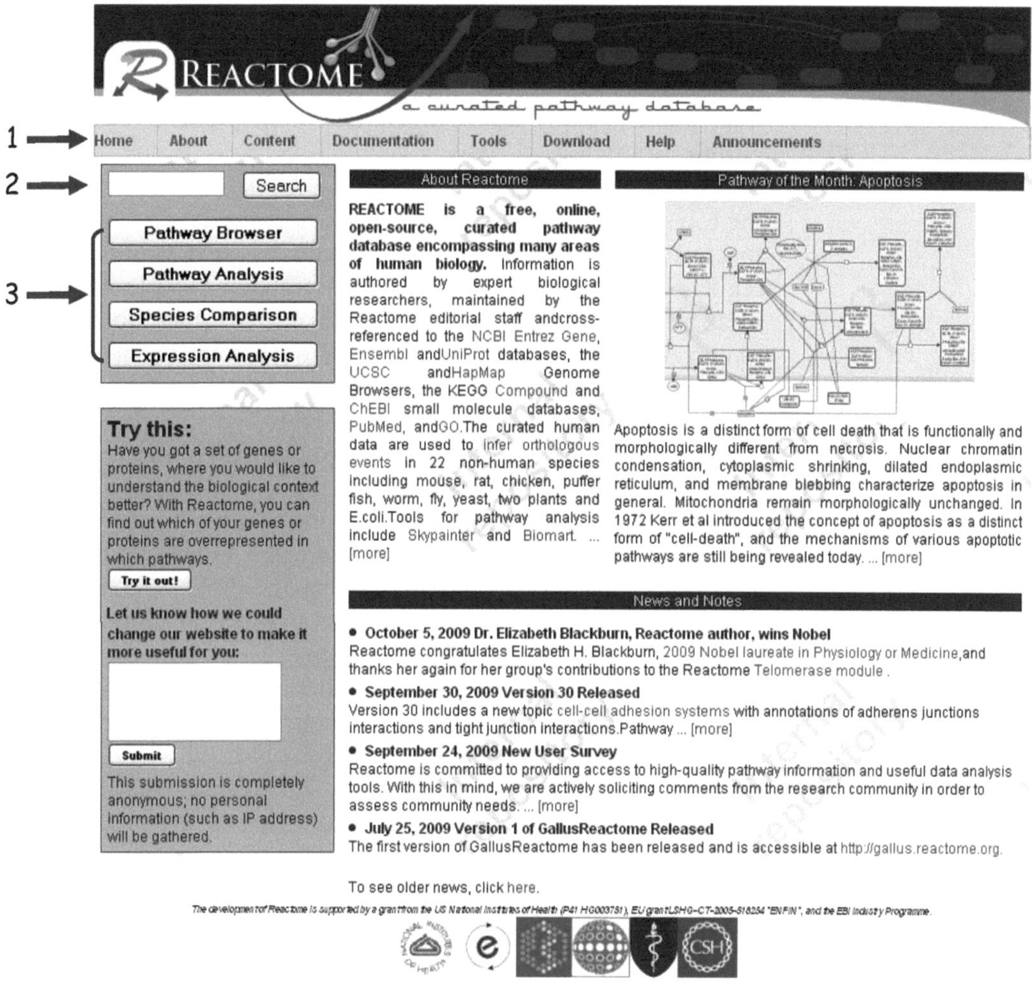

Fig. 2. The Reactome web site (www.reactome.org) home page. A menu bar at the top (1) and buttons on the left side (3) provide access to tools for browsing and analyzing Reactome data and for superimposing user-generated expression data on Reactome pathways. A simple search function for querying the database is also available (2).

A top menu bar (Fig. 2, label 1) provides access via its "Contents" tab to a detailed list of the pathways in Reactome and their digital object identifiers (DOIs), a specification of the data schema, and a calendar showing topics under development; via its "Documentation" tab to detailed descriptions of several aspects of the project and a users' guide, and via its "Download" tab to the complete Reactome software and the complete Reactome dataset in MySQL, BioPAX level 2, and BioPAX level 3 formats. Data for individual reactions and pathways can also be downloaded as described below.

Simple and advanced search tools function as an index of Reactome data. A simple search tool on the left side of the home page (Fig. 2, label 2) allows the user to enter a word or phrase and retrieve a list of all matching instances in the database.

For example, a search for a named protein will return lists of all modified and variously located forms of the protein, all complexes of which any form of the protein is a component, and all events (reactions and pathways) in which the protein participates as input, output, or catalyst.

An advanced search tool is accessible via the "Tools" tab in the top menu bar. This tool allows searches for database instances that match up to four attributes specified as text strings or Perl regular expressions. For example, a search for complexes whose location (compartment) is nucleoplasm and that have unmodified Cyclin A ("with the exact phrase *only*") as a component and that also have any form of Cdc2 ("with the exact phrase") as a component returns a list of 48 complexes, each hyperlinked to a web page that provides attributes of the complex and links to all events in which the complex participates.

The "Pathway Browser" button on the left side of the home page (Fig. 2, label 3) provides access to an individual pathway. Clicking on the button opens a web page displaying all of the pathways and reactions in the database arranged in a hierarchy on the left (Fig. 3a). Clicking on the name of a pathway causes the pathway to open, revealing its component sub-pathways and reactions. In the example shown, a reaction contained in a sub-pathway of apoptosis, "TNF: TNF-R1 binds TRADD, TRAF2 and RIP Complex," has been chosen. The reaction and the steps in the event hierarchy leading back to apoptosis are highlighted. In the right pane of the web page, the physical entities and reactions connecting them are displayed as nodes and edges with a standardized iconography. The center of the chosen reaction is highlighted by a red box. Placing the mouse over an entity node or a reaction edge causes its name to pop up. Standard map tools in the upper left hand corner of the pane enable panning and zooming. The pane can be expanded to fill the whole screen with the toggles on its edges (Fig. 3a, label 1). As most pathways are too big to be viewed in a single screen, a thumbnail view is provided in the lower left hand corner of the pane (Fig. 3a, label 2). Partly closing the bottom of the view pane reveals a panel in which the attributes of the chosen physical entity or event are shown (Fig. 3b). In the case of an event, these attributes include a brief free-text description of the reaction. External data sources are

Fig. 3. A Reactome event page. The *left-hand panel* displays the entire dataset as a hierarchically organized set of pathways. Choosing a reaction causes it to be displayed as entities (nodes) connected via reaction edges to other entities in the pathway in the *right panel* (**a**). The chosen reaction is highlighted (*red box*). The display panel can be opened or closed using toggles (1) and a thumbnail view of the whole pathway is provided (2). Closing the *bottom* of the *right panel* reveals a text description of the pathway (**b**). Links are provided to external data sources (3) and to Reactome pages for physical entities (4). Clicking on the name of a physical entity (**c**) causes the pathway display to re-format, highlighting all instances of the chosen entity. The text panel changes to display attributes of the entity.

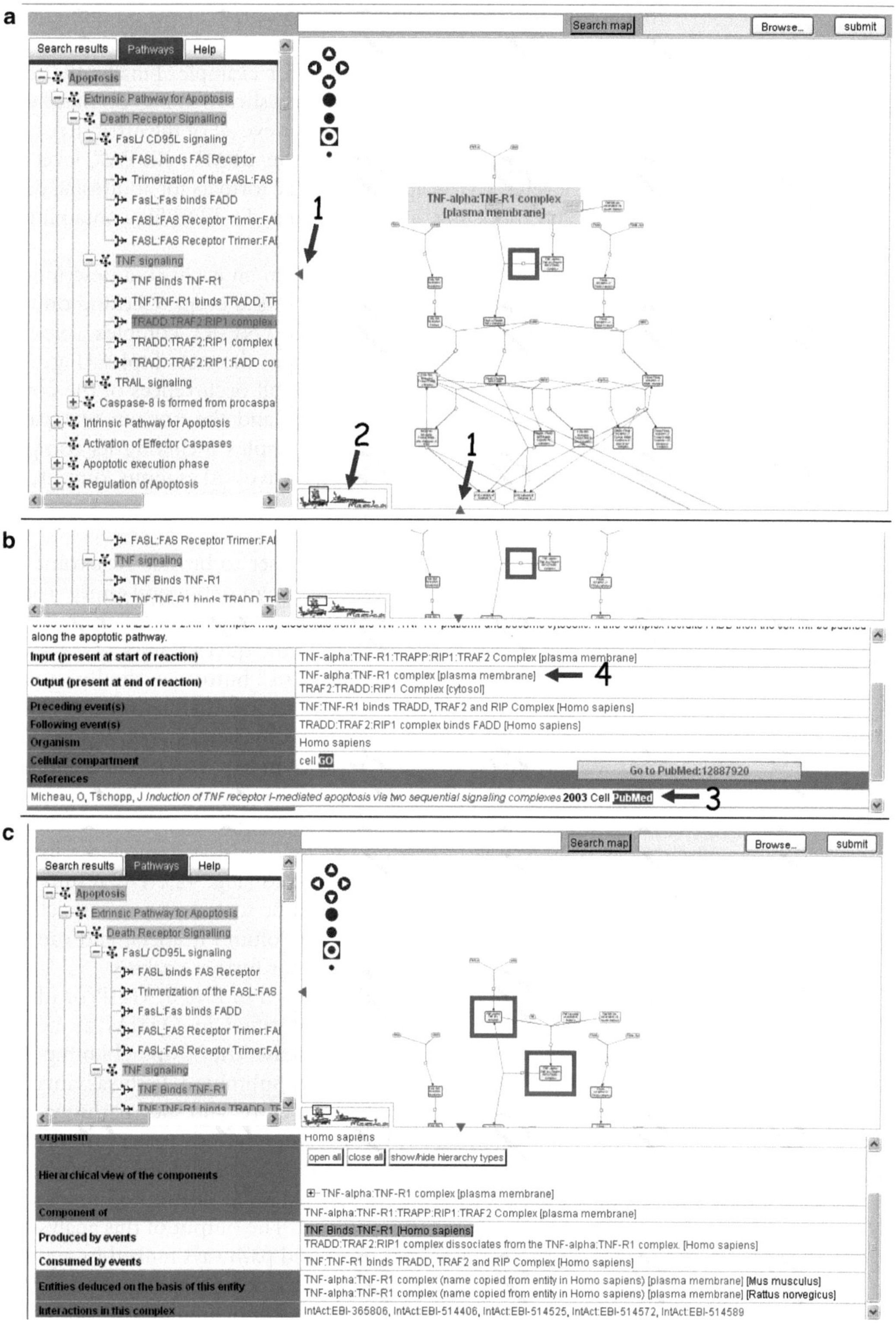
a
Search map
Browse...
submit
Search results
Pathways
Help
Apoptosis
Extrinsic Pathway for Apoptosis
Death Receptor Signalling
FasL/ CD95L signaling
FASL binds FAS Receptor
Trimerization of the FASL:FAS
FasL:Fas binds FADD
FASL:FAS Receptor Trimer:FA
FASL:FAS Receptor Trimer:FA
TNF signaling
TNF Binds TNF-R1
TNF:TNF-R1 binds TRADD, TR
TRADD:TRAF2:RIP1 complex
TRADD:TRAF2:RIP1 complex
TRADD:TRAF2:RIP1:FADD co
TRAIL signaling
Caspase-8 is formed from procaspa
Intrinsic Pathway for Apoptosis
Activation of Effector Caspases
Apoptotic execution phase
Regulation of Apoptosis
TNF-alpha:TNF-R1 complex [plasma membrane]
1
2
1
b
FASL:FAS Receptor Trimer:FA
TNF signaling
TNF Binds TNF-R1
along the apoptotic pathway.
Input (present at start of reaction)
TNF-alpha:TNF-R1:TRAPP:RIP1:TRAF2 Complex [plasma membrane]
Output (present at end of reaction)
TNF-alpha:TNF-R1 complex [plasma membrane]
TRAF2:TRADD:RIP1 Complex [cytosol]
4
Preceding event(s)
TNF:TNF-R1 binds TRADD, TRAF2 and RIP Complex [Homo sapiens]
Following event(s)
TRADD:TRAF2:RIP1 complex binds FADD [Homo sapiens]
Organism
Homo sapiens
Cellular compartment
cell GO
References
Go to PubMed:12887920
Micheau, O, Tschopp, J Induction of TNF receptor I-mediated apoptosis via two sequential signaling complexes 2003 Cell PubMed
3
c
Search map
Browse...
submit
Search results
Pathways
Help
Apoptosis
Extrinsic Pathway for Apoptosis
Death Receptor Signalling
FasL/ CD95L signaling
FASL binds FAS Receptor
Trimerization of the FASL:FAS
FasL:Fas binds FADD
FASL:FAS Receptor Trimer:FA
FASL:FAS Receptor Trimer:FA
TNF signaling
TNF Binds TNF-R1
Homo sapiens
Hierarchical view of the components
open all
close all
show/hide hierarchy types
TNF-alpha:TNF-R1 complex [plasma membrane]
Component of
TNF-alpha:TNF-R1:TRAPP:RIP1:TRAF2 Complex [plasma membrane]
Produced by events
TNF Binds TNF-R1 [Homo sapiens]
TRADD:TRAF2:RIP1 complex dissociates from the TNF-alpha:TNF-R1 complex. [Homo sapiens]
Consumed by events
TNF:TNF-R1 binds TRADD, TRAF2 and RIP Complex [Homo sapiens]
Entities deduced on the basis of this entity
TNF-alpha:TNF-R1 complex (name copied from entity in Homo sapiens) [plasma membrane] [Mus musculus]
TNF-alpha:TNF-R1 complex (name copied from entity in Homo sapiens) [plasma membrane] [Rattus norvegicus]
Interactions in this complex
IntAct:EBI-365806, IntAct:EBI-514406, IntAct:EBI-514525, IntAct:EBI-514572, IntAct:EBI-514589

hyperlinked to attribute values as appropriate: label 3 highlights a link to the PubMed entry for the literature reference on the basis of which the event was annotated, for example. Finally, the text panel for an event includes a menu not shown in Fig. 3b of download options that enable a user to retrieve all of the attributes of a single event (pathway or reaction) in SBML, BioPAX level 2, BioPAX level 3, Cytoscape, or Protégé formats for additional data analysis and modeling, and to download a PDF file containing a narrative text description of the event.

A user can browse seamlessly from an event to a description of a physical entity that participates in the event. Clicking on the participating complex "TNF-alpha: TNF-R1 complex (plasma membrane)" (Fig. 3b, label 4) changes the web display (Fig. 3c) so that in the pathway diagram pane all occurrences of the complex are highlighted with red boxes and the bottom text pane shows all of the attributes of the complex including its components and the reactions in which it is involved as input, output, or catalyst.

2.2. Pathway Analysis with the Reactome Knowledgebase

The tools described so far enable a user to browse Reactome to retrieve information about a specific event or physical entity. Tools are also provided to allow the data set to be queried more systematically and to search for patterns in the query results.

Clicking on the "Pathway Analysis" button on the Reactome home page (Fig. 2, label 3) opens a form that allows entry of a user-specified list of proteins, which can then be analyzed in various ways. A simple example is shown in Fig. 4. A user has entered the UniProt identifiers for six human proteins (the form also accepts EntrezGene, Ensembl, and Affymetrix protein identifiers; work is underway to expand this list) and has requested a list of the pathways in which each is involved (Fig. 4a). The results are returned as an HTML table that can be sorted on a user-specified column by selecting toggles in the column headers, and can be downloaded as a delimited text file for further analysis.

A more complex analysis is shown in Fig. 5. UniProt identifiers for all human proteins identified in OMIM as being associated with human genetic disease (ftp://ftp.ncbi.nih.gov/repository/OMIM/morbidmap) are submitted. Pathways involving these proteins are identified as before, and such pathways are ranked to highlight ones in which a significantly larger proportion of proteins are known to OMIM than would be expected by chance if proteins known to OMIM were distributed randomly over pathways curated in Reactome. The output of this analysis is displayed as a list of over-represented pathways ranked by significance of the protein – pathway association (Fig. 5a), as a HTML table that can also be downloaded as tab-delimited text listing the proteins associated with each pathway (Fig. 5b), and as an HTML list for each submitted protein of all of the reactions in which it is

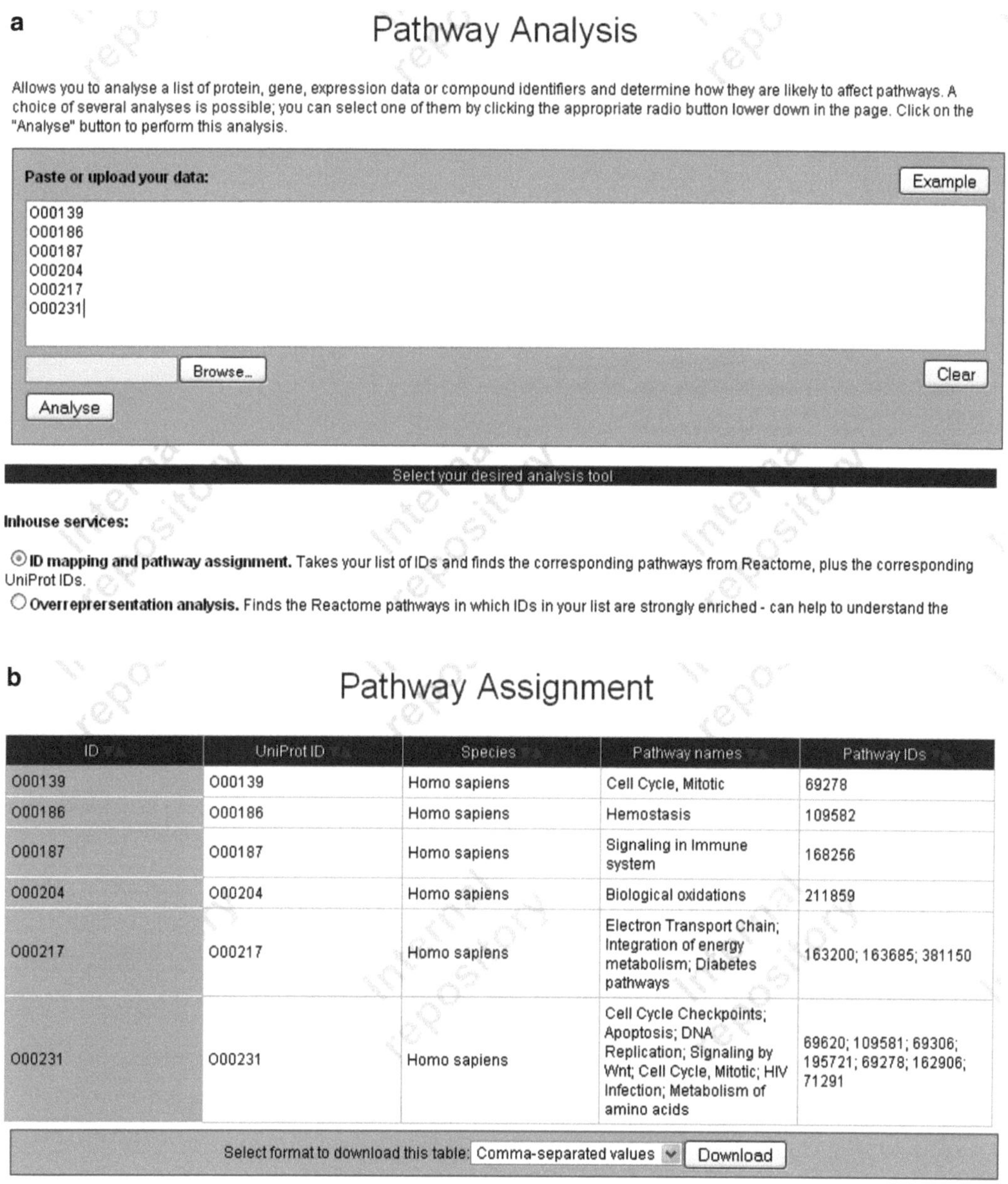

Fig. 4. Pathway analysis: a simple query to retrieve all reactions in which a user-specified list of proteins participates. The query set-up is shown in (**a**) and the results in (**b**).

involved (Fig. 5c). This approach provides a powerful means of searching for key perturbed biological processes in data sets such as those generated in high-throughput screens to identify proteins differentially expressed in response to a stress, or whose expression is altered in comparisons of a malignant tumor and the normal tissue from which it derives.

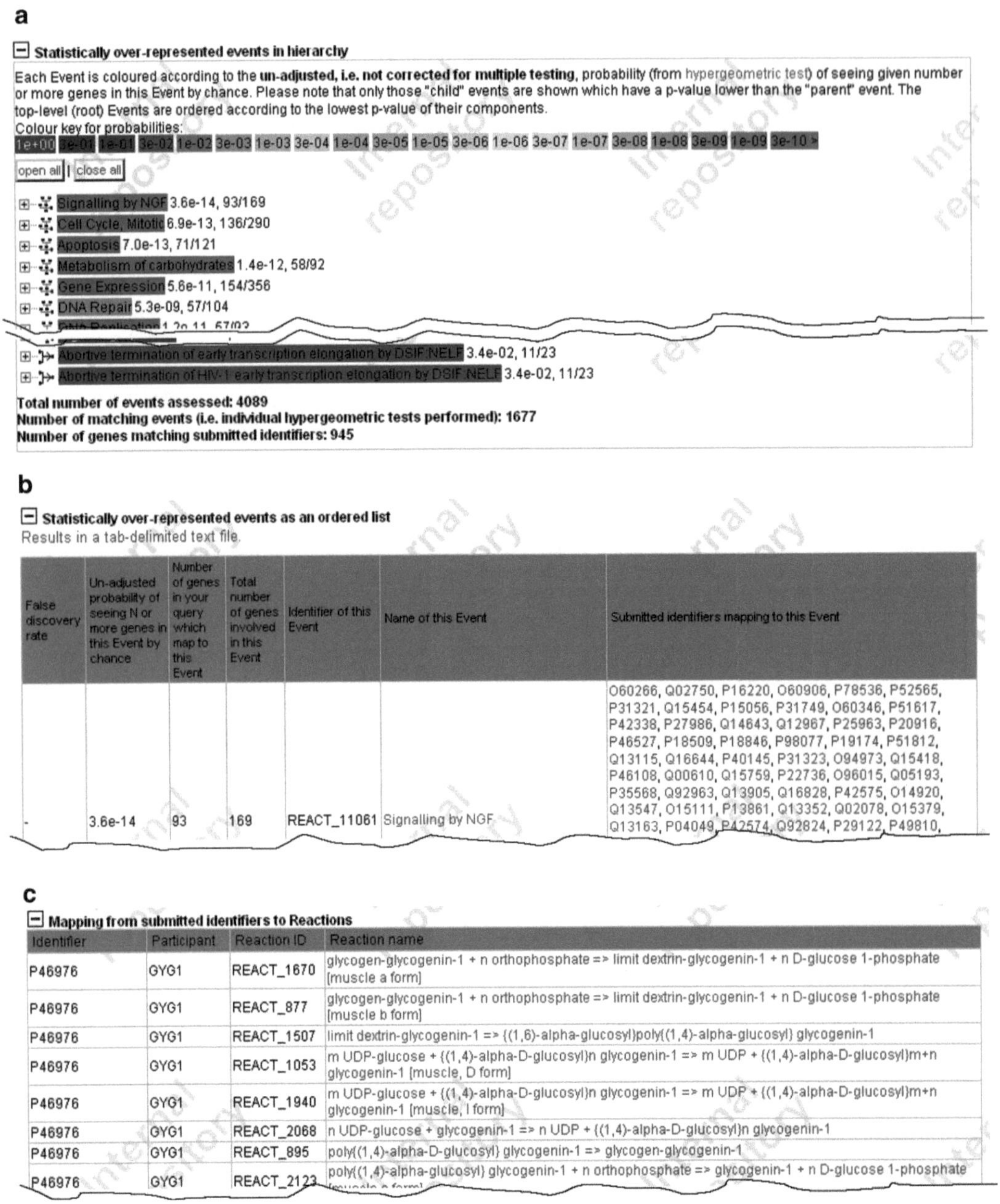

Fig. 5. Pathway analysis: results of a query to identify specific pathways within the entire Reactome dataset that involves significantly more of the proteins on a user-specified list than would be expected if proteins were distributed randomly over reactions. (**a**) A ranked list of over-represented pathways; (**b**) data for over-represented pathways in tabular form; (**c**) a list of reactions for each user-specified protein.

2.3. Electronic Inference of Nonhuman Reactions and Pathways

The Reactome Knowledgebase includes computationally inferred pathways and reactions in 20 nonhuman species, including *Mus musculus*, *Tetraodon nigroviridis*, *Drosophila melanogaster*, *Caenorhabditis elegans*, *Saccharomyces cerevisiae*, *Arabidopsis thaliana*, *Plasmodium falciparum*, and *Escherichia coli*. These species represent more than four billion years of evolution and span the main branches of life. We project the set of curated human reactions onto the genome of another species using OrthoMCL (http://orthomcl.cbil.upenn.edu/cgi-bin/OrthoMclWeb.cgi) protein similarity clusters (6). For each species, each human reaction is checked to determine whether all of its protein participants (inputs, outputs and catalysts) have at least one ortholog or recent paralog (OP) in the other species. In the case of protein complexes, we relax this requirement so that a complex is considered to be present in the other species if at least 75% of its protein components are present as OPs. For each reaction that meets these criteria, we create an equivalent reaction for the other species by replacing all human protein components with their corresponding OPs. For proteins with more than one OP in the other species, we create a "defined set" named "Homologues of..." that contains these OPs, and use this set as the corresponding component of the equivalent reaction. To order the inferred reactions, pathway instances are created corresponding to the ones that contain the human reactions from which the inferences were made. Inferred pathways may thus have large gaps in them relative to the homologous curated human ones.

The text panel of the web display for each human pathway and reaction includes a list of all such successfully inferred reactions, allowing access to specific inferred nonhuman reactions. To obtain all of the inferences for a species, click the "Species Comparison" button on the Reactome home page (Fig. 2, label 3). The new page that opens allows the user to choose a species from a drop-down menu and to retrieve a list of all pathways successfully inferred in that species as an HTML table (Fig. 6). Toggle buttons in the headers for each column allow the rows to be sorted. In the example shown, pathways inferred for chicken (*Gallus gallus*) are ranked according to the fraction of the human reactions in each pathway that was successfully projected on chicken. Each row contains a hyperlink to a web page for that pathway in the species chosen. Data can also be exported as delimited text.

2.4. BioMart: Using the Reactome Knowledgebase for Data Aggregation

The Reactome Knowledgebase can also be used for data mining and large-scale analysis of gene functions. Reactome Mart, accessed via the tools item on the top menu bar of the home page (Fig. 2, label 1) uses the BioMart query-oriented data-management system (7) to generate integrated queries across Reactome and other databases, including UniProt and Ensembl (http://www.ensembl.org/index.html). Several preformatted (canned) queries are available in the menu bar of the Reactome Mart tool. Fig. 7 shows the results of a query to identify all complexes of which several proteins, identified by their UniProt identifiers, are components and all other components of those complexes.

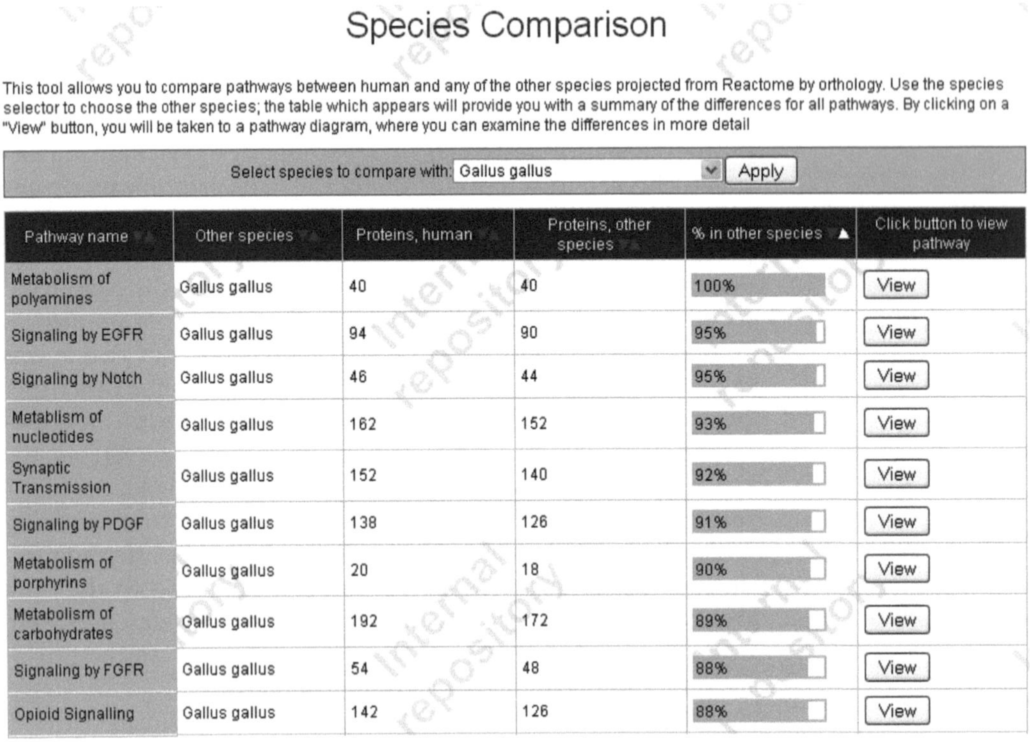

Pathway name	Other species	Proteins, human	Proteins, other species	% in other species	Click button to view pathway
Metabolism of polyamines	Gallus gallus	40	40	100%	View
Signaling by EGFR	Gallus gallus	94	90	95%	View
Signaling by Notch	Gallus gallus	46	44	95%	View
Metablism of nucleotides	Gallus gallus	162	152	93%	View
Synaptic Transmission	Gallus gallus	152	140	92%	View
Signaling by PDGF	Gallus gallus	138	126	91%	View
Metabolism of porphyrins	Gallus gallus	20	18	90%	View
Metabolism of carbohydrates	Gallus gallus	192	172	89%	View
Signaling by FGFR	Gallus gallus	54	48	88%	View
Opioid Signalling	Gallus gallus	142	126	88%	View

Fig. 6. An electronically inferred pathway dataset for a model organism (the chicken, *Gallus gallus*) generated from the human curated dataset. For each human pathway, the number of proteins involved in its human version is shown, together with the number of orthologous chicken proteins that could be identified and placed in a reaction context.

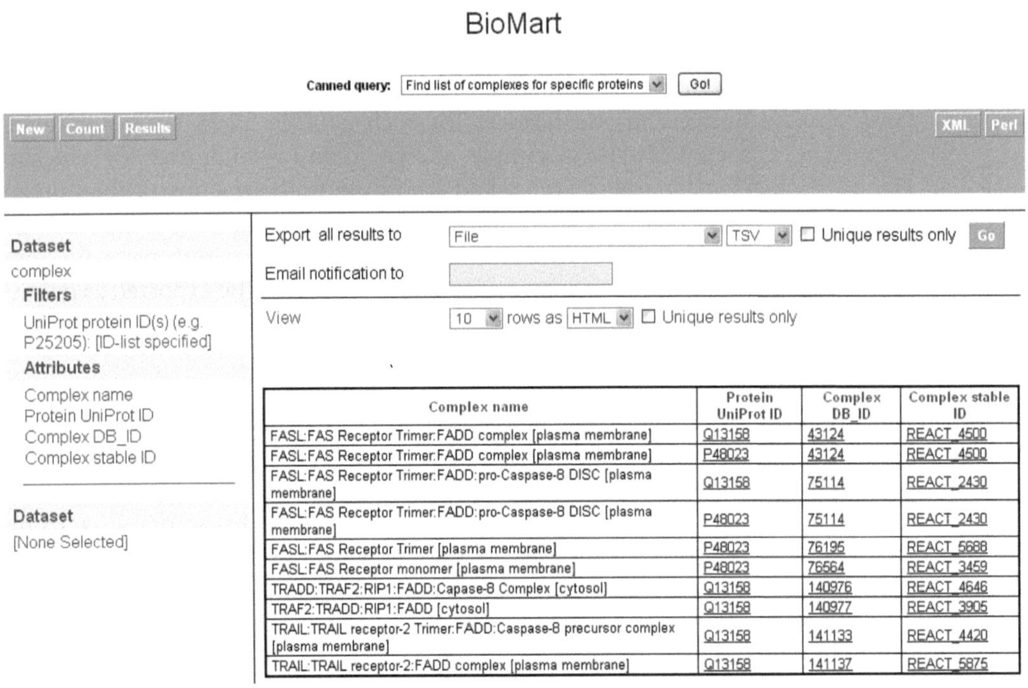

Complex name	Protein UniProt ID	Complex DB_ID	Complex stable ID
FASL:FAS Receptor Trimer:FADD complex [plasma membrane]	Q13158	43124	REACT_4500
FASL:FAS Receptor Trimer:FADD complex [plasma membrane]	P48023	43124	REACT_4500
FASL:FAS Receptor Trimer:FADD:pro-Caspase-8 DISC [plasma membrane]	Q13158	75114	REACT_2430
FASL:FAS Receptor Trimer:FADD:pro-Caspase-8 DISC [plasma membrane]	P48023	75114	REACT_2430
FASL:FAS Receptor Trimer [plasma membrane]	P48023	76195	REACT_5688
FASL:FAS Receptor monomer [plasma membrane]	P48023	76564	REACT_3459
TRADD:TRAF2:RIP1:FADD:Capase-8 Complex [cytosol]	Q13158	140976	REACT_4646
TRAF2:TRADD:RIP1:FADD [cytosol]	Q13158	140977	REACT_3905
TRAIL:TRAIL receptor-2 Trimer:FADD:Caspase-8 precursor complex [plasma membrane]	Q13158	141133	REACT_4420
TRAIL:TRAIL receptor-2:FADD complex [plasma membrane]	Q13158	141137	REACT_5875

Fig. 7. Results of a BioMart query to find all complexes that contain one or more of the proteins in a user-specified list.

Users can also define their own queries with the menus that are accessible by means of the highlighted terms and the boxes on the mart page. For example, a coupled search across the Reactome and Ensembl databases will retrieve a list of orthologs of the human proteins that are involved in a pathway, or will identify the Affymetrix IDs that are associated with genes in the selected Reactome pathways.

Acknowledgments

The data, data model, and data analysis tools described in this chapter are the product of the collaborative work of curators and software developers at the Ontario Institute for Cancer Research (Lincoln Stein, Michael Caudy, Marc Gillespie, Robin Haw, Bruce May, Guanming Wu), the European Bioinformatics Institute (Ewan Birney, Henning Hermjakob, David Croft, Phani Garapati, Bijay Jassal, Steven Jupe, Gavin O'Kelly, Esther Schmidt) and the NYU School of Medicine (Peter D'Eustachio, Shahana Mahajan, Lisa Matthews). We are grateful to the many scientists who collaborated with us as authors and reviewers to build the content of the knowledgebase. Grants from the National Human Genome Research Institute/NIH (P41 HG003751) and the European Union 6th Framework Programme (LSHG-CT-2005-518254 "ENFIN" support development of the Knowledgebase.

References

1. Ashburner, M., Ball, C. A., Blake, J. A., Botstein, D., Butler, H., Cherry. J. M., et al. (2000) Gene ontology: tool for the unification of biology. *Nat. Genet.* **25**, 25–29.
2. The Gene Ontology Consortium. (2010) The Gene Ontology in 2010: extensions and refinements. *Nucleic Acids Res.* **38**, D331–D335.
3. The UniProt Consortium. (2010) The Universal Protein Resource (UniProt) in 2010. *Nucleic Acids Res.* **38**, D142–D148
4. de Matos, P., Alcántara, R., Dekker, A., Ennis, M., Hastings, J., Haug, K., et al. (2010) Chemical Entities of Biological Interest: an update. *Nucleic Acids Res.* **38**, D249–D254.
5. Leinonen, R., Akhtar, R., Birney, E., Bonfield, J., Bower, L., Corbett, M., et al. (2010) Improvements to services at the European Nucleotide Archive. *Nucleic Acids Res.* **38**, D39–D45.
6. Li, L., Stoeckert, C. J. Jr., and Roos, D. S. (2003) OrthoMCL: identification of ortholog groups for eukaryotic genomes. *Genome Res.* **13**, 2178–2189.
7. Durinck, S., Moreau, Y., Kasprzyk, A., Davis, S., De Moor, B., Brazma, A., Huber, W. (2005) BioMart and Bioconductor: a powerful link between biological databases and microarray data analysis. *Bioinformatics* **21**, 3439–3940.

Chapter 5

eFIP: A Tool for Mining Functional Impact of Phosphorylation from Literature

Cecilia N. Arighi, Amy Y. Siu, Catalina O. Tudor, Jules A. Nchoutmboube, Cathy H. Wu, and Vijay K. Shanker

Abstract

Technologies and experimental strategies have improved dramatically in the field of genomics and proteomics facilitating analysis of cellular and biochemical processes, as well as of proteins networks. Based on numerous such analyses, there has been a significant increase of publications in life sciences and biomedicine. In this respect, knowledge bases are struggling to cope with the literature volume and they may not be able to capture in detail certain aspects of proteins and genes. One important aspect of proteins is their phosphorylated states and their implication in protein function and protein interacting networks. For this reason, we developed eFIP, a web-based tool, which aids scientists to find quickly abstracts mentioning phosphorylation of a given protein (including site and kinase), coupled with mentions of interactions and functional aspects of the protein. eFIP combines information provided by applications such as eGRAB, RLIMS-P, eGIFT and AIIAGMT, to rank abstracts mentioning phosphorylation, and to display the results in a highlighted and tabular format for a quick inspection. In this chapter, we present a case study of results returned by eFIP for the protein BAD, which is a key regulator of apoptosis that is posttranslationally modified by phosphorylation.

Key words: Text mining, BioNLP, Information extraction, Phosphorylation, Protein–protein interaction, PPI, Knowledge discovery

1. Introduction

There has been a general shift in paradigm from dedicating a lifetime's work to analyzing of a single protein to the analysis of cellular and biochemical processes and networks. This has been made possible by a dramatic improvement in technologies and experimental strategies in the fields of genomics and proteomics (1). Although bioinformatics tools have greatly assisted in data analysis, both protein identification and functional interpretation

Cathy H. Wu and Chuming Chen (eds.), *Bioinformatics for Comparative Proteomics*, Methods in Molecular Biology, vol. 694, DOI 10.1007/978-1-60761-977-2_5, © Springer Science+Business Media, LLC 2011

are still major bottlenecks (2). In this regard, public knowledge bases constitute a valuable source of such information, but the manual curation of experimentally determined biological events is slow compared to the rapid increase in the body of knowledge represented in the literature. Hence, literature still continues to be a primary source of biological data. Nevertheless, manually finding the relevant articles is not a trivial task, with issues ranging from the ambiguity of some names to the identification of those articles that contain the specific information of interest.

Fortunately, the text mining community has recognized in recent years the opportunities and challenges of natural language processing (NLP) in the biomedical field (3), and has developed a number of resources for providing access to information contained in life sciences and biomedical literature. Table 1 lists a sampling of freely-available tools that address the various BioNLP applications. In addition, there are a large number of papers discussing research and techniques for these applications. For an in-depth overview of these topics, please refer to review articles by Krallinger et al. (4) and Jensen et al. (5).

However, BioNLP tools are only useful if they are designed to meet real-life tasks (4). In fact, this has been one of the obstacles for the general adoption of BioNLP tools by biologists, because many of these applications perform individual tasks (like gene/protein mention, phosphorylation, or protein–protein interaction), thus providing only one piece of information, which in itself might not be enough to describe the biology. To address this issue, we have designed eFIP (*e*xtraction of *F*unctional *I*mpact of *P*hosphorylation), a system that combines several publicly available tools to allow identification of abstracts that contain protein phosphorylation mentions (including the site and the kinase), coupled with mentions of functional implications (such

Table 1
Biological applications and a sampling of available resources

Biological applications	Resources
Protein–protein interaction	iHOP, Chilibot, KinasePathway, PPI Finder, Protein Corral
Gene name recognition/mention/tagger	ABNER, AIIAGMT, ABGene, BANNER, BIGNER, GAPSCORE, KEX, LingPipe, SciMiner
Acronym expansion and disambiguation	Acromine, AcroTagger, ADAM, ALICE, ARGH, Biomedical Abbreviation
Protein sequence	Mutation Finder, MeInfoText, mSTRAP, MutationFinder, PepBank, RLIMS-P
Text-mining search aids	Anne O'Tate, e-LiSe, FABLE, GoPubMed, MedEvi, NextBio

as protein–protein interaction, function, process, localization, and disease). In addition, eFIP ranks these abstracts and presents the information in a user-friendly format for a quick inspection.

The rationale for performing this particular task relies on at least three aspects:

1. Phosphorylation is one of the most common protein post-translational modifications (PTMs). Phosphorylation of specific intracellular proteins/enzymes by protein kinases and dephosphorylation by phosphatases provides information of both activation and deactivation of critical cellular pathways, including regulatory mechanisms of metabolism, cell division, cell growth and differentiation (6).
2. Often protein phosphorylation has some functional impact. Proteins can be phosphorylated on different residues, leading to activation or down-regulation of their activity, alternative subcellular location, and binding partners. One such example is protein Smad2, whose phosphorylation state determines its interaction partners, its subcellular location, and its cofactor activity (7).
3. Currently, protein–protein interaction (PPI) data involving phosphorylated proteins is not yet well represented in the public databases. Thus, extracting this information is critical to the interpretation of PPI and prediction of the functional outcomes.

1.1. Goal of This Chapter

As mentioned before, interesting and important real-life tasks would require the combination of multiple individual tasks. A major focus of this chapter is to highlight how the combination of existing BioNLP tools can reveal some interesting biology about a protein. The specific goal is to describe eFIP, a tool that can assist a researcher in finding information in the literature about protein phosphorylation mentions that have some biological implication, such as PPI, localization, function, and disease.

1.2. The Approach

The BioNLP tasks behind eFIP include (1) document retrieval – selection of relevant scientific publications, and gene name disambiguation (eGRAB); (2) text mining – detection of functional terms (eGIFT); (3) information extraction – identification of substrate, phosphorylation sites, and kinase (RLIMSP); (4) protein–protein interaction identification (PPI module) and gene name recognition (AIIAGMT); and (5) document and sentence ranking – integration of text mining results with ranking and summarization (eFIP's ranking module) (Fig. 1).

For details regarding each individual tool mentioned here, please refer to Subheading 2. In Subheading 3, we will provide the user with a protocol to find relevant articles using the protein BAD as an example.

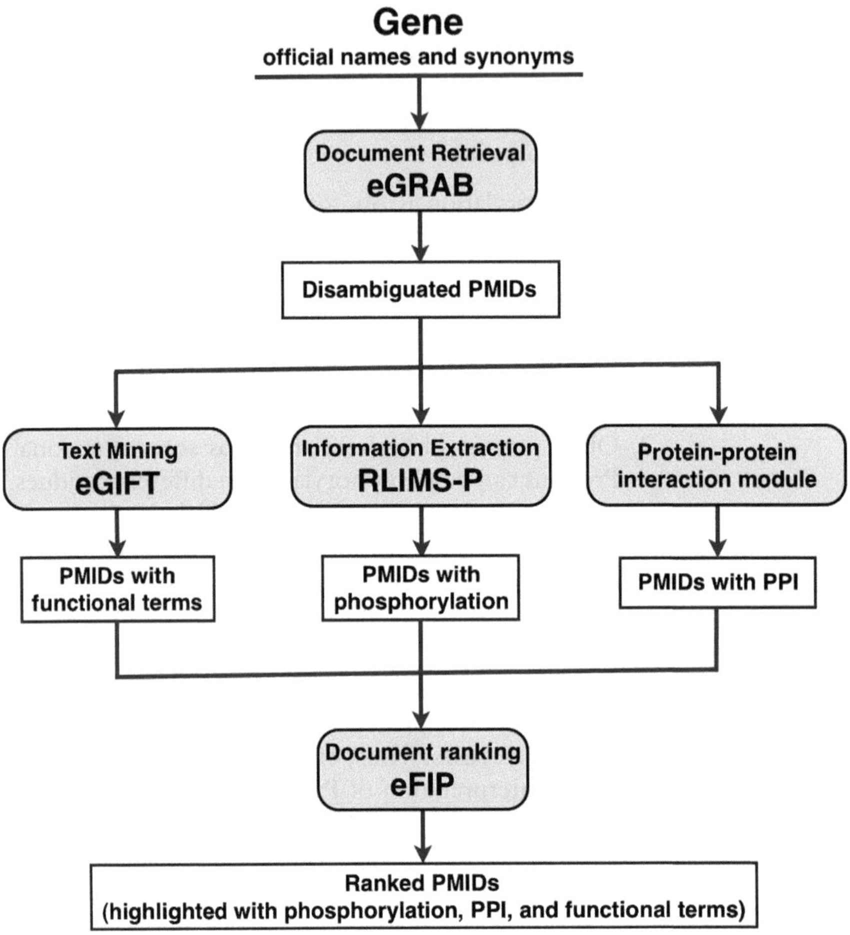

Fig. 1. General pipeline of BioNLP tasks, including specific tools used in our approach. The protein–protein interaction module includes the gene name recognition tool (AIIAGMT).

2. Materials

In this section, we briefly describe the tools depicted in Fig. 1.

2.1. Extractor of Gene-Related Abstracts

Extractor of Gene-Related ABstracts (eGRAB) is used to gather the literature for a given gene/protein. To retrieve all Medline abstracts relevant to a given gene/protein requires expanding the PubMed search query with all the synonyms of the gene/protein, as this is often mentioned in text by short names (acronyms and abbreviations) and gene symbols, with or without the accompanying long names. Searching short names and abbreviations is challenging as these names tend to be highly ambiguous, resulting in the retrieval of many irrelevant documents. Although augmenting the query using NOT operators, to disallow irrelevant expansions of the short names, may help in some cases with document retrieval, it does not circumvent the problem altogether. Short forms can be mentioned in text without the accompanying long form, thus making it impossible to automatically detect the relevance of the text based solely on the query.

For example, consider protein Carbamoyl-phosphate synthetase 1, whose short names are CPS1 and CPSI. The latter could also be an abbreviation for "cancer prevention study I," "chronic prostatis symptom index," and "chronic pain sleep inventory". Equally ambiguous are non abbreviated short names. The task of disambiguating words with multiple senses dates back to Bruce and Wiebe (8) and Yarowsky (9), who proposed a word sense disambiguation (WSD) technique for English words with multiple definitions (e.g., "bank" in the context of "river," and "bank" in the context of "financial institution").

eGRAB starts by gathering all possible names and synonyms of a gene/protein from knowledge bases of genes and proteins (such as Entrez Gene, Uniprot, or BioThesaurus), searches PubMed using these names, and returns a set of disambiguated Medline abstracts to serve as the gene's literature. This technique filters potentially irrelevant documents that mention the gene names in some other context, by creating language models for all the senses and assigning the closest sense to an ambiguous name. Similar methods have been described for disambiguating biomedical abbreviations by taking into consideration the context in which the abbreviations occur (10–13).

2.2. Extracting Genic Information from Text

Extracting Genic Information from Text (eGIFT) (14, 15) is a new, freely available online tool (http://biotm.cis.udel.edu/eGIFT/), which aims to link genes/proteins to key descriptors. The user can search for the gene/protein of interest and see its concepts grouped in categories: processes and functions, diseases, cellular components, motifs/domains, taxons, drugs, and genes. In eGIFT these concepts are extracted from the gene's literature when they are statistically more frequent in this set of abstracts, as compared to abstracts about genes in general. For example, given the protein BAD and its literature identified by eGRAB, eGIFT focuses on the abstracts that are mainly about BAD, and identify concepts, such as "apoptosis," "cell death," and "dephosphorylation" as highly relevant to this gene. Although different in the overall approach, scoring formula, redundancy detection, multi-word concept retrieval, and evaluation technique, eGIFT can be compared with methods described by Andrade and Valencia (16), XplorMed (17, 18), Liu et al. (19), and Shatkay and Wilbur (20).

2.3. Rule-Based Literature Mining System for Protein Phosphorylation

Rule-based LIterature Mining System for Protein Phosphorylation (RLIMS-P) (21, 22) is a system designed for extracting protein phosphorylation information from MEDLINE abstracts. Its unique features, which distinguishes it from other BioNLP systems, include the extraction of information about protein phosphorylation, along with the three objects involved in this process – the protein kinase, the phosphorylated protein (substrate), and the phosphorylation site (residue/position being phosphorylated). RLIMS-P employs techniques to combine information found in different sentences, because rarely are the three objects (kinase, substrate, and site) found in the same sentence. For this, RLIMS-P utilizes extraction rules that cover a wide range of patterns, including some specialized terms used only with phosphorylation. RLIMS-P was benchmarked using PIR annotated literature data from iProLINK (21). The online tool is available at http://www.proteininformationresource.org/pirwww/iprolink/rlimsp.shtml.

2.4. PPI Module

The PPI module is an internal implementation designed to detect mentions of PPI in text. This tool extracts text fragments, or text evidence, that explicitly describe a type of PPI (such as binding and dissociation), as well as the interacting partners. The primary engine of this tool is an extensive set of rules specialized to detect patterns of PPI mentions (manuscript in preparation).

The interacting partners identified are further sent to AIIAGMT, a gene/protein mention tool (described in more detail in the next sub-section), to confirm whether they are genuine protein mentions. Consider the sample phrase "several proapoptotic proteins commonly become associated with 14-3-3." "14-3-3" is a protein, whereas "several proapoptotic proteins" prompts the need to further identify the actual proteins (Bad and FOXO3a) that interact with 14-3-3. Our PPI module can be compared to other systems that also extract text evidence of PPI from literature, such as PIE (23), BIOSMILE (24, 25), Chilibot (26) and iHOP (27).

2.5. AIIAGMT

As mentioned previously in this chapter, genes and proteins often have many synonyms that come in short and long forms. To aid the PPI module to confirm whether an interacting partner in a PPI mention is indeed a protein, we employ AIIAGMT (28). AIIAGMT is a gene/protein mention tagger that detects all the proteins mentioned in some given text. The tool ranked second in the BioCreative II competition (29) for the gene mention task (F-score of 87.21) (30). Other systems that also extract gene and protein mentions from text are ABGene (31), BIGNER (32), GAPSCORE (33), T2K Gene Tagger (34), and LingPipe (35).

2.6. eFIP's Ranking Module

eFIP ranks abstracts mentioning a given protein based on three features: phosphorylation, functional terms, and proteins with which the given protein interacts. Because our main goal is to find information about a particular protein when it is in its phosphorylated state, we disregard abstracts that do not contain phosphorylation information. The next step is to distinguish the set of abstracts that mention a phosphorylation site for the given protein from the set of abstracts that mention only that the protein is phosphorylated. We rank the former set higher than the latter. Within these sets, a second ranking is performed, based on the following criteria (1) highly ranked are abstracts that include all three features, mentioned in one or two consecutive sentences; (2) following these are abstracts mentioning phosphorylation together with one other feature, in one or two consecutive sentences. When the features are found in the same sentence these abstracts are ranked higher than when they are found in two consecutive ones. Intuitively, the closer the two pieces of information, the higher the likelihood that they are related. We also consider the confidence level of rules or patterns matched for the PPI. For instance, "protein A binds to protein B" strongly indicates a PPI, whereas "the colocalization of proteins C and D" may suggest, but does not imply, a physical interaction. Some examples of the types of sentences mentioned above are depicted in Fig. 2. Based on our ranking, PMID:15161349 (A) would rank higher than PMID:12049737 (B).

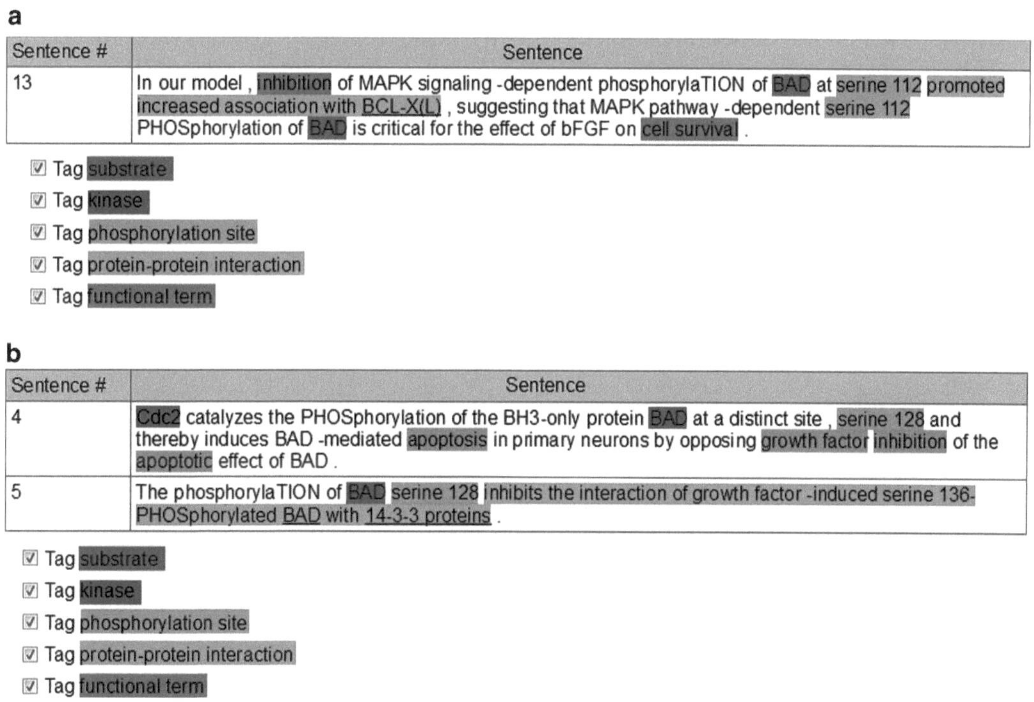

a

Sentence #	Sentence
13	In our model , inhibition of MAPK signaling -dependent phosphorylaTION of BAD at serine 112 promoted increased association with BCL-X(L) , suggesting that MAPK pathway -dependent serine 112 PHOSphorylation of BAD is critical for the effect of bFGF on cell survival .

- Tag substrate
- Tag kinase
- Tag phosphorylation site
- Tag protein-protein interaction
- Tag functional term

b

Sentence #	Sentence
4	Cdc2 catalyzes the PHOSphorylation of the BH3-only protein BAD at a distinct site , serine 128 and thereby induces BAD -mediated apoptosis in primary neurons by opposing growth factor inhibition of the apoptotic effect of BAD .
5	The phosphorylaTION of BAD serine 128 inhibits the interaction of growth factor -induced serine 136-PHOSphorylated BAD with 14-3-3 proteins .

- Tag substrate
- Tag kinase
- Tag phosphorylation site
- Tag protein-protein interaction
- Tag functional term

Fig. 2. Examples of sentences with different co-occurrence of ranked features. (**a**) Co-occurrence of the three features in one sentence (sentence 13); (**b**) Co-occurrence of phosphorylation and functional terms (sentences 4 and 5, respectively).

3. Methods

We present a use case on abstracts for the protein BAD (Bcl2-associated agonist of cell death). This protein is a key regulator of apoptosis that is posttranslationally modified by phosphorylation, which, in turn, defines BAD's binding partners and localization, as well as its function as an antiapoptotic or proapoptotic molecule. Ideally, we want to find papers about BAD that describe, together, phosphorylation and its functional consequence. Typically, we would start by searching PubMed using the protein/gene names (including/excluding its synonyms), coupled with phosphory* fuzzy search to retrieve abstracts that mention the given protein and its phosphorylation. For example, we might search using the following query (BAD AND phosphory*), which retrieves 1,050 papers. However, based on this search, some irrelevant abstracts may be retrieved (e.g., PMID: 8755886, where BAD is mentioned as an adjective). This example reflects the ambiguity problem mentioned before. From the list of abstracts obtained, we then need to check manually those for which phosphorylation has some implication on BAD biology. As an alternative to this approach, we present eFIP, a system that allows, in one step, document retrieval, disambiguation of names, and extraction of information.

eFIP combines information that is output by tools described in Subheading 2. Initially, eGRAB gathers abstracts specific to the gene/protein. These abstracts are input to (1) eGIFT, which mines, from this set of abstracts, terms that are highly related to the given gene/protein (e.g., "apoptosis" and "cell survival" for protein BAD); (2) RLIMS-P, which detects protein phosphorylation information from these abstracts; and (3) PPI module, which identifies interacting proteins. eFIP uses this information to rank abstracts mentioning a given protein of interest. However, these detailed steps are hidden from the user. eFIP combines these tools and requires only the following steps from its users:

3.1. Accessing eFIP's Website at http://biotm.cis.udel.edu/eFIP/

The search for a gene/protein is initiated from the Search eFIP link. Here, the gene/protein name or part of the name can be entered in the search box, and results are displayed for the search. For example, word BAD can be entered in the search box, and only one result is obtained for gene BAD. However, if a partial name is entered, such as bcl2 (initial part of one of BAD's name), many results are retrieved. In this case selecting the gene corresponding to BAD is required (Fig. 3).

3.2. Inspecting the Result Page

1. The primary result page contains the following information (Fig. 4):
   (a) Names, synonyms and statistics: The result page shows the names and synonyms used for retrieving the articles.

Fig. 3. eFIP search page. The screenshot shows the list of possible gene/protein names when using bcl2 as a query. The user needs to select BAD to inspect its specific literature.

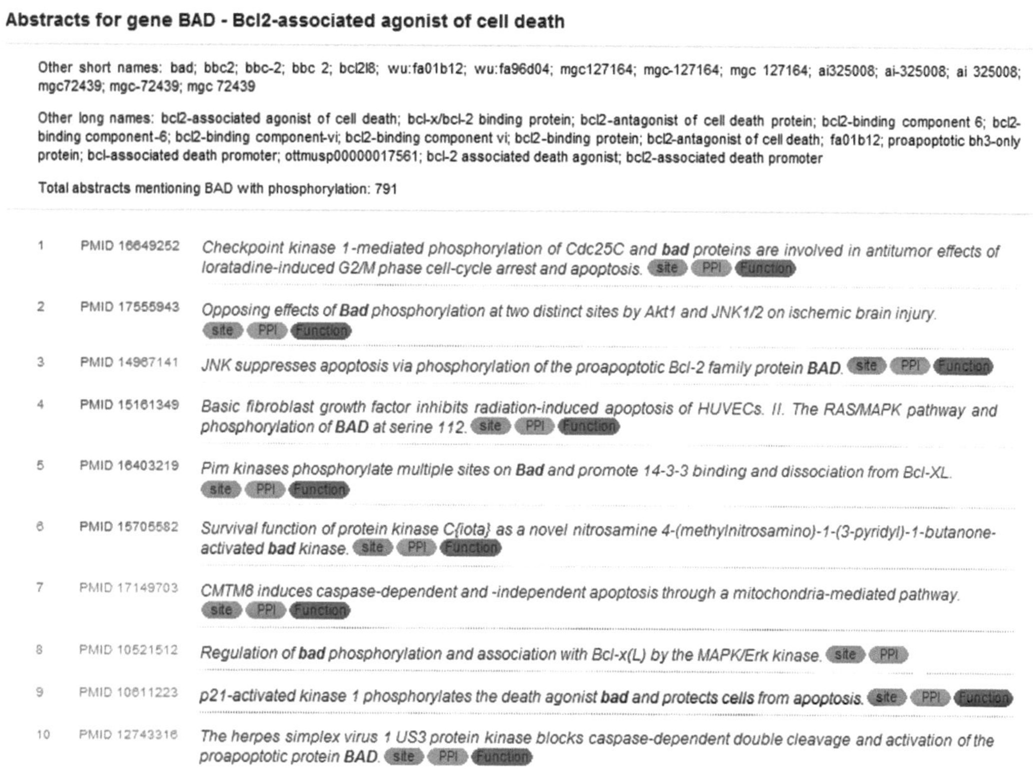

Fig. 4. Result page for the protein BAD.

It also shows the number of articles that contain phosphorylation mentions as evaluated by the RLIMS-P tool (791 in BAD's case). Note that the number of total articles disambiguated by eGRAB is 1,331.

(b) Ranked PMIDs, along with the information content of the abstract, are listed. Because all the abstracts have phosphorylation mentions by default, only the PPI and/or functional feature labels are displayed. Note that based on our ranking criteria, the first set of abstracts displayed are those that mention phosphorylation site information (206 abstracts).

2. Selecting a PMID leads to the abstract page (Fig. 5).

This page contains the summary table, with information extracted for phosphorylation and the predicted impact on function. We emphasize predicted here, because BioNLP tools are intended to assist the user by pointing to articles or sentences that are more likely to have the information needed. However, there is always a need to check the correctness of the information. The summary table, displayed on this page, consists of three main columns. The first column shows the number of the sentence that contains the evidence, thus facilitating its quick location within the abstract. The second column contains the phosphorylation information, as provided by RLIMS-P tool. Three different types of information are listed here: the substrate, the site, and the kinase. The third column provides information about the impact on phosphorylation. Here, we list the functional terms and/or interaction information provided by eGIFT and the PPI module, respectively. In this column,

PMID 10837486 for gene BAD - Bcl2-associated agonist of cell death

Predicted impact of phosphorylation:

Sentence #	Phosphorylation			Impact
	Substrate	Site	Kinase	
1	BAD	Ser-155	RSK1	regulates BAD/Bcl-XL interaction regulates cell survival
3,4	BAD	Ser-112 and Ser-136	N/A	promotes binding of BAD to 14-3-3 proteins
6	BAD	Ser-155	RSK1	blocking the binding of BAD to Bcl-XL
7	BAD	both Ser-112 and Ser-155	RSK1	rescues BAD -mediated cell death

☑ Tag substrate
☑ Tag kinase
☑ Tag phosphorylation site
☑ Tag protein-protein interaction
☑ Tag functional term

Text of title and abstract:

Sentence #	Sentence
1	TI - BAD Ser-155 PHOSphorylation regulates BAD/Bcl-XL interaction and cell survival .
2	AB - The BH3 domain of BAD mediates its death-promoting activities via heterodimerization to the Bcl-XL family of death regulators .
3	Growth and survival factors inhibit the death-promoting activity of BAD by stimulating PHOSphorylation at multiple sites including Ser-112 and Ser-136 .
4	PHOSphorylation at these sites promotes binding of BAD to 14-3-3 proteins , sequestering BAD away from the mitochondrial membrane where it dimerizes with Bcl-XL to exert its killing effects .
5	We report here that the phosphorylaTION of BAD at Ser-155 within the BH3 domain is a second PHOSphorylation -dependent mechanism that inhibits the death-promoting activity of BAD .
6	Protein kinase A , RSK1 and survival factor signaling stimulate PHOSphorylation of BAD at Ser-155 , blocking the binding of BAD to Bcl-XL .
7	RSK1 phosphorylaTEs BAD at both Ser-112 and Ser-155 and rescues BAD -mediated cell death in a manner dependent upon PHOSphorylation at both sites .

Fig. 5. Summary table and highlighted information for PMID 10837486. The different features are color coded.

we also include action words (e.g., regulates, promotes, blocks), present in the text, to point to the modification or to the influence on the meaning of the functional term. These action words, provided by the PPI module, provide a more accurate result. Listed below the table is the corresponding abstract, with highlighted information. Note that each type of information has a distinct color, and for each color there is a dark and a light version, to give different confidence levels to the prediction (the dark color hints to a higher likelihood of the prediction). At the bottom of the abstract, you can select which information to include in the highlighting.

4. Discussion

Using protein BAD and the information displayed in eFIP for this protein, we show in Fig. 6 the different phosphorylated forms of BAD, their functions, and their implication in PPI. The information

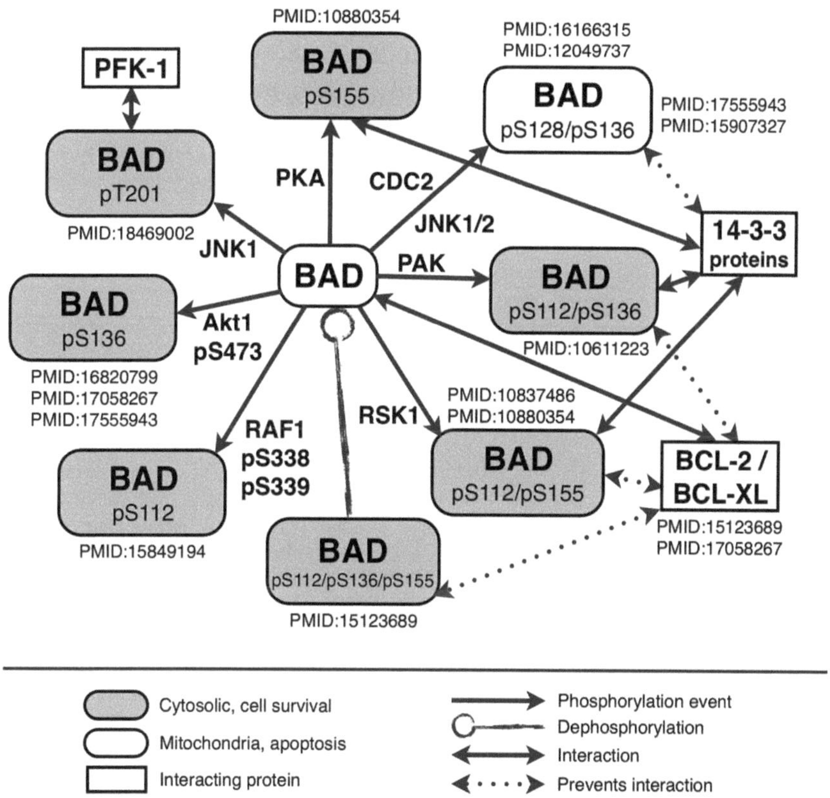

Fig. 6. Representation of different forms of phosphorylated BAD based on eFIP's results (only a subset is shown). Note that from information listed by eFIP, we are able to represent the impact (cytosolic vs. mitochondria; apoptosis vs. cell survival) for the different forms of BAD. Moreover, the kinases, which accompany the phosphorylation arrows, help to link BAD to pathways. Whenever available, the phosphorylation state of the kinase is extracted and displayed here, as in the case of RAF1 p338/pS339.

depicted here is extracted from a subset of the highest ranked abstracts, as provided by eFIP. The rich information from the eFIP text mining tool uncovers interesting facts about BAD (1) BAD is a common hub for several pathways to regulate apoptosis, as evidenced by the various kinases that are able to phosphorylate this protein; (2) BAD has specific partners for its distinct phosphorylated forms; and (3) phosphorylation on BAD may have two opposing effects: apoptosis (through phosphorylation at Ser128) and cell survival (phosphorylation on other residues), which is mainly dictated by the association/disassociation to 14-3-3 proteins and BCL-2/BCL-XL proteins. This example highlights the importance of detecting more than just the phosphorylation mention. The phosphorylation site, as well as the kinase that links to the pathway, are important aspects in understanding the regulation of BAD. The majority of abstracts describing BAD focus on BAD's interaction with apoptotic and antiapoptotic proteins. However, in this figure, we also point to an example in which phosphorylated BAD (Thr-201) leads to binding to phosphofructokinase (PFK-1), and the subsequent activation of glycolysis (a pathway that is key to cell survival).

Thus, we show that eFIP provides the means to find the most relevant papers about BAD phosphorylation, interaction partners, and its functions. Based on the literature data collected from eFIP for BAD protein, it is possible to predict, for example, how the regulation or inhibition of a certain pathway may affect the cell fate.

References

1. Preisinger, C., von Kriegsheim, A., Matallanas, D., and Kolch, W. (2008) Proteomics and phosphoproteomics for the mapping of cellular signalling networks. *Proteomics* **8**, 4402–4415.
2. Huang, H., Hu, Z. Z., Arighi, C., and Wu, C. H. (2007) Integration of bioinformatics resources for functional analysis of gene expression and proteomic data. *Front Biosci* **12**, 5071–5088.
3. Hirschman, L., Park, J. C., Tsujii J., Wong, L., and Wu, C. H. (2002) Accomplishments and challenges in literature data mining for biology. *Bioinformatics* **18**, 1553–1561.
4. Krallinger, M., Morgan, A., Smith, L., Leitner, F., Tanabe, L., Wilbur, J., Hirschman, L., and Valencia, A. (2008) Evaluation of text-mining systems for biology: overview of the second BioCreative community challenge. *Genome Biol* **9**, S1.
5. Jensen, L. J., Saric, J., and Bork, P. (2006) Literature mining for the biologist: from information retrieval to biological discovery. *Nat Rev Genet* **7**, 119–129.
6. Salih, E. (2005) Phosphoproteomics by mass spectrometry and classical protein chemistry approaches. *Mass Spectrom Rev* **24**, 828–846.
7. Wicks, S. J., Lui, S., Abdel-Wahab, N., Mason, R. M., and Chantry, A. (2000) Inactivation of smad-transforming growth factor beta signaling by Ca(2+)-calmodulin-dependent protein kinase II. *Mol Cell Biol* **20**, 8103–8111.
8. Bruce, R., and Wiebe, J. (1994) Word-sense disambiguation using decomposable models. In: *Proceedings of the 32nd Annual Meeting on ACL* 139–146.
9. Yarowsky, D. (1995) Unsupervised word sense disambiguation rivaling supervised methods. In: *Proceedings of the 33rd Annual Meeting on ACL* 189–196.
10. Pakhomov, S. (2001) Semi-supervised maximum entropy based approach to acronym and abbreviation normalization in texts.

In: *Proceedings of 40th Annual Meeting on ACL* 2001.

11. Yu, Z., Tsuruoka, Y., and Tsujii, J. (2003) Automatic resolution of ambiguous abbreviations in biomedical texts using support vector machines and one sense per discourse hypothesis. In: *SIGIR'03 Workshop on Text Analysis and Search for Bioinformatics.*
12. Gaudan, S., Kirsch, H., and Rebholz-Schuhmann, D. (2005) Resolving abbreviations to their senses in Medline. *Bioinformatics* **21**, 3658–3664.
13. Stevenson, M., Guo, Y., Amri, A. A., and Gaizauskas, R. (2009) Disambiguation of biomedical abbreviations. In: *Proceedings of the BioNLP 2009 Workshop, ACL* 71–79.
14. Tudor, C. O., Vijay-Shanker, K., and Schmidt, C. J. (2008) Mining the biomedical literature for genic information. In: *Proceedings of Workshop on Current Trends in BioNLP, ACL* 28–29.
15. Tudor, C. O., Schmidt, C. J., and Vijay-Shanker, K. (2008) Mining for gene-related key terms: where do we find them? In: *Proceedings of the Third International Symposium on Semantic Mining in Biomedicine (SMBM)* 157–160.
16. Andrade, M. A., and Valencia, A. (1998) Automatic extraction of keywords from scientific text: application to the knowledge domain of protein families. *Bioinformatics* **14**, 600–607.
17. Perez-Iratxeta, C., Keer, H. S., Bork, P., and Andrade, M. A. (2002) Computing fuzzy associations for the analysis of biomedical literature. *BioTechniques* **32**, 1380–1385.
18. Perez-Iratxeta, C., Perez, A. J., Bork, P., and Andrade, M. A. (2003) Update on XplorMed: a web server for exploring scientific literature. *Nucleic Acid Res* **31**, 3866–3868.
19. Liu, Y., Brandon, M., Navathe, S., Dingledine, R., and Ciliax, B. J. (2004) Text mining functional keywords associated with genes. *MedInfo* 292–296.
20. Shatkay, H., and Wilbur, W. J. (2000): Finding themes in medline documents: probabilistic similarity search. In: *Proceedings of the Seventh IEEE Advances in Digital Libraries (ADL'00)* 183–192.
21. Hu, Z. Z., Narayanaswamy, M., Ravikumar, K. E., Vijay-Shanker, K., and Wu, C. H. (2005) Literature mining and database annotation of protein phosphorylation using a rule-based system. *Bioinformatics* **21**, 2759–2765.
22. Narayanaswamy, M., Ravikumar, K. E., and Vijay-Shanker, K. (2005) Beyond the clause: extraction of phosphorylation information from Medline abstracts. *Bioinformatics* **21** Suppl 1, i319–i327.
23. Kim, S., Shin, S. Y., Lee, I. H., Kim, S. J., Sriram, R., and Zhang, B. T. (2008) PIE: an online prediction system for protein–protein interactions from text. *Nucleic Acids Res* **36**, W411–W415.
24. Dai, H. J., Huang, C. H., Lin, R. T., Tsai, R. T., and Hsu, W. L. (2008) BIOSMILE web search: a web application for annotating biomedical entities and relations. *Nucleic Acids Res* **36**, W390–W398.
25. Tsai, R. T. H., Chou, W. C., Su, Y. S., Lin, Y. C., Sung, C. L., Dai, H. J., Yeh, I. T. H., Ku, W., Sung, T. Y., and Hsu, W. L. (2007) BIOSMILE: a semantic role labeling system for biomedical verbs using a maximum-entropy model with automatically generated template features. *BMC Bioinformatics* **8**, 325.
26. Chen, H., and Sharp, B. M. (2004) Content-rich biological network constructed by mining PubMed abstracts. *BMC Bioinformatics* **5**, 147.
27. Hoffmann, R., and Valencia, A. (2005) Implementing the iHOP concept for navigation of biomedical literature. *Bioinformatics* **21**, ii252–ii258.
28. Hsu, C. N., Chang, Y. M., Kuo, C. J., Lin, Y. S., Huang, H. S., and Chung, I. F. (2008) Integrating high dimensional bi-directional parsing models for gene mention tagging. *Bioinformatics* **24**, i286–i294.
29. Morgan, A. A., Lu, Z., Wang, X., Cohen, A. M., Fluck, J., Ruch, P., Divoli, A., Fundel, K., Leaman, R., Hakenberg, J., Sun, C., Liu, H. H., Torres, R., Krauthammer, M., Lau, W. W., Liu, H., Hsu, C. N., Schuemie, M., Cohen, K. B., and Hirschman, L. (2008) Overview of BioCreative II gene normalization. *Genome Biol* **9** Suppl 2, S3.
30. URL:http://www.bcsp1.iis.sinica.edu.tw:8080/aiiagmt/.
31. Tanabe, L., and Wilbur, W. J. (2004) Tagging gene and protein names in biomedical text. *Bioinformatics* **20**, 216–225.
32. Li, Y., Lin, H., and Yang, Z. (2009) Incorporating rich background knowledge for gene named entity classification and recognition. *BMC Bioinformatics* **10**, 223.
33. Chang, J. T., Schütze, H., and Altman, R. B. (2004) GAPSCORE: finding gene and protein names one word at a time. *Bioinformatics* **20**, 216–225.
34. URL: http://www.bioinformatics.org/~hyy/textknowledge/genetag.php.
35. URL: http://www.alias-i.com/lingpipe/.

Chapter 6

A Tutorial on Protein Ontology Resources for Proteomic Studies

Cecilia N. Arighi

Abstract

The protein ontology (PRO) is designed as a formal and well-principled open biomedical ontologies (OBO) foundry ontology for proteins. The components of PRO extend from the classification of proteins, on the basis of evolutionary relationships at the full-length level, to the representation of the multiple protein forms of a gene, such as those resulting from alternative splicing, cleavage and/or post-translational modifications, and protein complexes. As an ontology, PRO differs from a database in that it provides description about the protein types and their relationships. In addition, the representation of specific protein types, such as a phosphorylated protein form, allows precise definition of objects in pathways, complexes, or in disease modeling. This is useful for proteomics studies where isoforms and modified forms must be differentiated, and for biological pathway/network representation where the cascade of events often depends on a specific protein modification. PRO is manually curated starting with content derived from scientific literature. Only annotation with experimental evidence is included, and is in the form of relationship to other ontologies. In this tutorial, you will learn how to use the PRO resources to gain information about proteins of interest, such as finding conserved isoforms (ortho-isoforms), and different modified forms and their attributes. In addition, it will provide some details on how you can contribute to the ontology via the rapid annotation interface RACE-PRO.

Key words: Biomedical ontology, Protein ontology, Community annotation, Protein

1. Introduction

Biomedical ontologies have emerged as critical tools in genomic and proteomic research where complex data in disparate resources need to be integrated. In this context, gene ontology (GO) (1) has become the common language to describe biological processes, protein function and localization. Protein or peptides detected in proteomic experiments are usually mapped to database entries, followed by data mining for GO terms and other data with the aim of characterizing the proteomic products (2).

Cathy H. Wu and Chuming Chen (eds.), *Bioinformatics for Comparative Proteomics*, Methods in Molecular Biology, vol. 694, DOI 10.1007/978-1-60761-977-2_6,

However, there are some issues in capturing scientific knowledge based on the current infrastructure in that most sequence and organism databases provide gene-centric organization: one entry for one gene or canonical gene product. But in reality, many protein forms may derive from a single gene as a result of alternative splicing and/or subsequent posttranslational modifications. These various protein forms may have different properties. Therefore, the functional annotation of a protein may represent composite annotation of several protein forms, which may lead to noisy data mining results, and eventually to misinterpretation of data mining results. This missing infrastructure may also affect interoperability since some of the databases need to represent this level of granularity and create these objects independently, adding complexity to data integration.

The protein ontology (PRO) (3, 4) is an OBO Foundry ontology that describes the different protein forms and their relationships in order to provide the appropriate framework for tackling the above-mentioned problems. PRO provides a means to refer to a specific protein object and append the corresponding annotations. This means that, for example, posttranslationally modified and unmodified forms of a given protein are two distinct objects in the ontology. Figure 1 shows a schematic representation of the ontology, which is organized in different levels (see Note 1) that can be grouped into four main categories (in decreasing hierarchical order):

1. *Family*: a PRO term at this level refers to proteins that can trace back to a common ancestor over the entire length of the protein. The leaf-most nodes at this level are usually families comprising paralogous sets of gene products (of a single or multiple organisms). In Fig. 2, PRO:000000676 is an example of this level. Note that the hierarchy in the ontology (Fig. 2a) reflects the evolutionary relationship of this group (Fig. 2b), HCN1-4 are paralogs that belong to the same homeomorphic family (full-length sequence similarity and have common domain architecture); therefore, in the ontology they are all under the same parent node (PRO:000000676).

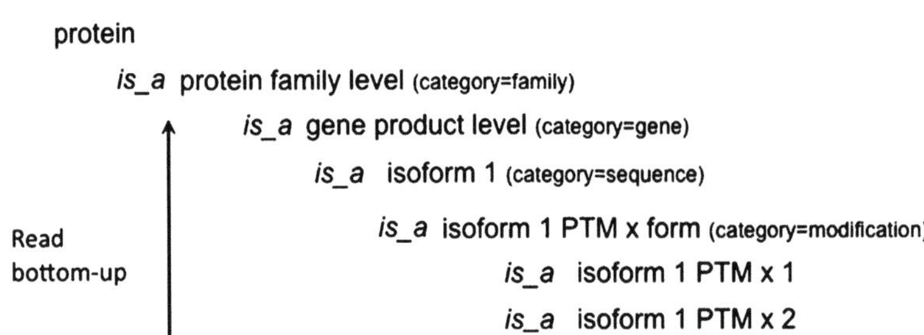

Fig. 1. PRO hierarchical organization. The ontology is read from bottom-up. *PTM* post-translational modification, *x* type of modification (such as acetylation, phosphorylation).

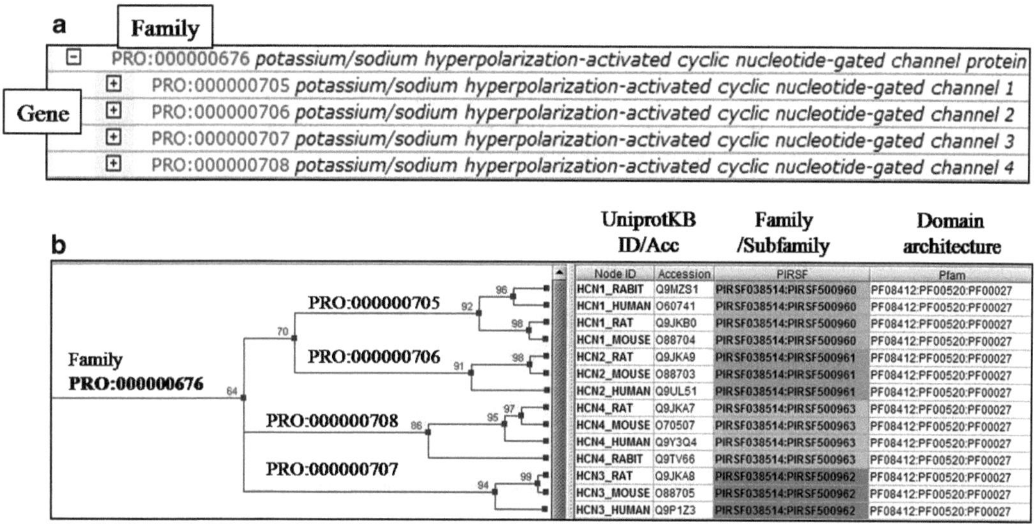

Fig. 2. Family category reflects the evolution of full-length proteins. (**a**) PRO ontology terms for the potassium/sodium hyperpolarization-activated cyclic nucleotide-gated channel protein. The family and gene product levels are shown. (**b**) *Left panel:* neighbor-joining tree showing the evolutionary relation of some representative proteins of the HCN1-HCN4 genes. The PRO IDs of each class is shown. *Right panel:* display of the corresponding database identifiers for: protein (UniProtKB), family (PIRSF (17)), and domain (Pfam).

2. *Gene*: a PRO term at this level refers to the protein products of a distinct gene. A single term at the gene-level distinction collects the protein products of a subset of orthologs for that gene (the subset that is so closely related that its members are considered the same gene). From the example depicted in Fig. 2 the HCN1 gene product (PRO:000000705) would include the proteins of the rat, mouse, rabbit and human HCN1 genes.

3. *Sequence*: a PRO term at this level refers to the protein products with a distinct sequence upon initial translation. The sequence differences can arise from different alleles of a given gene, from splice variants of a given RNA, or from alternative initiation and ribosomal frame shifting during translation. One can think of this as a mature mRNA-level distinction. Similarly to the gene product level, this level collects the protein products of a subset of orthologous splice variants for that gene, and we call them ortho-isoforms. Figure 3a shows an example of two nodes at the sequence level, PRO:000003420 and PRO:000003423, corresponding to isoform 1 (p75) and isoform 2 (p52) derived from gene LEDG. In this case literature is the data source for these protein forms. Figure 3b depicts the experimentally determined LEDG gene products (protein known as PC4 and SFRS1-interacting protein) based on the PMID:18708362 (5). Note that, although the experimental data displayed is from human, the article also describes

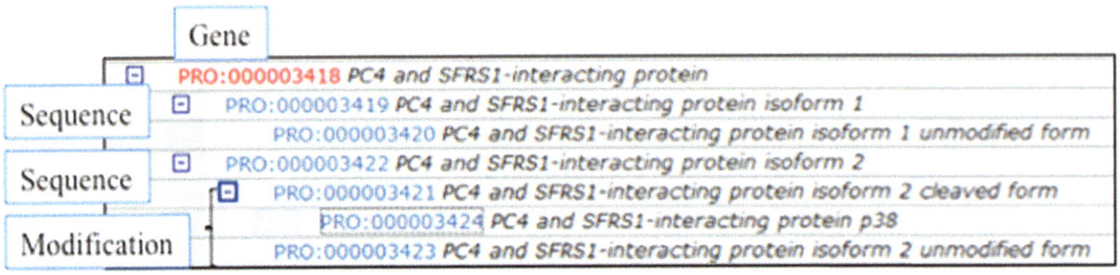

Fig. 3. Protein ontology to describe protein forms. (**a**) PRO ontology terms for the PC4 and SFRS1-interacting protein (derived from LEDG gene) depicting the isoforms, and modified forms. (**b**) Literature is the source for PRO forms; the scheme shows the different protein forms derived from the LEDG gene as described in a given article.

the existence of these isoforms in mouse, so the human and mouse p75 isoforms will be both described by the PRO:000003420 term.

4. *Modification*: a PRO term at this level refers to the protein products derived from a single mRNA species that differ because of some change (or lack thereof) that occurs after the initiation of translation (co- and posttranslational). This includes sequence differences due to cleavage and chemical changes to one or more amino acid residues. Figure 3a shows an example of the cleaved version (p38) of isoform 2 (p52) of the LEDG gene. This level represents ortho-modified forms, the presence of posttranslational modifications on equivalent residues in ortho-isoforms.

1.1. Relevance

We have previously described the various states of proteins involved in the TGF-beta signaling pathway (4), and also in the intrinsic apoptotic pathway (6). In the latter case, one key regulator of apoptosis is Bcl2 antagonist of cell death (Bad, PRO:000002184), whose phosphorylation state determines whether the cell fate is apoptosis or survival. It is generally stated that the BAD unphosphorylated form activates apoptosis and that the phosphorylated form of BAD leads to cell survival. However, the ontology shows that there are at least six distinct phosphorylated forms, which can be phosphorylated via activation of various kinases, such as AKT1, MAPK8 (JNK1), PKA, and CDC2. Although phosphorylation by the first three leads to interaction with the 14-3-3 proteins and cell survival, the outcome of the phosphorylation by CDC2 is the opposite, leading to translocation to the mitochondria an activation of apoptosis. This knowledge is key for the correct interpretation of proteomic results.

Therefore in this tutorial, you will learn how to use the PRO resources to gather this type of information about your protein(s) of interest.

2. Materials

The PRO website is accessible at http://pir.georgetown.edu/pro/pro.shtml.

2.1. Download

The ontology (pro.obo), the annotation (PAF.txt), and mappings to external databases can be downloaded from the ftp site at ftp://ftp.pir.georgetown.edu/databases/ontology/pro_obo/. This chapter is based on Release 8.0 v1. The ontology is also available in OBO and OWL formats through the OBO Foundry (7) and Bioportal (8). For general documentation please see http://pir.georgetown.edu/pro/pro_dcmtt.shtml.

2.2. PRO Files

The pro.obo file is in OBO 1.2 format and can be opened with OBO Edit 2.0 (9). This file displays a version information block, followed by a stanza of information about each term. Each stanza in the obo file is preceded by [Term] and it is composed of an ID, a name, synonyms (optional), a definition, comment (optional), cross-reference (optional) and relationship to other terms (see example below).

```
format-version: 1.2
date: 15:12:2009 13:48
saved-by: cecilia
auto-generated-by: OBO-Edit 2.0
default-namespace: pro
remark: release: 8.0, version 1

[Term]
id: PRO:000000003
name: HLH DNA-binding protein inhibitor
def: "A protein with a core domain composition consisting of a Helix-loop-helix DNA-binding domain (PF00010) (HLH), common to the basic HLH family of transcription factors, but lacking the DNA binding domain to the consensus E box response element (CANNTG). By binding to basic HLH transcription factors, proteins in this class regulate gene expression." [PRO:CNA]
comment: Category=family.
synonym: "DNA-binding protein inhibitor ID" EXACT []
synonym: "ID protein" RELATED []
xref: PIRSF:PIRSF005808
is_a: PRO:000000001 ! protein
```

The annotations to PRO terms are distributed in the PAF.txt file. To facilitate interoperability to the best extent, this tab delimited file follows the structure of the gene ontology association

(GAF) file. Please read the README file and the PAF guidelines. pdf in the ftp site to learn about the structure of this file. PRO terms are annotated with relation to other ontologies or databases. Currently in use: Gene ontology (GO) to describe processes, function and localization; Sequence ontology (SO) (10) to describe protein features; PSI-MOD (11) to describe protein modifications; MIM (12) to describe disease states; and Pfam (13) to describe domain composition.

2.3. Link to PRO

Use the persistent URL: http://purl.obolibrary.org/obo/PRO_xxxxxxxxx, where PRO_xxxxxxxxx is the corresponding PRO ID with an underscore (_) instead of semicolon (:). Example: link to PRO:000000447 would be http://purl.obolibrary.org/obo/PRO_000000447

3. Methods

3.1. PRO Homepage

The PRO homepage (http://pir.georgetown.edu/pro/pro.shtml) (Fig. 4) is the starting point to navigate through the protein ontology resources. The menu on the left side links to several documents and information pages, as well as to the ftp download page. The functionalities in the homepage include the subheadings: "PRO Browser," "PRO Entry Retrieval," "Text Search," and "Annotation."

3.1.1. PRO Browser

The browser is used to explore the hierarchical structure of the ontology (Fig. 5). The icons with a plus and minus signs allow expanding and collapsing nodes, respectively. Next to these icons is a PRO ID, which links to the corresponding entry report, followed by the term name. Unless otherwise stated the implicit relation between nodes is *is_a*.

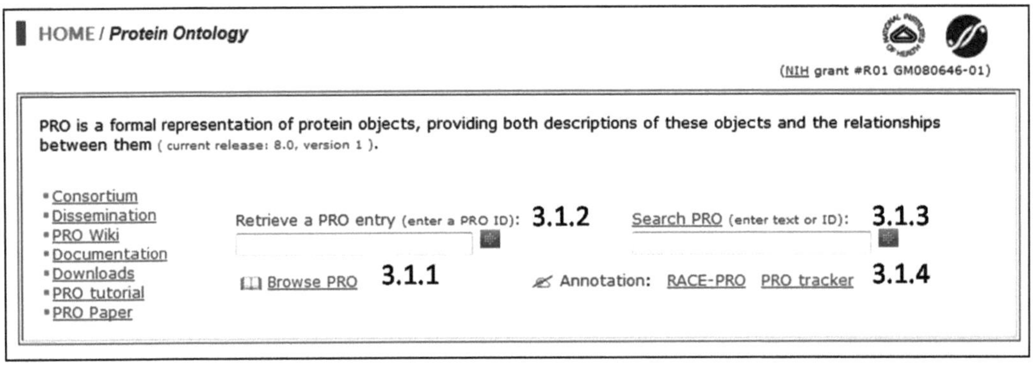

Fig. 4. PRO homepage (partial snapshot). The left menu links to documentation and downloads, whereas the *right part* displays the current functionalities.

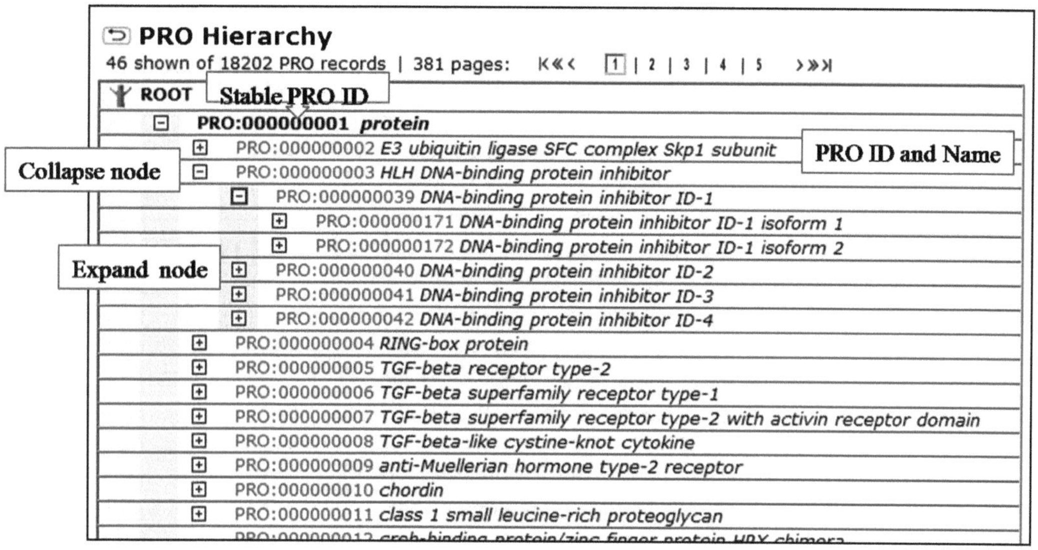

Fig. 5. The PRO browser shows the ontology hierarchy. Use icons to expand/collapse nodes, or select an ID to go to the PRO entry view.

3.1.2. PRO Entry

The PRO entry provides an integrated report about the ontology and annotation available for a given PRO term. If you know the PRO ID you can use the "retrieve PRO entry" box in the homepage. Alternatively, you can open an entry by clicking on the PRO ID in any other page (search, browser, etc.). The entry report contains four sections (Fig. 6):

(a) *Ontology information:* this section displays the information from the ontology about a term (source: the pro.obo file). You can link to the parent node, to the hierarchy, and find the definition and synonyms of the term, among other things.

(b) *Information about the entities that were use to create the PRO entry:* this section lists the sequences, in the case where category corresponds to gene, sequence or modification, for which some experimental information exists. Taxon information as well as PSI-MOD ID and modification sites are indicated when applicable. In many cases, the modifications sites are unknown and therefore only the PSI-MOD ID is listed. For cleaved products, the protein region is indicated and is underlined in the displayed sequence (Fig. 6b). In the case of category corresponding to family, this section provides a cross-reference to the database that is the source of the class.

(c) *Synonymous mappings:* this section contains mappings to external databases that link to protein forms as described in the given class (information source: mapping files). This is the case for Reactome (14) entry REACT_13251which

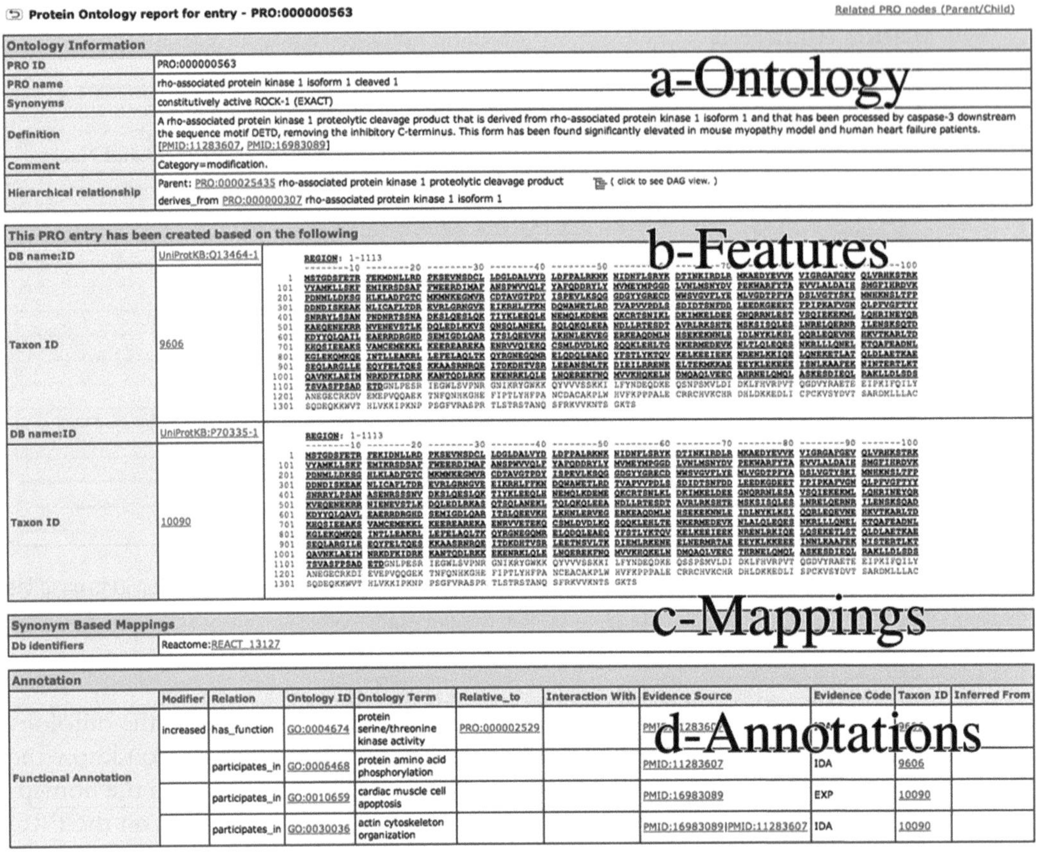

Fig. 6. Sample PRO entry report. The different sections are indicated and explained in detail in the text.

represents the human constitutive active form of ROCK-1 (Fig. 6c).

(d) *Annotation:* This section shows the annotation of the term with the different ontologies (source: PAF file). These annotations were contributed by the PRO consortium group and by community annotators through submission of RACE-PRO annotations (see Subheading 3.1.4).

3.1.3. Searching PRO

The search can be performed by entering a keyword or ID in the text box provided in the homepage. For example, you could just type the name of the protein for which you want to find related terms. Alternatively, advanced text search is available by clicking on the Search PRO title above the search box on the home page. Advanced text search supports Boolean (AND, OR, NOT) searches, as well as null (not present)/null (present) searches with several field options (see Note 2). Figure 7 shows an example of advanced search, which should retrieve all PRO terms that are in the modification category and contain annotation for protein–protein interaction.

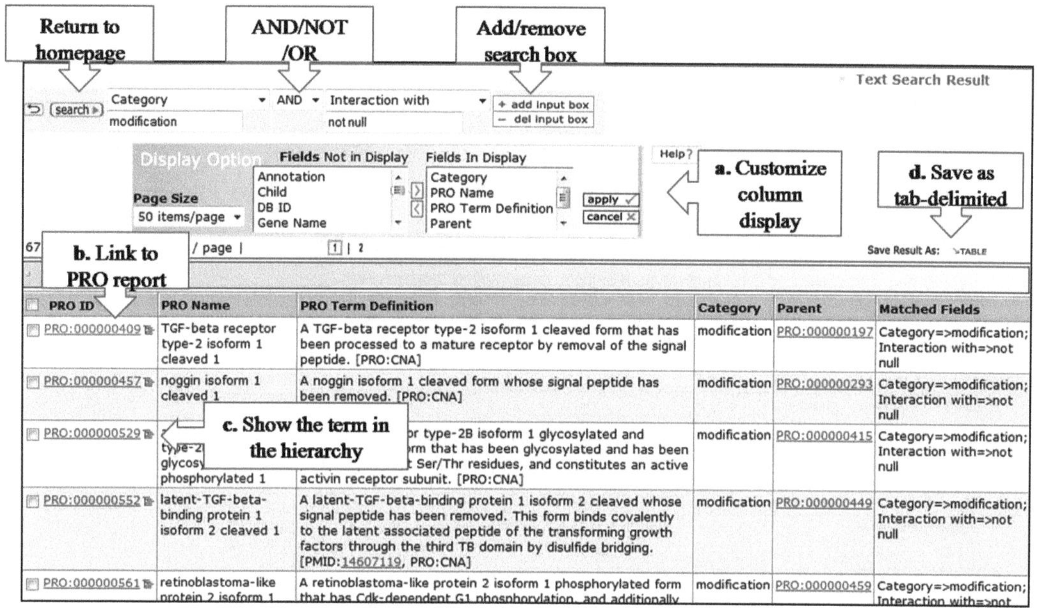

Fig. 7. Advance search and result table.

Results are shown in a table format with the following default columns (Fig. 7): the PRO ID, PRO name, PRO term definition, the category, the parent term ID, and the matched field. Some of the functionality in this page includes:

(a) *Display Option:* to customize result table by adding or removing columns. Use > to add or < to remove items from the list, but always select apply for the changes to take effect.

(b) *Link to PRO entry report:* the link is available by selecting the PRO ID

(c) *Link to hierarchical view:* the icon shows the term in the hierarchy, i.e., opens the browser.

(d) *Save:* the result table as a tab-delimited file.

3.1.4. Annotation

The annotation section is for community interaction. The PRO tracker should be used to request new terms or to change/comment on existing ones. The link is directed to an external page (sourceforge) where you will need to provide the details about the terms of interest. On the other hand, if you have the data and domain knowledge you can directly submit annotation via the rapid annotation interface RACE-PRO as described below.

3.1.4.1. Rapid Annotation Interface RACE-PRO

Follow a few simple steps and become an author of annotations in PRO. As an example of the procedure, the annotation pertinent for the cleaved product p38 from Fig. 3 is shown in Fig. 8. First fill your personal information. This information will not be

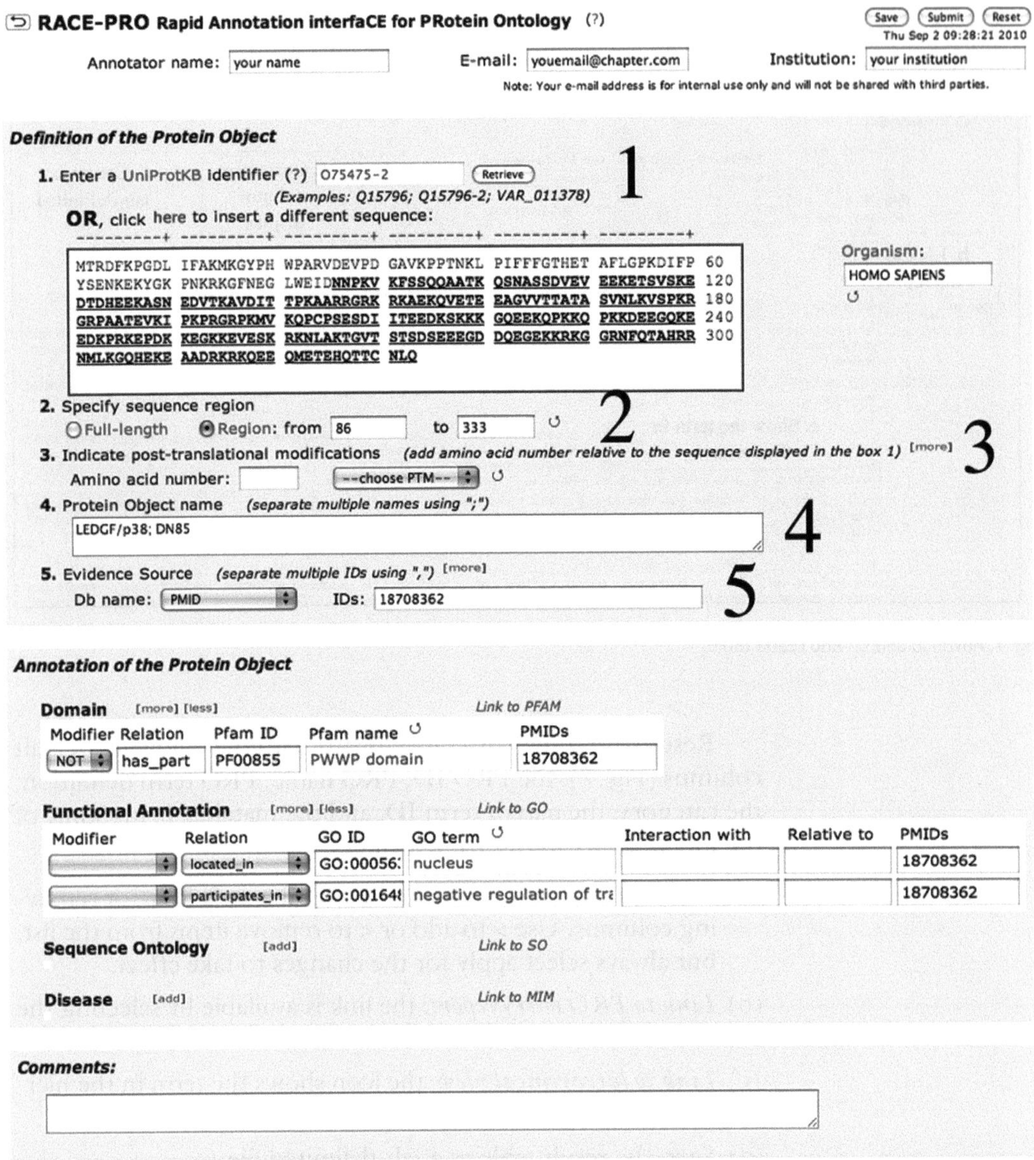

Fig. 8. RACE-PRO entry to describe the cleaved product (p38) shown in Fig. 3b.

distributed to any third party, but will only be used for saving your data and for communication purposes.

3.1.4.1.1. Definition of the Protein Object

This block allows you to enter all the information about a protein form along with the source of evidence. It is mandatory to add all the information relevant to this section whenever applicable.

1. *Retrieve the sequence:* if you use a UniProtKB identifier (15) and click "Retrieve," the sequence retrieved is formatted to show the residue numbers, and the organism box is automatically filled. You can use identifiers for isoforms (a UniProtKB

accession followed by a dash and a number) as in the example shown here. If you happen to have an identifier from a different database, you can use the ID mapping or batch retrieval services either from the PIR (16) or UniProt (17) websites to obtain the corresponding UniProtKB accession and retrieve the sequence – just be aware of which isoform or variant that you want to describe. Alternatively, you can paste a sequence, but in this case you will need to add the organism name (the link to NCBI taxonomy browser by clicking on the Organism title is provided as help).

2. *Protein region:* once the sequence is retrieved, you can select a subsequence in the cases where the protein form you are describing is not the full length, but a cleaved product or a fragment (as is the case of this example). After you do this, click on the circle arrow and the selected region will be underlined.
3. *Selecting the Modification*: If you need to describe a modification (or modifications), enter the residue number and the type of modification. If the modification is not in the list, use the "Other" option to add it. These terms will be later mapped to the corresponding PSI-MOD terms. If the modification site is unknown, please enter "?" in the residue number box. Use the [more] or the [less] to add or remove a modification line.
4. Be aware that the amino acid number should always refer to the sequence displayed in the sequence box. When clicking on the circle arrow, you will see the residues highlighted. Check that these are the ones expected. If there is no information about any posttranslational modification, then do not complete this line (as in the current example).
5. *Protein object name*: add names by which this object is referred to in the paper or source of data (separated by;). In the current example, both LEDG/p38 and DN85 are used to refer to the shorter cleaved form of LEDG isoform 2.
6. *Evidence source*: add the database (DB) that is the source of the annotation, in this case it is PubMed so we select as PMID. If the DB is not listed use the "Other" option and provide it. In the ID box you can add many IDs for a given DB separated by comma. Use the [more] or [less] to add or remove DB lines.

3.1.4.1.2. Annotation of the Protein Object

Only annotation from experimental data that is pertinent for the protein form (and species) described in the previous section should be added. There are three types of annotation that are based on different databases/ontologies: domain (Pfam), GO, and disease (MIM). If the paper describes the existence of a protein form with no associated properties, then do not fill this section.

All the information about the different columns in the table is described in the PAF guidelines. But below are some clarifications:

1. Modifiers: used to modify a relation between a PRO term and another term. It includes the GO qualifiers NOT, contributes to plus increased, decreased, and altered (to be used with the relative to column) e.g., "NOT has part PF00085 PWWP domain" is used because LEDG/p85 lacks this domain as determined in the paper, although it is present in the full length form.
2. Relation to the specific annotation. For some databases/ ontologies there is a single relation possible and therefore it is already displayed, for GO we use three depending on the ontology used. Example: located in is used for GO component for subcellular locations, whereas participates in is used for GO biological processes.
3. Add ID for the specific database/ontology. If you need to search use the "link to.." link. If you enter the ID, the name autofills. Example: The paper shows in Fig. 5 that the p38 interferes with the transactivation potential of the full-length protein. Also the same figure shows the nuclear subcellular localization of this protein form. Then we can search for both GO terms in AMIGO and add the IDs to the annotation table.
4. The "Interaction with" column is used with the GO term "protein binding" to indicate to the binding partner. Please add the corresponding UniProtKB Acc and/or PRO ID. Examples with "Interaction with" column are found in any of the entry annotations from the PRO terms listed in Fig. 6.7.
5. The "Relative to" column is used only in conjunction with modifiers of the type increased, decreased and altered. In this column add the reference protein to which the protein form is being compared to. Either provide its UniProtKB Acc, its PRO ID, or its name. The annotation for rho-associated protein kinase 1 isoform 1 cleaved 1 in Fig. 6.6 has one such example: increased has_function GO:0004674 Relative to PRO:000002529 (rho-associated protein kinase 1 isoform 1 unmodified form).

3.1.4.1.3. Comment Section

Just add any comment that clarifies any of the content.

3.1.4.1.4. Saving/ Submitting the Annotation

These options are found in the right upper corner of the RACE-PRO form. The save option allows saving the data in case you have not finished and need to complete the annotation later. When you save you are given a REF number; you can insert this number in the UniProtKB identifier box to retrieve your entry. Submit is used when you are done with the entry. You will still have the same reference number. Please keep it for tracking purposes.

3.1.4.1.5. What Happens Next?

An editor from the PRO team will review the entry and send you back comments/suggestions. Then the corresponding PRO term is generated along with the annotations. These will have the corresponding source attribution.

3.2. Conclusion

The PRO website can be used to retrieve information about the various protein forms derived from a given gene and to learn about their relationships. The integrated information for each form can be viewed in the entry report that collects information about the ontology and annotation (whenever available), and also provides mappings to external databases. This website constitutes a highly valuable resource providing a landscape of protein diversity and associated properties that is relevant for proteomics analysis.

4. Notes

1. Recently PRO has been funded to include protein complexes, so be aware that the structure of the framework may look slightly different in the future but the definition of each of the existing levels should not change. In addition, the PRO ID will soon change from PRO: to PR: to avoid confusion with other existing database identifiers.
2. Some search tips:
   (a) If you want to retrieve all the entries from a given category, for example, all the nodes for gene product level, then search selecting the category field and type gene. Search for category has the following options: family, gene, sequence, and modification.
   (b) Some of the search fields are of the type null/not null. This is the case for the ortho-isoform and ortho-modified form. So if you are interested in retrieving the ortho-isoform entries, please select as a search field ortho-isoform and type not null.
   (c) The specifics about what are the options for the DB ID, Modifiers and relations fields are listed in the PAF guidelines (*see* Subheading 2).

Acknowledgments

PRO Consortium participants: Protein Information Resource, The Jackson Laboratory, Reactome, and the New York State Center of Excellence in Bioinformatics and Life Sciences. PRO is funded by NIH grant #R01 GM080646-01.

References

1. The Gene Ontology Consortium. (2000) Gene ontology: tool for the unification of biology. *Nat. Genet.* **25**, 25–29.
2. Li, D., Li, J-Q., Ouyang, S-G., Wang, J., Xu, X., Zhu, Y-P., He, F-C. (2005) An integrated strategy for functional analysis in large-scale proteomic research by gene ontology. *Prog. Biochem. Biophys.* **32**, 1026–1029.
3. Natale D., Arighi C., Barker W.C., Blake J., Chang T., et al. (2007) Framework for a Protein Ontology. *BMC Bioinformatics* **8** (Suppl 9), S1.
4. Arighi, C.N., Liu, H., Natale, D.A., Barker, W.C., Drabkin, H., Blake, J.A., Smith, B., Wu, C.H. (2009) TGF-beta signaling proteins and the Protein Ontology. *BMC Bioinformatics* **10** (Suppl 5), S3.
5. Brown-Bryan, T.A., Leoh, L.S., Ganapathy, V., Pacheco, F.J., Mediavilla-Varela, M., Filippova, M., Linkhart, T.A., Gijsbers, R., Debyser, Z., Casiano, C.A. (2008) Alternative splicing and caspase-mediated cleavage generate antagonistic variants of the stress oncoprotein LEDGF/p75. *Mol. Cancer Res.* **6**, 1293–1307.
6. Nchoutmboube, J., Arighi, C.N., and Wu, C.H. (2009) Data integration and literature mining for the curation of protein forms in the protein ontology (PRO). *BIBM09, IEEE International Conference on Bioinformatics & Biomedicine*, Washington, DC.
7. URL: http://www.obofoundry.org/.
8. URL: http://bioportal.bioontology.org/.
9. Day-Richter, J., Harris, M.A., Haendel, M.; Gene Ontology OBO-Edit Working Group, Lewis, S. (2007) OBO-Edit – an ontology editor for biologists *Bioinformatics* **23**, 2198–2200.
10. Eilbeck, K., Lewis, S.E., Mungall, C.J., Yandell, M., Stein, L., Durbin, R., Ashburner, M. (2005) The Sequence Ontology: a tool for the unification of genome annotations *Genome Biol* **6**, R44.
11. URL: http://psidev.sourceforge.net/mod/.
12. URL: http://www.ncbi.nlm.nih.gov/sites/entrez?db=omim.
13. Finn, R.D., Mistry J., Schuster-Bockler, B., Griffiths-Jones, S., Hollich, V., et al. (2006) Pfam: clans, web tools and services. *Nucleic Acids Res.* **34**, D247–D251.
14. Vastrik, I., D'Eustachio, P., Schmidt, E., Joshi-Tope, G., Gopinath, G., et al. (2007) Reactome: a knowledge base of biologic pathways and processes. *Genome Biol.* **8**, R39.
15. UniProt Consortium. (2010) The Universal Protein Resource (UniProt) in 2010. *Nucleic Acids Res.* **38**, D142–D148.
16. URL: http://proteininformationresource.org/pirwww/search/idmapping.shtml.
17. Wu, C.H., Nikolskaya, A., Huang, H., Yeh, L-S., Natale, D.A., Vinayaka, C.R., Hu, Z., Mazumder, R., Kumar, S., Kourtesis, P., Ledley, R.S., Suzek, B.E., Arminski, L., Chen, Y., Zhang, J., Cardenas, J.L., Chung, S., Castro-Alvear, J., Dinkov, G., Barker, W.C. (2004) PIRSF family classification system at the Protein Information Resource. *Nucleic Acids Res.* **32**, D112–D114.

Chapter 7

Structure-Guided Rule-Based Annotation of Protein Functional Sites in UniProt Knowledgebase

Sona Vasudevan, C.R. Vinayaka, Darren A. Natale, Hongzhan Huang, Robel Y. Kahsay, and Cathy H. Wu

Abstract

The rapid growth of protein sequence databases has necessitated the development of methods to computationally derive annotation for uncharacterized entries. Most such methods focus on "global" annotation, such as molecular function or biological process. Methods to supply high-accuracy "local" annotation to functional sites based on structural information at the level of individual amino acids are relatively rare. In this chapter we will describe a method we have developed for annotation of functional residues within experimentally-uncharacterized proteins that relies on position-specific site annotation rules (PIR Site Rules) derived from structural and experimental information. These PIR Site Rules are manually defined to allow for conditional propagation of annotation. Each rule specifies a tripartite set of conditions whereby candidates for annotation must pass a whole-protein classification test (that is, have end-to-end match to a whole-protein-based HMM), match a site-specific profile HMM and, finally, match functionally and structurally characterized residues of a template. Positive matches trigger the appropriate annotation for active site residues, binding site residues, modified residues, or other functionally important amino acids. The strict criteria used in this process have rendered high-confidence annotation suitable for UniProtKB/Swiss-Prot features.

Key words: PIR Site-rules, Functional sites, Functional annotation, PIR, Features

1. Introduction

Success in high-throughput genome sequencing projects and structural genomic initiatives has put tremendous pressure on development of computational approaches for large-scale functional annotation. There are over ten million sequences deposited in public databases. However, experimental characterization of these proteins lags far behind, indicating a need for the development of systematic approaches for reliable error-free transfer of functional

Cathy H. Wu and Chuming Chen (eds.), *Bioinformatics for Comparative Proteomics*, Methods in Molecular Biology, vol. 694, DOI 10.1007/978-1-60761-977-2_7, © Springer Science+Business Media, LLC 2011

annotations from the characterized proteins to the remaining set that lack any functional information. This realization, not surprisingly, has given rise to many tools, methodologies and servers that are aimed at function prediction and propagation of functional information (1–4). Most attempt to annotate certain global properties based on conservation of sequence and/or gene order within defined protein families (5–9).

It is generally recognized that, with respect to family-based annotation, greater specificity yields greater accuracy and confidence. Assuming that each family comprises proteins that share end-to-end similarity (homeomorphicity), specificity can be increased based on a number of different parameters. Simplest of these is classification (family vs. subfamily). For example, annotation specificity is increased if a family of insulin-like growth factor binding proteins is subdivided into subfamilies one through six, each representing a different subtype that could be applied to a potential member. Another specificity discriminator is taxonomy. In some cases the activities of same-family proteins are known or suspected to differ in one branch of the taxonomic tree but, more typically, differing nomenclatures are used. An example of the latter is the Sec preprotein translocase, which in eukaryotes is known as Sec61, but in prokaryotes is known as SecY.

Yet another basis for discriminating between different possible annotations is the presence of specific amino acid residues at specific positions in the sequence. This could have important implications for binding specificity, or in determining the likelihood that a given protein is competent to catalyze an enzymatic reaction. It is not always possible to distinguish between these possibilities based on the typical parameters used to create protein families; alternative methods are needed.

The UniProt Consortium aims to provide high quality annotation to proteins based both on experiment and on confident prediction (10). Here we describe an approach that addresses the need for residue-specific discrimination as implemented in the PIR Site Rule system. The method described in this chapter requires that proteins be classified based on both evolutionary relatedness and homeomorphicity. As mentioned earlier we use the PIRSF protein classification system, a hierarchical classification of whole proteins designed to accurately propagate biological and biochemical information from proteins with known experimental characterization to those without (11). However, although the hierarchical classification allows for increased specificity in some cases, global sequence similarity does not afford the fine-tuning of annotation that, instead, can be provided using site-specific rules. We call the site-specific rules that make use of the PIRSF classification system "PIR Site Rules" (PIRSR). We take a three-step approach to determining whether or not annotation is appropriately given. Briefly, these are: (1) determine if the protein belongs to a family that contains proteins related to one with the supposed activity; (2) determine if the protein contains the conserved regions found in proteins known to have the supposed activity; and (3) determine if the protein contains the precise amino acids required for the supposed activity. Using condition-based rules, the approach combines information from sequence, structure, domains, motifs, and common ancestry to both make predictions of global function and to provide annotation (herein called "features") to individual amino acids.

2. Materials

1. Curated homeomorphic protein families (PIRSFs).
2. Three-dimensional structural representative from each family.
3. Structure guided multiple sequence alignment for each PIRSF.
4. Site-rule system (Site-HMM alignments and propagation).

3. Methods

3.1. Site-Types for Annotation

Primary annotations (to individual residues) include the UniProt feature keys ACT_SITE (amino acid/s involved in the activity of an enzyme), BINDING (binding site for any chemical group such as coenzyme, prosthetic group, substrate etc.), DISULFID (disulfide bond), METAL (binding site for a metal ion), MOD_RES (posttranslational modification of a residue), MOTIF (short sequence motif of biological interest), NP_BIND (extent of a nucleotide phosphate-binding region), REGION (extent of a region of interest in the sequence), SITE (any interesting single amino acid, that is not defined by another feature key), MUTAGEN (amino acid that has been experimentally altered by mutagenesis) and CARBOHYD (glycosylation site). These are annotated manually for the template and then the corresponding annotations are specified in the rule. The dependent UniProtKB fields – such as keywords (KW) and comments (CC), are also annotated in the template and specified in the rule. For example, an entry with protease active sites will get the keyword "Protease," and the name given to the protein should reflect that activity (or at least not conflict with it). Full curation effort is given to every line within the scope of a rule, including references, keywords, comments, and cross-references to PDB.

3.2. Site Rule Definition

PIR site rules are defined starting with PIRSF families that contain at least one known 3D structure with experimentally verified site information in published scientific literature. As references for site selection, a few protein structure resources are used, including the binding site information from PDB (12), LIGPLOT interactions in PDBSum (13), and enzyme active sites (catalytic residues) documented in Catalytic Residue Dataset (14) and Catalytic Site Atlas (15). The rules are manually defined following extensive scientific literature review to determine site residues (confirmed by site-directed mutagenesis or other experimental data) and mechanism of action. Each PIR site rule consists of the rule ID, template sequence (a representative sequence with known 3D structure), rule condition and feature for propagation (denoting site feature to be propagated if the entire rule condition is tested true). The rules are PIRSF-specific and there may be more than one site rule for a PIRSF family. Each family can have as many site rules as there are site types (ACT_SITE, BINDING, etc.). That is, every site type has a single rule dedicated to it for each family, and this rule will be used to annotate every appropriate residue of that type. A residue can be annotated by more than one rule. For example, an active-site cysteine can be also involved in disulfide bonding (see thioredoxin example below). However, if a residue is involved in both binding to a ligand and mediates catalysis, and if supporting experimental literature data are available for its role in catalysis, this residue is annotated under the rule defining ACT_SITE and not BINDING. Under some circumstances multiple residues are permissible at a given position. For example, if a particular position is found to be an Asp in 50% of the cases and a Glu in the remaining half, this flexibility is noted in the rule.

3.3. Site Rule Curation

PIR Site Rules are created for a given PIRSF homeomorphic protein family using a dedicated editing system. The editor enables the curator to input the various feature information (derived from the chosen structural template) that is desired to propagate to a typical UniProtKB/Swiss-Prot entry; specifically, the annotation fields FT (feature), CC (comment), and KW (keyword). Appropriate syntax and controlled vocabulary are used for site description and evidence attribution. The controlled vocabulary for annotating UniProtKB sequence feature lines include terms for feature type (such as "ACT_SITE" for catalytic residues, BINDING for binding site residues and METAL for binding of metal ions) and corresponding terms for feature descriptions (such as "nucleophile" or "proton acceptor"). The interface allows for an edited structure-guided alignment to be imported as the basis for the Site HMM (see below). Criteria used to curate a Site Rule are described in later sections.

3.4. Site Rule Match Conditions

3.4.1. Family HMM

Rule construction begins with a curated PIRSF, the process of which is described elsewhere (6, 11). The important steps for Site Rule purposes include membership verification – which includes a size-based filter – and selection of sequences for family HMM construction (used for future recruitment). For each family, a set of seed sequences is chosen for multiple alignment based on evolutionary distance, with more-distant members preferred. The alignments are not edited. The family HMM is generated based on the seed-sequences alignment using hmmbuild and hmmcalibrate – both part of the HMMER package (16) – using default parameters. Protein sequences that score better than the HMM cut-off (set to that of the lowest-scoring true member) and whose length does not deviate beyond what is normal for the family are recruited automatically.

3.4.2. Site HMM

The family HMM described in the previous section, optimized for a different purpose, may not be suitable as a discriminator for a particular site of interest. Therefore a second "site HMM" is created for this purpose. The "seeds" for the site HMM always include a template that serves as a source of information for rule-based annotation. Structure-guided manual editing of the alignment is done after visual inspection using an alignment editor, and score thresholds are manually set if necessary to verify that the residues of interest in the template are conserved among the aligned sequences. Conservation in this context means that any difference between the template and aligned sequence at critical positions is either known to be allowed, or that such difference is a plausible chemical and structural replacement. Any sequence containing a nonconserved residue in important positions is removed from the alignment. In certain cases, some regions from the alignment are removed. The removed regions usually include largely nonconserved residues, or partially conserved residues that are not important to define the site. The remaining conserved regions of the alignment covering the propagatable residues are concatenated to form the site specific alignment. The site HMM is then generated based on this site specific alignment using HMMER. The site-specific HMM is thus much more focused on the propagatable residues than the original full-length family HMM. Using the site-specific HMM also increases the signal to noise ratio compared to the original family HMM (data not shown).

3.4.3. Site Match

The final match condition that must be defined is an enumeration of the amino acids of interest. Included are those amino acids that are directly involved in catalysis, binding, bonding, or modification and those that contribute to activity. The sequence of each member of the appropriate family that passes the previous two filters will be checked by aligning both the template and target

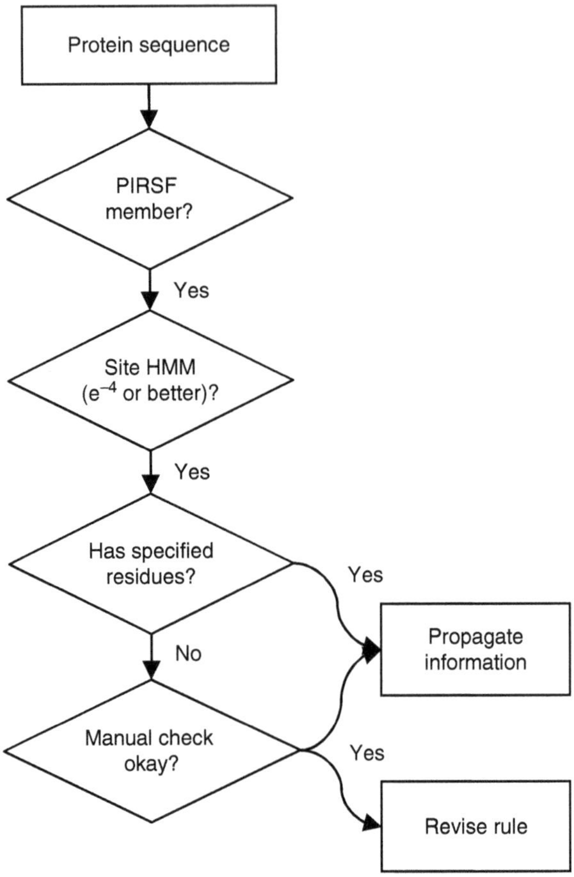

Fig. 1. PIR Site rule propagation pipeline.

sequences to the profile HMM using hmmalign. Target residues that match those defined in the rule are eligible for propagation of information, if so specified. Conservative substitutions are allowed after manual inspection. The substitutions that are generally allowed are Asp/Glu, Lys/Arg, Tyr/Phe, and Ser/Thr. Additional substitutions may be allowed if supported by experimental evidence or acceptable by curator judgment.

3.5. Rule Propagation

In order to apply the rules to UniProtKB, each rule must specify both the feature annotation (given to the appropriate amino acids) and any annotation that is logically dependent on features in a typical UniProtKB/Swiss-Prot entry. Feature information on specific residues is propagated only to those proteins that pass all the conditions described in the previous sections (see flowchart in Fig. 1). Those that do not pass the conditions are deemed suspect and are subjected to manual scrutiny. Failures prompt a review of the rule itself to determine if revision is warranted.

```
           10        20        30        40
            |         |         |         |
P0AA25  IHLTDDSFDTDVLKADGAILVDFWAEWCGPCKMIAPILDE
Q9X2T1  VNVTDASFEQDVLKADGPVLVDYWAEWCGPCKMIAPVLDE
Q9K8A8  VNVTDQTFAQETSEG--LVLADFWAPWCGPCKMIAPVLEE
Q9S386  KEITDATFEQETSEG--LVLTDFWATWCGPCRMVAPVLEE
P09857  VHVTDDSFEEEVXKSPDPVLVDYWADWCGPCKMXAPVXDE
Q97IU3  QEINDKSFVNVISNSKKVVVVDFWATWCEPCKMIAPILEE
P56430  IELTEENFESTIKKG--VALVDFWAPWCGPCKMLSPVIDE
P57653  IELTDQNFEEQVLNSKSFFLVDFWAQWCNPCKILAPILEE
Q9CF37  YNITDATFDEETKEG--LVLIDFWATWCGPCRMQAPILEQ
Q97EM7  KEINESIFDEEIKTSGEPVIVDFWAPWCGPCKMLGPIIDE
Q9ZEH4  VKVTDADFDSKVESG--VQLVDFWATWCGPCKMIAPVLEE
P51225  SQVTDASFKQEVINNDLPVLVDFWAPWCGPCRMVSPVVDA
Q8ZAD9  IHLSDDSFDTDVLKASGLVLVDFWAEWCGPCKMIAPILDE
Q7M0Y9  KDINDSNFQEEVKAG--TVVVDFWAAWCGPCKMLGPVIDE
P14949  VKATDQSFSAETSEG--VVLADFWAPWCGPCKMIAPVLEE
Q9KV51  LQLTDDGFENDVIKAAGPVLVDFWAEWCGPCKMIAPILDE
P43785  LHINDADFESVVVNSDIPILLDFWAPWCGPCKMIAPVLDE
            .  :           . :.:: :: ::..  :. .
PIRSR000077-1                ^     !  !
PIRSR000077-2                ^     ^!!^
PIRSR000077-3                !     ^  ^
PIRSR000077-4                      !  !
```

Fig. 2. Partial alignment of seed sequences for PIRSF000077 Site HMM. *Single dots* indicate highly conserved residues, whereas *double dots* indicate perfectly conserved residues. For each rule, residues subject to propagation are indicated by exclamation points, whereas additional checked residues are indicated by *carets*.

3.6. Strict Site Rule Propagation Criteria

A PIR Site Rule is only applied to sequences from the corresponding PIRSF family. Sequences must therefore pass the family HMM by computational means, or be assigned to the family manually. The sequences that match the site specific HMM with an *e*-value better than the 10^{-4} cutoff are then automatically checked for functional residues by aligning the potential target sequence with the rule-specified template sequence and verifying the presence of the rule-specified residues (Fig. 1). To avoid false positives, site features are only propagated automatically if all site residues match perfectly in the conserved region. When propagation is warranted, the position (within the target sequence) and identity of the appropriate residues are noted, along with the propagatable information specified for those residues.

Potential functional sites missing one or more residues or containing conservative substitutions not already noted by the rule are only annotated after careful manual analysis on a case by case basis. For example, if a rule specifies an Asp at a given position and a particular sequence contains a Glu at that position, the sequence fails the automatic propagation, but annotation might be allowed after considering the available structural or mutational evidence. Manual review is also triggered when a sequence fails to reach the cutoff *e*-value.

3.7. Rule Evidence Attribution

Associated with the rule-based automated annotation is evidence tagging that distinguishes experimentally-verified from computationally-predicted annotation (17). In the new UniProtKB evidence attribution system, all protein annotation will be attributed for the data source, the types of evidence and methods for annotation.

3.8. System Implementation

To facilitate Site Rule curation and propagation, a specialized system has been developed. This system includes three components: the backend Oracle database, the underlying application procedures and the curator interface. The underlying procedures connect the alignment and HMM programs for rule curation and propagation. The curator web interface is built using Perl/Javascript.

3.9. Site Rule Creation Criteria

Not every family is amenable to Site Rule creation. Members of families containing at least one member with residue-specific experimental characterization can be annotated with both local and global information. Suitable experimental data include kinetic, mutation, spectroscopic and structural data. Furthermore, the data should define active, binding, metal, or some other functional site. Lacking this, local annotation can still be provided if a member has a solved structure with a bound ligand, or has a solved structure that can be aligned to other structures with residue-specific annotation.

4. Case Studies

4.1. Case Study 1: Thioredoxin

The thioredoxin family (PIRSF000077) has over 1,000 members found in archaea, prokaryotes, eukaryotes, and viruses. There are four Site Rules covering this family: PIRSR000077-1, PIRSR000077-2, PIRSR000077-3, and PIRSR000077-4. Each uses the *Escherichia coli* strain K12 thioredoxin (UniProtKB accession P0AA25) as a template.

4.1.1. Underlying Biology

Thioredoxins regulate other enzymes by reducing their disulfide bonds. Two vicinal cysteines are involved in this process. Cys33 of reduced thioredoxin (lacking a Cys33-Cys36 disulfide bridge) is exposed to the solvent and loses the proton to act as a nucleophile. Cys36 is buried and hence cannot initially act as nucleophile. Nucleophilic attack of Cys33 on the substrate disulfide bond results in the formation of a mixed disulfide intermediate. A conformational change then exposes Cys36, allowing Asp27 to deprotonate it (18). The nucleophilic Cys36 attacks the mixed disulfide intermediate to produce reduced enzyme and an oxidized thioredoxin (with a Cys33-Cys36 disulfide bridge) (19). Gly34 and Pro35 modulate the redox potential of the active site disulfide bond (20). By varying either of these two residues, optimal redox potential can be obtained.

Table 1
Site Rules for PIRSF000077 (thioredoxin)[a]

Rule ID	Feature key	Checked residues	Propagated features/annotated residues		PubMed ID
1	ACT_SITE	D27, C33, C36	C33 C36	Nucleophile	2181145
2	SITE	D27, C33, G34, P35, C36	G34 P35	Contributes to redox potential value	9099998 10489448
3	SITE	D27, C33, C36	D27	Deprotonates active site C-terminal Cys	9374473 11563970
4	DISULFID	C33, C36	C33-C36	Redox-active	2181145

[a]All rules use the *E. coli* thioredoxin (UniProtKB accession P0AA25, PDB accession 2TRX:A) as a template, and include the Catalytic Site Atlas (15) entry 2TRX as a reference. Rule IDs are prepended by PIRSR000077 and a *dash*

4.1.2. Rule Construction

The underlying biology of thioredoxin activity indicates four types of functions (features) attributable to specific residues. Accordingly, four rules were constructed (Table 1). The feature key for site rule 1 (PIRSR000077-1) is ACT_SITE, indicating that the residues specified under "propagated features" are directly involved in catalytic function, specifically, as nucleophiles (18). Site rule 2 (PIRSR000077-2) and site rule 3 (PIRSR000077-3) uses the feature key SITE, used for residues with critical functions that do not fit any of the other specific UniProt feature keys. The variation corresponding to Gly34 and Pro35 of the template in various organisms is allowed as this is the mechanism to modulate the redox potential. PIRSR000077-2 indicates the contribution of these two residues to the redox potential of the disulfide bond, whereas PIRSR000077-3 indicates the deprotonating function of Asp27. The feature key DISULFID is used for PIRSR0000077-4, specifying the disulfide bonds between Cys33 and Cys36 and indicating that they are redox active. Note that, for rules one through three, other residues are checked in addition to those eligible for propagation. This reflects the underlying biology. For example, Cys36 specified in rule PIRSR000077-1 will fail to act as nucleophile if Asp27 is substituted by Asn. Note also that not every highly-conserved residue is checked – only those pertinent to the annotation (Fig. 2).

4.1.3. Propagation

There are 152 members of PIRSF000077 in the reviewed (Swiss-Prot) section of UniProtKB, (currently, Site Rules are being applied only to this section). All 152 members passed the Site HMM test with an *e*-value of 10^{-4} or better. They then were tested for the presence of the two cysteines (Cys33 and

Cys36) and the aspartic acid (Asp27) by hmmalign. One-hundred forty-two members passed this test. The remaining ten had both the cysteines, but not the aspartic acid, where instead they had either a His, Glu, or Tyr in the corresponding position. The function of Asp27 in the template is to deprotonate the active site Cys36 by accepting a proton. This aspartic acid can be conservatively substituted by Glu, but not by residues such as His or Tyr; the higher pKa values of the latter two mean they are likely to be already protonated and therefore cannot act as proton acceptors. Six of the ten mismatches have the conservative Glu substitution, and are therefore eligible for the annotation, bringing the total to 148. To indicate that these annotations are based on sequence similarity to the template, a "By similarity" tag is added to each of the annotations. However, proteins that have experimental data on the individual sites will not have this tag.

4.2. Case Study 2: Oxygen-Independent Coproporphyrinogen III Oxidase

The so-called "radical SAM enzymes," which generate radical species by reductive cleavage of *S*-adenosyl-L-methionine (SAM), catalyze a diverse set of reaction types, including methylation, isomerization, sulfur transfer, and anaerobic oxidation (21). One such enzyme is the oxygen-independent coproporphyrinogen III oxidase (PIRSF000167). The family contains a total of 1,400 members. There are four Site Rules covering this family: PIRSR000167-1, PIRSR000167-2, PIRSR000167-3, and PIRSR000167-4. Each uses the *E. coli* enzyme (UniProtKB accession P32131) as a template.

4.2.1. Underlying Biology

During the biosynthesis of heme and chlorophyll, coproporphyrinogen III oxidase converts coproporphyrinogen III to protoporphyrinogen IX by oxidatively decarboxylating the propionate side chains of rings A and B to their corresponding vinyl groups. Two unrelated enzymes – oxygen-dependent coproporphyrinogen III oxidase (encoded by the HemF gene), and the oxygen-independent coproporphyrinogen III oxidase (encoded by the HemN gene)–catalyze these reactions. The HemN family members contain a conserved iron–sulfur cluster (4Fe-4S) binding motif (CxxxCxxC). In its reduced state, the cluster transfers a single electron to SAM, resulting in its reductive cleavage to methionine and a 5′-deoxyadenosyl radical. This highly reactive oxidizing radical abstracts a hydrogen atom from the β-C atom of the propionate side chain of coproporphyrinogen III. These steps occur twice – once for each bound SAM (22). Between each round, the oxidized cluster (created during the reduction of SAM), is once again reduced.

Table 2
Site Rules for PIRSF000167 (oxygen-independent coproporphyrinogen III oxidase)[a]

Rule ID	Feature key	Checked residues	Propagated features/annotated residues		PubMed ID
1	BINDING	Y56, G112, E/D145	Y56 G112 E/D145	*S*-adenosyl-L-methionine 1	14633981
2	METAL	C62, C66, C69	C62 C66 C69	Iron–sulfur (4Fe-4S-S-AdoMet)	14633981
3	BINDING	F/Y68, G113, T/S114,Q172, R184, D209	F/Y68 G113 T/S114 Q172 R184 D209	*S*-adenosyl-L-methionine 2	14633981

[a]All rules use the *E. coli* oxygen-independent coproporphyrinogen III oxidase (UniProtKB accession P32131, PDB accession 1OLT) as a template, and include the Catalytic Site Atlas entry 1OLT as a reference. *Slashes* indicate allowed conservative substitutions

4.2.2. Rule Construction

The template structure (1OLT) presents the oxygen-independent coproporphyrinogen III oxidase bound to two SAM molecules. There is also available experimental mutagenesis data for this protein. Based on the data, four site rules covering three different feature keys have been created (Table 2). Site-Rules 1 and 3, with feature key BINDING, are used to annotate the residues involved in binding to the two SAM molecules (SAM1 and SAM2). Site Rule 2, with feature key METAL, is used to annotate the cysteine residues involved in coordinating the iron–sulfur cluster. The structure-guided alignment used for creating the rules is given in Fig. 3.

4.2.3. Propagation

Of the 29 members of PIRSF000167 that are in the Swiss-Prot section of UniProtKB, all passed the Site HMM test with an *e*-value of 10^{-8} or better. They were then tested for the presence of the binding site residues for the two SAM molecules (SAM1: Tyr56, Gly112, and Glu/Asp145; SAM2: Phe/Tyr68, Gly113, Thr/Ser114, Gln172, Arg184, Asp209) and for the iron–sulfur cluster residues (Cys62, Cys66, and Cys69) by hmmalign. Twenty-eight passed the criteria defined by the three Site Rules. The one failure is a sequence fragment.

```
                       60        70            90             110       120       130  140       150       160       170
                        |         |  \/         |      \/       |         |         | \/ |         |         |         |
1OLT             LSLYVHIPFCHKLCYFCGCNKHKADQYLDALEQEIVHRAGRHVSQLHWGGGTPTYLNKAQISRLMKLLREADAEISIEVDPREIELDVLDHLRAEGFNRLSMGVQDFN
O25824           MILYIHIPFCENKCGYCAFNSGLKEEYTQALCLDLKHALDEPIESVFIGGGTPNTLSVKAFERIFESIYQLDCEITTEANPELITKAWCQGLKGLGINRLSLGVQSFR
O34162           LSLYLHIPFCENICYYCGCNKGRSARYVNYLGREMALVARRQVLQSHWGGGTPTFLDPGEMRRVMALLHEAEGEHSIEIDPRRVDHARMALLAELGFNRVSLGVQDFD
O67418           KALYVHVPFCSYKCPYCDFLSVSVRDYLSALKREINLY-DVKLETLYFGGGTPTLLKPEELGEIIEEIDKGVKEITVECNPETYRKKEFEELKKLGVNRISVGVQSFT
O83216           ASLYVHIPFCAQRCAYCDFYSQPCPHFIDRLLQDVALQRVQGWQTVYMGGGTPSLLAPQDIRHFCVALRA-IQEFTLEVNPEDVTEEFLCACAEGGVNRLSLGVQSLR
O84055           LALYIHIPFCSKKCHYCSFYTELMRSYCEAVIKEGLKKLSHYIDTVFFGGGTPSLVPPALIQDILVALEAHATEITLEANPENLSLEYIQALALTSINRISIGVQTFN
P43899           LSLYIHIPWCVQKCPYCDFNSIPEQDYIYHLLQDLQADLQRKLHSIFIGGGTPSLFSAESIAYLLKEIKKDNIEITLEANPGTVEAERFKGYVSAGIMRISMGIQSFN
P54304           KSAYIHIPFCEHICHYCDFNKQPVDEYLNALEQEMINTIQPDLKTIFIGGGTPTSLSEEQLKKLMDMINRSDSEFAVDANPDDLSAEKLKILKEAGVNRLSFGVQTFE
P57615           ISLYIHIPWCLKKCGYCDFYSIPENKYIEHLLRDFERDLNRNINTIFIGGGTPSLLKNTSIKNLLNGIKKKNIEISIEANPKTLEYQNFIQYKNSGINRFSLGIQTFN
P71756           FGVYLHVPFCLTRCGYCDFNTVSPDRWLLALRAELELAAAPTVHTVYVGGGTPSLLGGERLATLLDMVRDPDAEVSTEANPESTWPEFFATIRAAGYTRVSLGMQSVA
P73245           TAAYIHIPFCRQRCFYCDFPIGWVGEYVEAVCREIAGQQGQPLQTVFFGGGTPSLLPITGLEKILLAVDQPDAEISIEIDPGTFDQVQLQGYKNLGINRFSLGVQAFQ
P95651           ISVYIHVPFCERLCWFCACPTAPVEAYVGTLLQELELVKGVKAGRLHWGGGTPTILSPELIHKLAQAIKAEDYEFSVEIDPMMVDEPKIRALSEEGMNRASIGIQDFT
Q5HVJ0           MHFYIHIPFCESKCNYCAFTSDYEKAYFKALKEDIVFQLSNQIKTLFIGGGTPSCVNAYNYEDIFKILYPKNVEISCEANPNSATLNWLKNMKNLGVNRISFGVQSFH
Q5N1A7           IAAYLHIPFCRRRCFYCDFPIGMIRQYVDAICQEIAATPGPALQTVFFGGGTPSLLSPNQLDRILQTLDRATAEISIEMDPASFDRSQAIATRQLGFNRVSIGAQAFQ
Q5NII4           IGIYIHFPWCVRKCPYCDFNSYLSEQYYQKLIADFDTHLAREVISIFIGGGTPSLFKAEYLDKVLQHITKINCEITLEMNPGTVERGSILSYQNIGINRISLGVQSFQ
Q5NMS2           ISIYLHVPFCQQLCHYCGCHTEIITAYVDTLVKEIETVGHLKIGHLSWGGGTPSIIGIEGIRRVMKALQREDMIHAMELDPRLLDAEFAQNLFDIGINRTSLGVQDFN
Q64PH0           AGIYIHVPFCKTRCIYCDFYSEWKGRYIEALCKELEMRYGKPIETLYFGGGTPSQLDEKDFRKVFDTVRRNCHEITLEANPDDLCPEYLQMLSELPFNRISMGIQTFD
Q64V17           LSFYLHIPFCRHLCHYCGCNSEIIESYVEALHQEIDLILDRPIAQIHYGGGSPTAIPVALIKELNAHLLSDRPEIAIECHPGYLSEKDWLQLTECGFNRLSIGVQDFN
Q6AQC8           SSLYLHIPFCRSKCRYCAFNSHLFVRYIAALKRELSFLFGRPLDTVFFGGGTPTVLGAAGLVEILDHCQQKQAEISTEANPETVSAADLILLRRGGFNRISFGVQSLS
Q8UF01           ISLYLHIPYCDRLCWFCACHTEPIAVYLEALKQEIAAVGDAVVSAVHFGGGSPTMLKPDDMVSLMDCLRRLDPEISVEMDPNDLDDSRYDALAAVGMSRASLGVQDFD
Q8YDS9           TSLYLHIPFCRSMCWYCGCHTQPILDYLDVLHKEIEMIASFNLGEIHFGGGTPTIIQPDELVALMAALRDGELNAAVEIDPRRMTPEMAAALAYSGITRASLGVQSFD
Q8YEV6           FGIYVHWPFCLAKCPYCDFNSVDQARYGAAFRREMETLRPRTVTSIFLGGGTPSLMQPETVGALLDGIASPDIEMTLEANPTSVEADRFRGYRAAGVNRVSLGIQALN
Q8YG68           VSLYIHVPYCSQICHYCGCNADVIENFVNALLGEIRTVSRPRVAHLHWGGGTPSILNAGQFQTVLAAIRTLSMEHAIELDPRTVTPLLAKTLAEMGVNRASLGVQDVD
Q92J50           LSIYIHWPFCVSKCPYCDFNSD-HNQWLKSYETEIEYFKNKYIKSIFFGGGTPSLMNPVIVEGIINKISTNQTEITLETNPTSFETEKFKAFKAVGINRISIGVQSLK
Q97JJ5           LGLYIHIPFCKSKCLYCDFPSDYMIPYAKALCTEIEKASNKVFSSIFIGGGTPSFLSIEALNILCDALKKKDVEFTVEGNPNSFTEKKLMVFKDMGVNRLSIGLQACQ
Q9A8E3           LSVYVHVPFCRRLCWYCGCHTERVGTYADVLAREIDLTAHDGLAHLHFGGGSPNALSPSDFKRLVAQLKTPGAEIAAELDPGMLSDTFVDATGEAGVTRVSLGVQTFD
Q9ABS5           VGIYVHWPYCARICPYCDFNVDEQRALADAIVADLEAQRERRLLSIFFGGGTPSLMDPAQVARVIETAKRDDLEISLEANPTDAETDRFAALAAAGVQRLSLGVQALD
Q9RY22           RHLYVHVPFCPSICPYCDFHVGLVERYLERLEQEAAELAV-DLDTVYLGGGTPSFLRDHELRALVESVQR-RAENTLEINPGTVSTERAALWRSLGFDRASVGVQSLD
Q9X0P9           LAVYVHVPFCKSKCVYCDFYS-KFDEYFSHLLREIDLYEESEIKTVYFGGGTPSVVPPSFLKMVLEKLERTPDEITIEVNPESVETEKLKIYKQIGINRISLGVQACD
Q9Z717           LSLYFHIPFCQSMCLYCGCSVDIVEAYINTLIQEMKLVVRPQVSRIHFGGGTPSRLSRELFTLLFDHIHKHAEEIAIEVDPRSLRNEKADFFQNVGFNRVSLGVQDTQ
Q9ZLH0           LSLYTHLPFCRSACYFCACSVEKKVRYISYLKKELALLKNREVAQFHYGGGTPTFFSPPQLDEITQSIQEQDIEMSCEIDPRHFTKEHMQTLFDRGFNRLSFGVQDFD
                    : : :.:   : .:                               :::.:                      . .  :                 : : : :
PIRSR000167-1       !                                             !                           !
PIRSR000167-2             !   !  !
PIRSR000167-3                  !                                  !!                                                    !
```

Fig. 3. Partial alignment of seed sequences for PIRSF000167 Site HMM. *Dots* and exclamation points are as given for Fig. 2. Numbering is for the template sequence, and accounts for deleted columns (positions of which are indicated by *backward* and *forward slashes*). Two positions of interest for PIRSR000167-3 are beyond the sequence boundary, and therefore are not shown.

5. Discussion

Homology-based annotation transfer – rooted on the idea that proteins that evolved from a common ancestor will have similar function – is by far the most common and widely accepted method for annotation transfer. Although this method generally works for close homologs with sequence similarities greater than 50–60%, an error-rate as high as 30% has been reported (23, 24). One contributor to this error is the reliance on global similarities without accounting for local differences in sequence and without accounting for structural or physico-chemical constraints when such local differences occur. That is, not all substitutions are permissible in all contexts.

The main purpose of PIR Site Rules therefore is to afford a mechanism that allows for high-confidence annotation of both global and local properties of a protein that is supported by three-dimensional structure information. Accordingly, our three-step approach is designed to rigorously determine whether or not predicted annotation is appropriate. Each step performs a different type of certainty check. The importance of all three steps is illustrated by thioredoxin, which requires a CxxC motif. Such a motif occurs in a multitude of proteins, including some zinc-finger transcription factors. It would therefore be ill-advised to annotate every protein with this motif as "thioredoxin." However, if this motif is found in a member of a family related to thioredoxins (first criterion) and furthermore, contains the conserved regions found in known thioredoxins (second criterion), there is some confidence that the member is a real, functional thioredoxin. Conversely, a member of this family that lacks this motif (third criterion) would not actually function as a thioredoxin.

On average, each protein family that serves as a basis for Site Rules has two-to-three rules that potentially apply a total of three-to-four feature lines to each of about 80 reviewed members, and the system has currently added over 18,000 features to over 5,000 entries (data not shown). The propagated information is expected to facilitate experiments by providing potential targets for mutation studies. Indeed, there are several documented examples where mutation in a ligand binding site has led to disease states (25–27). To further extend the reach of Site Rules, we are integrating them into the UniProtKB rule-based annotation propagation system (UniRule) and will apply them to the non-reviewed (TrEMBL) section of the UniProt Knowledgebase in the future, thereby meeting the demand for a confident method for automated annotation of protein sequence features.

Acknowledgments

Thanks are given to our colleagues at the Protein Information Resource (PIR) and to our UniProt collaborators at the Swiss Institute of Bioinformatics (SIB) and European Bioinformatics Institute (EBI) for their support and fruitful discussions. This work was funded under a grant to the UniProt Consortium. UniProt is supported by the National Institutes of Health, grant number: 5U01HG02712-05.

References

1. Date, S.V. (2007) Estimating protein function using protein-protein relationships. *Methods Mol Biol.* **408**,109–127.
2. Glaser, F., Pupko, T., Paz, I., Bell, R.E., Bechor-Shental, D., Martz, E., and Ben-Tal, N. (2003) ConSurf: identification of functional regions in proteins by surface-mapping of phylogenetic information. *Bioinformatics.* **19**, 163–164.
3. Laskowski, R.A., Watson, J.D., and Thornton, J.M. (2005) ProFunc: a server for predicting protein function from 3D structure. *Nucleic Acids Res.* **33**, W89–W93.
4. Standley, D.M., Toh, H., and Nakamura, H. (2008) Functional annotation by sequence-weighted structure alignments: Statistical analysis and case studies from the Protein 3000 structural genomics project in Japan. *Proteins.* **72**, 1333–1351.
5. Tatusov, R.L., Fedorova, N.D., Jackson, J.D., Jacobs, A.R., Kiryutin, B., Koonin, E.V., Krylov, D.M., Mazumder, R., Mekhedov, S.L., Nikolskaya, A.N., Rao, B.S., Smirnov, S., Sverdlov, A.V., Vasudevan, S., Wolf, Y.I., Yin, J.J., and Natale, D.A. (2003) The COG database: an updated version includes eukaryotes. *BMC Bioinformatics.* **4**, 41.
6. Nikolskaya, A.N., Arighi, C.N., Huang, H., Barker, W.C., and Wu, C.H. (2006) PIRSF family classification system for protein functional and evolutionary analysis. *Evol Bioinform Online.* **2**, 197–209.
7. Aziz, R.K., Bartels, D., Best, A.A., DeJongh, M., Disz, T., Edwards, R.A., Formsma, K., Gerdes, S., Glass, E.M., Kubal, M., Meyer, F., Olsen, G.J., Olson, R., Osterman, A.L., Overbeek, R.A., McNeil, L.K., Paarmann, D., Paczian, T., Parrello, B., Pusch, G.D., Reich, C., Stevens, R., Vassieva, O., Vonstein, V., Wilke, A., and Zagnitko, O. (2008) The RAST Server: rapid annotations using subsystems technology. *BMC Genomics.* **9**, 75.
8. Hunter, S., Apweiler, R., Attwood, T.K., Bairoch, A., Bateman, A., Binns, D., Bork, P., Das, U., Daugherty, L., Duquenne, L., Finn, R.D., Gough, J., Haft, D, Hulo, N., Kahn, D., Kelly, E., Laugraud, A., Letunic, I., Lonsdale, D., Lopez, R., Madera, M., Maslen, J., McAnulla, C., McDowall, J., Mistry, J., Mitchell, A., Mulder, N., Natale, D., Orengo, C., Quinn, A.F., Selengut, J.D., Sigrist, C.J., Thimma, M., Thomas, P.D., Valentin, F., Wilson, D., Wu, C.H., and Yeats, C. (2009) InterPro: the integrative protein signature database. *Nucleic Acids Res.* **37**, D211–D215.
9. Lima, T., Auchincloss, A.H., Coudert, E., Keller, G., Michoud, K., Rivoire, C., Bulliard, V., de Castro, E., Lachaize, C., Baratin, D., Phan, I., Bougueleret, L., and Bairoch, A. (2009) HAMAP: a database of completely sequenced microbial proteome sets and manually curated microbial protein families in UniProtKB/Swiss-Prot. *Nucleic Acids Res.* **37**, D471–D478.
10. UniProt Consortium. (2008) The Universal Protein Resource (UniProt). *Nucleic Acids Res.* **36**, D190–D195.
11. Wu, C.H., Nikolskaya, A., Huang, H., Yeh, L-S., Natale, D.A., Vinayaka, C.R., Hu, Z-Z., Mazumder, R., Kumar, S., Kourtesis, P., Ledley, R.S., Suzek, B.E., Arminski, L., Chen, Y., Zhang, J., Cardenas, J.L., Chung, S., Castro-Alvear, J., Dinkov, G., and Barker, W.C. (2004) PIRSF: family classification system at the Protein Information Resource. *Nucleic Acids Res.* **32**, D112–D114.
12. Bourne, P.E., Westbrook, J., and Berman, H.M. (2004) The Protein Data Bank and lessons in data management. *Brief Bioinform.* **5**, 23–30.
13. Laskowski, R.A. (2001) PDBsum: summaries and analyses of PDB structures. *Nucleic Acids Res.* **29**, 221–222.

14. Bartlett, G.J., Porter, C.T., Borkakoti, N., and Thornton, J.M. (2002) Analysis of catalytic residues in enzyme active sites. *J Mol Biol.* **324**, 105–121.
15. Porter, C.T., Bartlett, G.J., and Thornton, J.M. (2004) The Catalytic Site Atlas: a resource of catalytic sites and residues identified in enzymes using structural data. *Nucleic Acids Res.* **32**, D129–D133.
16. Eddy, S.R. (1998) Profile hidden Markov models. *Bioinformatics.* **14**, 755–763.
17. Wu, C.H., Huang, H., Yeh, L.S., and Barker, W.C. (2003) Protein family classification and functional annotation. *Comput Biol Chem.* **27**, 37–47.
18. LeMaster, D.M., Springer, P.A., and Unkefer, C.J. (1997) The role of the buried aspartate of *Escherichia coli* thioredoxin in the activation of the mixed disulfide intermediate. *J Biol Chem.* **272**, 29998–30001.
19. Katti, S.K., LeMaster, D.M., and Eklund, H. (1990) Crystal structure of thioredoxin from *Escherichia coli* at 1.68 A resolution. *J Mol Biol.* **212**, 167–184.
20. Chivers, P.T., Prehoda, K.E., and Raines, R.T. (1997) The CXXC motif: a rheostat in the active site. *Biochemistry.* **36**, 4061–4066.
21. Frey, P.A., Hegeman, A.D., and Ruzicka, F.J. (2008) The Radical SAM Superfamily. *Crit Rev Biochem Mol Biol.* **43**, 63–88.
22. Layer, G., Grage, K., Teschner, T., Schünemann, V., Breckau, D., Masoumi, A., Jahn, M., Heathcote, P., Trautwein, A.X., Jahn, D. (2005) Radical S-adenosylmethionine enzyme coproporphyrinogen III oxidase HemN: functional features of the [4Fe-4S] cluster and the two bound S-adenosyl-L-methionines. *J Biol Chem.* **280**, 29038–29046.
23. Bork, P., and Koonin, E.V. (1998) Predicting functions from protein sequences--where are the bottlenecks? *Nat Genet.* **18**, 313–318.
24. Devos, D., and Valencia, A. (2001) Intrinsic errors in genome annotation. *Trends Genet.* **17**, 429–431.
25. Astner, I., Schulze, J.O., van den Heuvel, J, Jahn, D., Schubert, W.D., and Heinz, D.W. (2005) Crystal structure of 5-aminolevulinate synthase, the first enzyme of heme biosynthesis, and its link to XLSA in humans. *EMBO J.* **24**, 3166–3177.
26. Janosik, M., Oliveriusova, J., Janosikova, B., Sokolova, J., Kraus, E., Kraus, J.P., and Kozich, V. (2001) Impaired heme binding and aggregation of mutant cystathionine beta-synthase subunits in homocystinuria. *Am J Hum Genet.* **68**, 1506–1513.
27. Nakazawa, T., Takai, T., Hatanaka, H., Mizuuchi, E., Nagamune, T., Okumura, K., and Ogawa, H. (2005) Multiple-mutation at a potential ligand-binding region decreased allergenicity of a mite allergen Der f 2 without disrupting global structure. *FEBS Lett.* **579**, 1988–1994.

Part II

Proteomic Bioinformatics

Chapter 8

Modeling Mass Spectrometry-Based Protein Analysis

Jan Eriksson and David Fenyö

Abstract

The success of mass spectrometry based proteomics depends on efficient methods for data analysis. These methods require a detailed understanding of the information value of the data. Here, we describe how the information value can be elucidated by performing simulations using synthetic data.

Key words: Protein identification, Simulations, Synthetic mass spectra, Significance testing, Value of information, Peptide mass fingerprinting, Tandem mass spectrometry

1. Introduction

Mass spectrometry based proteomics is a method of choice for identifying, characterizing, and quantifying proteins. Proteomics samples are often complex and the range of protein amounts is typically large ($>10^6$), whereas the dynamic range of mass spectrometers is limited ($<10^3$) (1). Because of this mismatch, it is necessary to process the protein samples so that the protein mixture that reaches the mass spectrometer at any given time is much less complex. This is often achieved by first separating the proteins, followed by digestion, and separation of the peptides. The peptides are subsequently analyzed in the mass spectrometer.

With mass spectrometry, it is possible to measure the mass and the intensity of peptide ions and their fragments. To identify proteins and to characterize their posttranslational modifications, the mass measurements are used (2–4) and sometimes to lesser degree the intensity measurements can also be used (5, 6). For quantification, the intensity measurements can be used, but only if the intensity scale is calibrated for each peptide, because the intensity of a peptide ion signal depends strongly on its sequence.

Cathy H. Wu and Chuming Chen (eds.), *Bioinformatics for Comparative Proteomics*, Methods in Molecular Biology, vol. 694, DOI 10.1007/978-1-60761-977-2_8, © Springer Science+Business Media, LLC 2011

The two most common types of analysis are peptide mass fingerprinting and tandem mass spectrometry. In both these approaches, the proteins are digested with an enzyme having high digestion specificity (usually trypsin) prior to the mass spectrometric analysis. The digestion results in mixtures of proteolytic peptides. In peptide mass fingerprinting the mass spectrometer detects ions of the proteolytic peptides and measures their respective mass. The mass of a proteolytic peptide is typically not unique (7) and therefore observation of several proteolytic peptides from a single protein is needed to generate a peptide mass fingerprint that is useful for protein identification. The peptide mass fingerprinting approach is usually used for samples where the protein of interest can be purified quite well, because peptide ion signals from different proteins can interfere with each other in an individual mass spectrum and the inclusion of mass values of peptides from more than one protein reduces the specificity of the peptide mass fingerprint. In tandem mass spectrometry, individual proteolytic peptide ion species are isolated in the mass spectrometer and are subjected to fragmentation. The masses of the proteolytic peptides and their fragments are measured, making it more applicable to complex mixtures, because a large amount of information is obtained for each peptide and the interference from peptides originating from other proteins is reduced.

Here we describe a few methods for generating synthetic mass spectra, including peptide mass fingerprints and tandem mass spectra. We also give a few examples of how these synthetic mass spectra can be used to better understand the dependence of the value of information in mass spectra on the nature and accuracy of the measurements.

2. Methods

2.1. Peptide Mass Fingerprinting

In peptide mass fingerprinting, protein identification is achieved by comparing the experimentally obtained peptide mass fingerprint to masses calculated from theoretical proteolytic digests of protein sequences from a sequence collection. Each sequence in the collection that has some extent of matching with the experimental peptide mass fingerprint is given a score, the statistical significance of the high scoring matches is tested, and the statistically significant proteins are reported. The statistical significance is tested by generating a distribution of scores for false and random matches. The score of the high-scoring proteins are then compared to the distribution of scores for false and random matches, and the significance level of the match is calculated. The distribution of scores for false and random matches can be obtained by direct calculations (8), by collecting statistics during the search (9, 10), or by simulations using random synthetic peptide mass fingerprints (11). Here we describe a method for generation of synthetic random peptide mass fingerprints to obtain a distribution of scores for false and random identification that can be used to test the significance of protein identification results (11) (Fig. 1):

1. Analyze the experimental data to obtain information about the parameter space that the synthetic random peptide mass fingerprints should cover, including number of peaks, intensity distribution, mass distribution, and mass accuracy.
2. Select a protein sequence collection, digest it with the enzyme used in the experiment, and calculate the masses of the proteolytic peptides.
3. Randomly pick a set of masses from the proteolytic peptide masses of the sequence collection according to the distributions obtained from the analysis of experimental data, and making sure that no more than one peptide is picked from each protein (see Note 1).
4. Add a mass error sampled from the expected error distribution.
5. Assign intensities to each mass (see Note 2).
6. Search the protein sequence collection and record the highest score.
7. Repeat steps 3–6 until sufficient statistics are obtained, and construct a distribution of scores for false and random identifications.

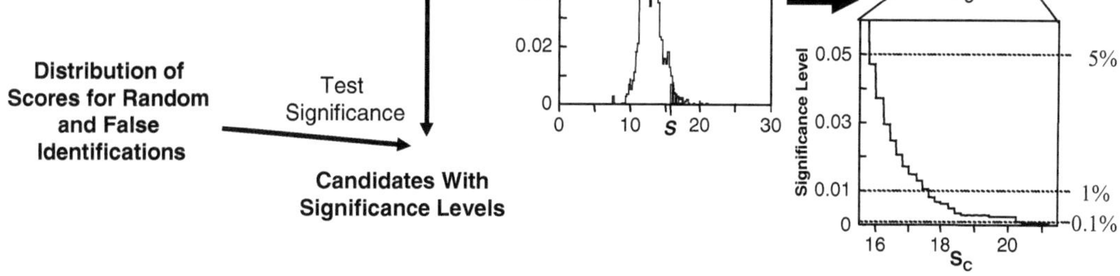

Fig. 1. *Left panel*: The principle of significance testing utilizing the distribution of scores for random and false identifications. *Right panel*: Detailed view of a simulated score distribution for random and false identifications (adapted from (11)).

8. Use the score distribution generated in step 7 to convert the scores from the search with the experimental data to a significance level.

For investigating other aspects of protein identification, it is useful to construct nonrandom peptide mass fingerprints. This can be achieved by modifying step 3:

3a. Select one or more proteins.

3b. For each of the selected proteins, pick a few peptides (see Note 3).

3c. Add background peaks by randomly picking a set of masses from the entire set of proteolytic peptide masses of the sequence collection according to the distributions obtained from the analysis of experimental data, and making sure that no more than one peptide is picked from each protein.

These nonrandom synthetic peptide mass fingerprints can be used to for example improve or compare algorithms, and investigate the effect of search parameters including mass accuracy, enzyme specificity, number missed cleavage sites, and size of sequence collection searched (8, 12). Nonrandom synthetic peptide mass fingerprints have also been used to investigate the potential of identifying complex mixtures of proteins by peptide mass fingerprinting (13). It was concluded that mass fingerprinting could be applied to complex mixtures of a few hundred proteins, if the mass accuracy and the dynamic range of the measurement are sufficient (Fig. 2).

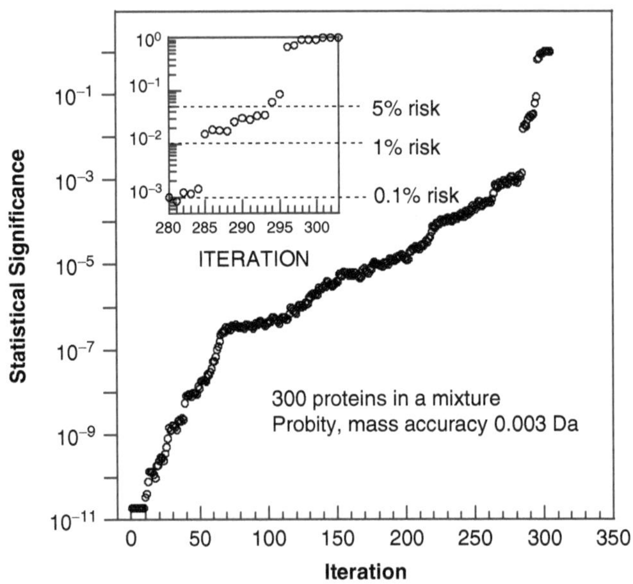

Fig. 2. The statistical significance of proteins identified by peptide mass fingerprinting in a mixture of 300 proteins using an iterative method. The inset displays a magnified portion of the graph for the 280–300th protein identified (Source: ref. 13).

In most practical cases, however, the dynamic range of the measurement is severely limiting and only a few proteins can be identified by peptide mass fingerprinting (14).

2.2. Tandem Mass Spectrometry

The method of choice for complex protein mixtures is to search sequence collections using the observed mass of an intact individual peptide ion species together with the masses of the fragment ions observed upon inducing fragmentation of the peptide in the mass spectrometer. This method requires much lower sequence coverage, and in some cases, even one peptide can be sufficient to identify a protein. Synthetic peptide tandem mass spectra can be generated by the following method:

1. Analyze the experimental data to obtain information about the parameter space of interest (see Note 4 and Fig. 3).
2. Select a protein sequence collection and digest it with the enzyme used in the experiment.
3. Randomly pick a peptide and calculate the peptide mass.
4. Add to the peptide mass an error sampled from the expected error distribution.
5. Calculate the mass of all expected fragment ions.
6. Randomly pick a set of fragment ion masses (Fig. 3a, b).
7. Add to the fragment ion masses an error sampled from the expected error distribution.
8. Assign intensities to each fragment ion mass sampled from the expected error distribution (Fig. 3e).
9. Add background ions by randomly picking peptides that have similar mass as the peptide in step 3, and randomly picking one fragment ion mass from each (Fig. 3c, d).
10. Add to the background masses an error sampled from the expected error distribution.
11. Assign intensities to background fragment ions sampled from the expected intensity distribution (Fig. 3f).
12. Search the protein sequence collection and record the highest score.
13. Repeat steps 6–12 until sufficient statistics are obtained.
14. Repeat steps 3–13 to cover the desired parameter space.

Random synthetic tandem mass spectra can be constructed by skipping steps 3–8 above. These random synthetic tandem mass spectra can be used for significance testing in a similar way as for peptide mass fingerprinting (15).

Nonrandom synthetic tandem mass spectra can, for example, be used to answer the question: How many fragment ions are needed for identification? By generating nonrandom synthetic

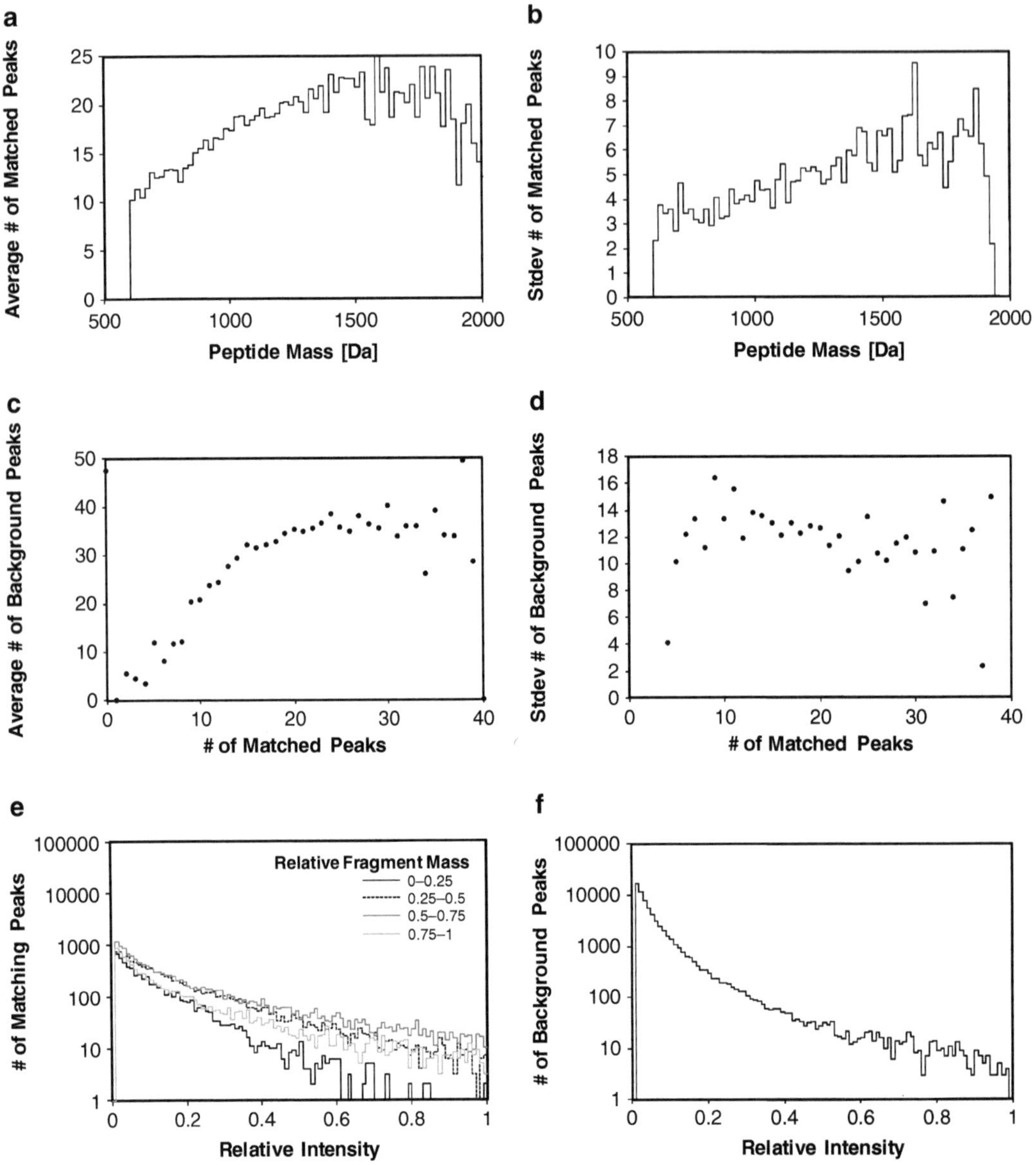

Fig. 3. Properties of tandem mass spectra with significant matches to a dataset acquired with an LTQ-Orbitrap (Thermo Fisher, San Jose, CA): (**a**) the average number and (**b**) the standard deviation of peaks matching the sequence as a function of peptide mass; (**c**) the average number and (**d**) the standard deviation of background peaks as a function of the number of peaks matching the sequence; (**e**) the intensity distribution of matching peaks; and (**f**) the intensity distribution of background peaks.

tandem mass spectra containing varying amounts of sequence information the number of matching fragments needed for identification can be determined (see Note 5 and Fig. 4). In this way it is possible to investigate how many fragment ions are needed for identification depending on the precursor mass, precursor and fragment mass errors, background levels, and modification states (16).

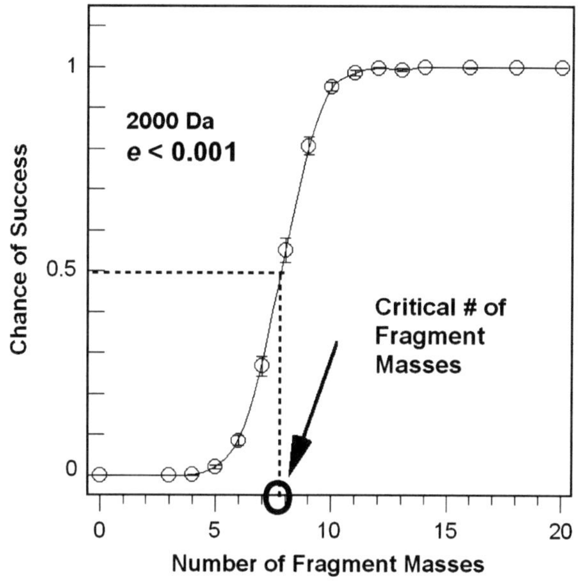

Fig. 4. The chance of success of identification, i.e., the fraction of the spectra that yield a true result and an *e*-value below a desired threshold, as a function of the number of fragment masses in the spectra. Each data point represents the mean value with standard error of the results for 50 randomly selected peptides and with 20 different randomly generated spectra from each peptide. The chance of success is low for few matching fragment and high for many matching fragments. The critical number of fragment masses is defined as the number of fragment masses that yield a 50% chance of success.

3. Notes

1. The distribution of peptide masses is far from uniform, because peptides contain only a few different types of atoms, and it is, therefore, important to use actual peptide masses in simulations. The distribution of peptide masses consists of peaks with centroids approximately 1 Da apart, and regions in between the peaks that are devoid of peptide masses. Using a uniform mass distribution would therefore result in unrealistic synthetic peptide mass fingerprints.
2. The intensities are often set to the same value for all masses. Alternatively, an intensity distribution derived from experimental data can be used.
3. The number of peptides to pick can for example be determined by selecting a target coverage for the proteins, and then randomly picking peptides until that coverage is reached.
4. An example of the kind of information that can be extracted from experiments is shown in Fig. 3. First the data acquired

on an LTQ-Orbitrap was searched using X! Tandem and all peptides with expectation value $<10^{-3}$ were used to characterize the data set. The average and the standard deviation of the number of ions that match the peptide sequence first increases with mass, and at masses above 1,500 Da the average saturates (Fig. 3a, b). The average number of background peaks increases with the number of matching peaks up to about 15 matching peaks, and then saturates (Fig. 3c). The standard deviation of the number of background peaks is constant within the uncertainty of the measurement (Fig. 3d). The matching peaks dominate at high intensity, but even though the majority of peaks with low relative intensity are background (<20% of the base peak), there are still a considerable number of low-intensity peaks that match the sequence (Fig. 3e, f).

5. Tryptic peptides were randomly selected from a proteome, and a set of fragment mass spectra was generated for each selected peptide assuming that they were unmodified or phosphorylated. These fragment mass spectra were constructed by randomly selecting fragment ions, and the number of fragments selected was varied over a wide range. The fragment mass spectra were searched against the proteome using X! Tandem and the probability of successful peptide identification was obtained as a function of the number of fragment ions in the spectra. From these curves, the critical number of fragment masses was derived for a given experimental condition, i.e., the number of fragment masses needed for successfully identifying half of the peptides.

Acknowledgments

This work was supported by funding provided by the National Institutes of Health Grants CA126485, DE018385, NS050276, RR00862 and RR022220, the Carl Trygger foundation, and the Swedish research council.

References

1. Eriksson J, Fenyo D. (2007) Improving the success rate of proteome analysis by modeling protein-abundance distributions and experimental designs. *Nat Biotechnol* **25**, 651–655.
2. Henzel WJ, Billeci TM, Stults JT, Wong SC, Grimley C, Watanabe C. (1993) Identifying proteins from two-dimensional gels by molecular mass searching of peptide fragments in protein sequence databases. *Proc Natl Acad Sci U S A* **90**, 5011–5015.
3. Mann M, Wilm M. (1994) Error-tolerant identification of peptides in sequence databases by peptide sequence tags. *Anal Chem* **66**, 4390–4399.
4. Eng JK, McCormack AL, Yates JR. (1994) An approach to correlate mass spectral data with amino acid sequences in a protein database. *J Am Soc Mass Spectrom* **5**, 976.

5. Craig R, Cortens JC, Fenyo D, Beavis RC. (2006) Using annotated peptide mass spectrum libraries for protein identification. *J Proteome Res* **5**, 1843–1849.
6. Lam H, Deutsch EW, Eddes JS, Eng JK, King N, Stein SE, Aebersold R. (2007) Development and validation of a spectral library searching method for peptide identification from MS/MS. *Proteomics* 7, 655–667.
7. Fenyo D, Qin J, Chait BT. (1998) Protein identification using mass spectrometric information. *Electrophoresis* **19**, 998–1005.
8. Eriksson J, Fenyo D. (2004) Probity, a protein identification algorithm with accurate assignment of the statistical significance of the results. *J Proteome Res* **3**, 32–36.
9. Field HI, Fenyo D, Beavis RC. (2002) RADARS, a bioinformatics solution that automates proteome mass spectral analysis, optimises protein identification, and archives data in a relational database. *Proteomics* **2**, 36–47.
10. Fenyo D, Beavis RC. (2003) A method for assessing the statistical significance of mass spectrometry-based protein identifications using general scoring schemes. *Anal Chem* **75**, 768–774.
11. Eriksson J, Chait BT, Fenyo D. (2000) A statistical basis for testing the significance of mass spectrometric protein identification results. *Anal Chem* **72**, 999–1005.
12. Eriksson J, Fenyo D. (2004) The statistical significance of protein identification results as a function of the number of protein sequences searched. *J Proteome Res* **3**, 979–982.
13. Eriksson J, Fenyo D. (2005) Protein identification in complex mixtures. *J Proteome Res* **4**, 387–393.
14. Jensen ON, Podtelejnikov AV, Mann M. (1997) Identification of the components of simple protein mixtures by high accuracy peptide mass mapping and database searching. *Anal Chem* **69**, 4741–4750.
15. Eriksson J, Fenyo D. (2009) Peptide identification with direct computation of the significance level of the results. *Proceedings of the 57th ASMS Conference on Mass Spectrometry.*
16. Fenyo D, Ossipova E, Eriksson J. (2008) The peptide fragment mass information required to identify peptides and their post-translational modifications. *Proceedings of the 56th ASMS Conference on Mass Spectrometry.*

Chapter 9

Protein Identification from Tandem Mass Spectra by Database Searching

Nathan J. Edwards

Abstract

Protein identification from tandem mass spectra is one of the most versatile and widely used proteomics workflows, able to identify proteins, characterize post-translational modifications, and provide semi-quantitative measurements of relative protein abundance. This manuscript describes the concepts, prerequisites, and methods required to analyze a tandem mass spectrometry dataset in order to identify its proteins, by using a tandem mass spectrometry search engine to search protein sequence databases. The discussion includes instructions for extraction, preparation, and formatting of spectral datafiles; selection of appropriate search parameter settings; and basic interpretation of the results.

Key words: Protein identification, MS/MS spectra, Protein sequence databases, Peptide identification, Search engine

1. Introduction

The identification of proteins by tandem mass spectrometry is one of the most widely used techniques in mass-spectrometry based proteomics. It can be applied to purified protein samples containing just a few protein components, or to complex samples containing many proteins, such as those resulting from the analysis of a cell-line or clinical sample. Tandem mass spectrometry can be applied without prior knowledge of the proteins to be analyzed and can readily identify proteins, characterize post-translational modifications, and provide semi-quantitative measurements of relative abundance. Furthermore, the collection of tandem mass-spectra using modern mass-spectrometers, in conjunction with specific protein chemistry techniques, can be highly automated. This automation makes it possible to conduct high-throughput,

Cathy H. Wu and Chuming Chen (eds.), *Bioinformatics for Comparative Proteomics*, Methods in Molecular Biology, vol. 694, DOI 10.1007/978-1-60761-977-2_9, © Springer Science+Business Media, LLC 2011

comprehensive analyses of complex protein mixtures, generating thousands of tandem mass-spectra per sample.

The most common application of tandem mass spectrometry in proteomics workflows seeks to identify the proteins in a sample. Known as shotgun proteomics, by analogy with whole genome shotgun sequencing, the sample's proteins are solubulized and digested into short peptides of 10–20 amino acids using a proteolytic enzyme. The resulting peptide mixture is separated in time according to the peptides' physical and chemical properties using liquid chromatography and analyzed in real-time by the mass-spectrometer. As the peptides with similar physicochemical properties elute from the column, the mass spectrometer acquires survey scans to identify and select the most abundant peptide ions for analysis by tandem mass-spectrometry. Each selected peptide ion, also called the precursor ion, is fragmented in turn to acquire the product ion scan, also known as the MS/MS spectrum or the tandem mass-spectrum. Mass-spectrometers are typically configured to collect three to five tandem mass spectra for each survey scan, with each cycle, consisting of one survey scan and multiple tandem mass spectra, taking a few seconds. Over the course of a 2–4 h chromatography run, this shotgun proteomics workflow can collect thousands of MS/MS spectra representing the fragmented peptides of a sample's proteins. The automated acquisition of tandem mass spectra in conjunction with liquid chromatography is often shortened to LC-MS/MS.

Computer software called tandem mass-spectrometry search engines analyze these shotgun proteomics datasets to identify the sample's proteins. These search engines match the tandem mass spectra with peptide sequences from a protein sequence database and use the identified peptides to infer the protein content of the sample. This manuscript describes the concepts, prerequisites, and methods required to analyze a shotgun proteomics dataset using a tandem mass-spectrometry search engine. The discussion includes instructions for extraction, preparation, and formatting of spectral datafiles; selection of appropriate search parameter settings; and interpretation of the results.

For more background on these various techniques, we refer the reader to one of many excellent reviews (1–7).

This manuscript will focus on the analysis of a typical shotgun proteomics dataset acquired from a complex protein sample for the purposes of protein identification. The various experimental technologies of mass-spectrometry based proteomics are constantly changing, with improved instrument resolution, fragmentation analysis of larger peptides and proteins, new ionization and fragmentation technologies, and novel separation techniques and digest reagents. We will not attempt to cover these newer technologies, though we expect that this manuscript will provide a foundation for understanding how the software tools should be used for these datasets, too.

This manuscript will not directly address the analysis of datasets from proteomics quantitation workflows, except where these impact the settings used for protein identification from tandem mass spectra. This manuscript will also not attempt to describe, in explicit terms, instructions for any specific search engine, but will document the concepts and principles behind each of the common parameter settings so that the appropriate options can be selected for any search engine. Finally, we note that we do not address a variety of other techniques and software tools for protein identification from tandem mass spectra, notably the de novo (8–11) and hybrid sequence tag (12–14) approaches.

2. Concepts

In order to streamline the description of the methods to follow, we introduce some important mass spectrometry concepts and terminology that will be used throughout the manuscript.

2.1. Ionization

Mass spectrometry is an analytical technique that measures the mass of molecules and atoms (15). The molecules to be analyzed are transformed into charged, gas-phase *ions* which can be manipulated and detected by the mass spectrometer. The *ionization* of peptides for mass spectrometry is typically carried out using one of two technologies: *electrospray ionization* (*ESI*) or *matrix assisted laser desorbsion ionization* (*MALDI*). Peptides are given charge during ionization by protonation, the addition of one or more protons, with the peptides observed in shotgun proteomics experiments typically acquiring one proton when subject to MALDI ionization, and between one and four protons when subject to ESI ionization.

2.2. Mass-to-Charge Ratio

The mass-spectrometer's mass analyzer uses electrical, magnetic, and RF fields to separate the gas-phase ions in time or space before they are counted and detected. These fields manipulate the ions based on their *mass-to-charge ratio* (*m/z value*), rather than their mass; therefore the number of attached protons must be determined before the mass of an ion can be inferred. The number of protons acquired by a peptide or fragment ion is called its charge state. The X-axis of a mass spectrum, such as the example in Fig. 1, records observed ions' m/z values in atomic mass units, the approximate mass of a hydrogen atom, or equivalently, that of a proton or a neutron. The mass spectrometry community generally refers to atomic mass units as *Daltons* (*Da*). We point out that the protonation of a peptide affects not only its m/z value but also its mass, increasing the mass by 1 Da (approximately).

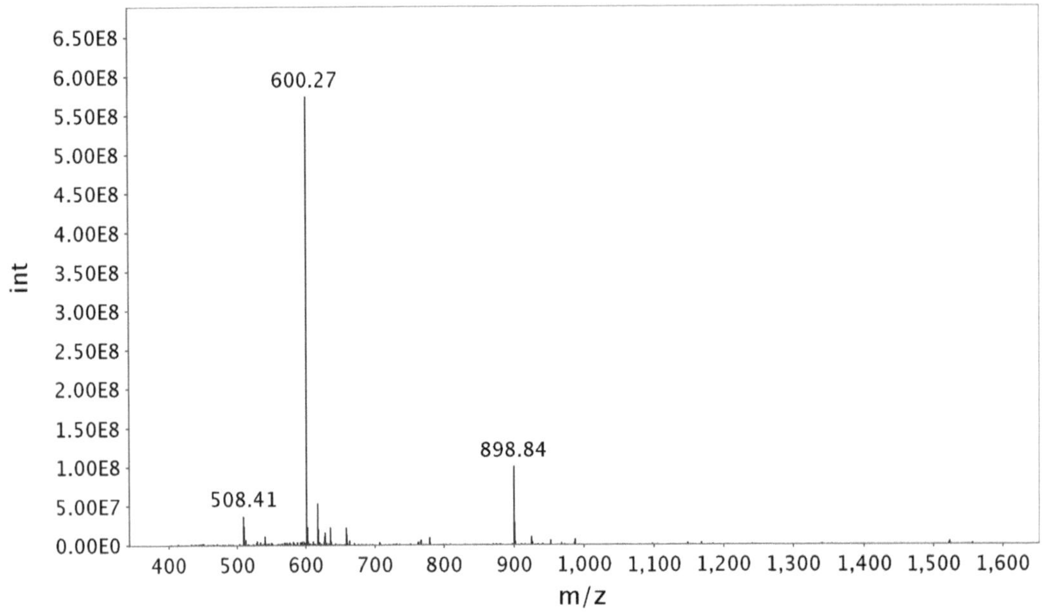

Fig. 1. Survey scan (#984) of spectra file raftflow37 from Peptide Atlas dataset raftflow.

2.3. Ion-Counts and Resolution

Mass-spectra are essentially histograms of *ion-counts*, with the *Y*-axis of a mass-spectrum representing a (relative) measure of the number of ions at a particular *m/z* value in arbitrary units, as shown in Fig. 1. The ion-counts of individual molecular ions are generally observed in a number of adjacent *m/z* bins, revealing a characteristic peak shape. A spectrum that samples each ions' peaks at many adjacent *m/z* values is called a *profile spectrum*. The ability of a mass spectrometer to distinguish two ions' *m/z* values, called *resolution*, depends on their true *m/z* value difference and the width of the peak shape observed in the mass spectrum. Figure 2 shows a schematic representation of the peak shapes of some peptide ions, demonstrating how the peaks of individual ions may overlap, or convolve, as they get close together. Low-resolution instruments may ultimately be unable to measure the *m/z* values of individual ions if they get too close, an effect which has important consequences for determining peptide ion charge states (see Subheading 2.5) and therefore mass.

2.4. Peak-Detection

During or after acquisition, profile spectra are analyzed using *peak detection* algorithms, to obtain *centroided spectra* or *peak lists*, in which each peak shape is integrated and summarized by *m/z* value and intensity. After centroiding, the peaks of Fig. 2 become impulses, without shape, and depending on the peak detection algorithm and instrument resolution, may represent the underlying, convolved peptide ions poorly. Figure 3 shows the centroided spectrum result of a simple apex-based peak detection

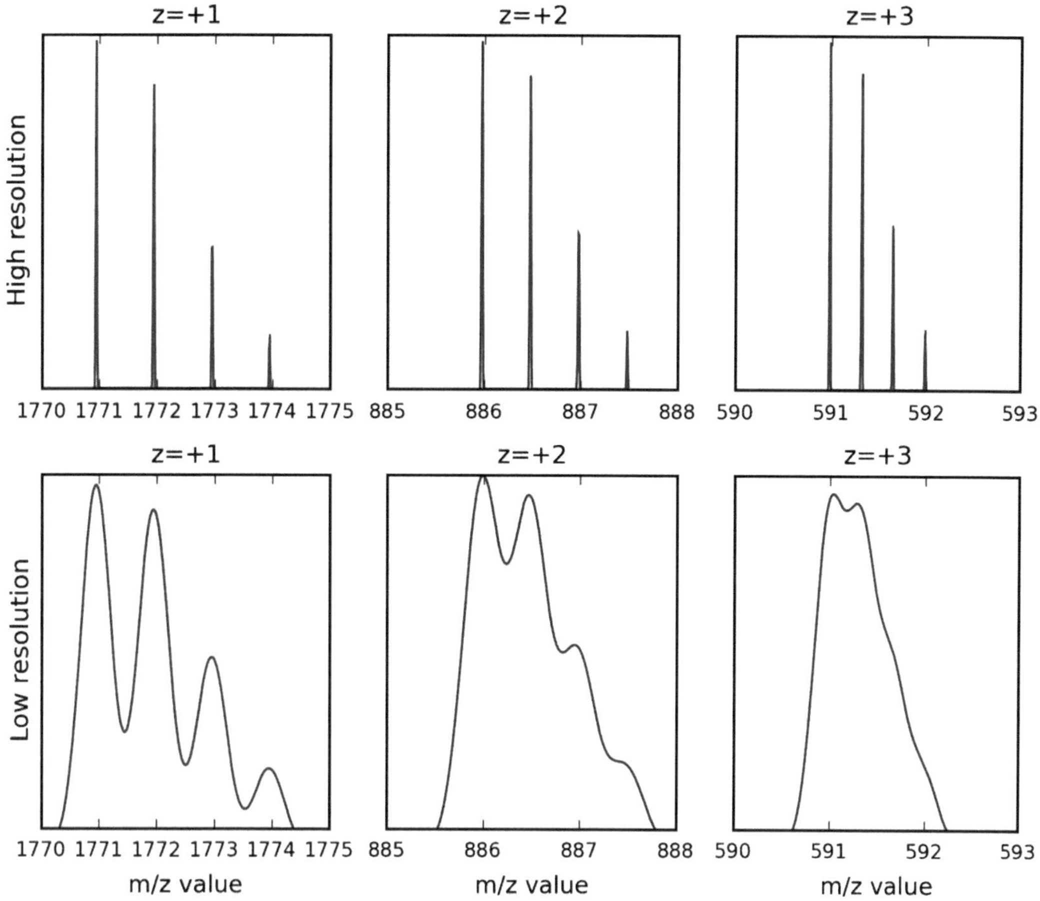

Fig. 2. Schematic isotope clusters for peptide AACLLPKLDELRDEGK (molecular weight 1769.93 Da), in charge state +1, +2, and +3, as might be observed in profile spectra from high (*top row*) and low (*bottom row*) resolution instruments.

algorithm applied to each of the schematic profile spectra of Fig. 2. While the low-resolution peak shapes for charge state +2 and +3 in this figure represent the convoluted shapes of four distinct peptide ions, most peak detection algorithms will output just two or three peaks in this region. Unsurprisingly, the results of peak detection on the high-resolution schematic profile spectra capture the peptide ions well.

2.5. Isotope Clusters and Charge State

Peptides, as naturally occurring organic molecules, contain elements such as carbon, that sometimes incorporate one or more extra neutrons in the nucleus. Approximately 1% of naturally occurring carbon is observed as the ^{13}C isotope rather than the more common ^{12}C isotope, and when a ^{13}C isotope is present in a peptide its molecular mass increases by 1 Da (approximately). In any proteomics sample, the masses observed for the many copies of a particular peptide are probabilistically distributed amongst those with no ^{13}C isotopes, one ^{13}C isotope, two ^{13}C isotopes, and

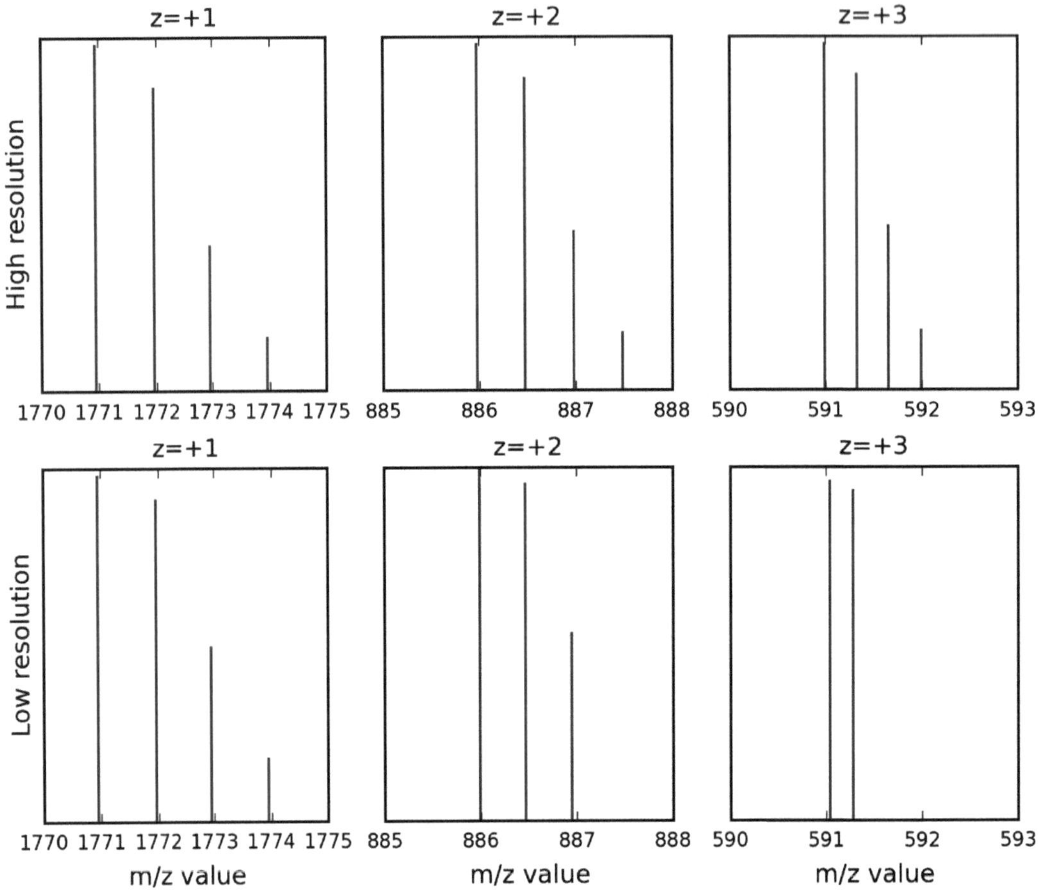

Fig. 3. Schematic isotope clusters for peptide AACLLPKLDELRDEGK (molecular weight 1769.93 Da) after peak detection, in charge state +1, +2, and +3, as might be observed in centroided spectra from high (*top row*) and low (*bottom row*) resolution instruments.

so on. As the mass of the peptide increases, the probability of incorporating ^{13}C isotopes increases too, so the relative likelihood of each of the peptide's *isotope peaks* changes with mass. Figure 2 shows a schematic isotope cluster for the peptide AACLLPKLDELRDEGK (molecular weight 1769.93 Da) in charge states +1, +2, and +3, as might be observed in a low-resolution and high-resolution mass spectrometer. Notice that the peaks of the +1, +2, and +3 charge state isotope clusters are separated by 1, $\frac{1}{2}$, and $\frac{1}{3}$ Da, making it possible to infer the charge state of the ion. Depending on the instrument's resolution, however, we may not be able to reliably determine the *m/z* value spacing of the isotope cluster peaks, before or after peak detection, in order to establish the charge state of the peptide ion.

2.6. Average and Monoisotopic Mass

Depending on the resolution of the instrument and the mass of the molecules being analyzed, the result of peak detection may reflect the weighted average of the *m/z* value of the individual isotope cluster peaks and the integration of the entire cluster. In this case, the observed *m/z* value represents the *average mass* of the peptide ion. With sufficient resolution or for small enough masses, a peak representing the monoisotopic mass of the peptide ion will be output. The *monoisotopic mass* is the mass of the peptide ion calculated using only the most abundant isotopic form of each element, which is the left-most peak of the isotope cluster for peptides in typical shotgun proteomics experiments. The calculated monoisotopic mass of peptide AACLLPKLDELRDEGK of Fig. 2 in charge state +1 is 1770.94, while the average mass is 1771.93.

2.7. Peptide Fragmentation Spectra

Peptide tandem mass spectra measure the *m/z* values of fragment ions formed by collisionally induced dissociation (CID), in which precursor ions break apart due to collisions with inert gas molecules under pressure. The precursor ions, selected by their *m/z* value, typically represent many copies of a particular peptide. When peptides break apart in CID, their protons are retained by one or more of the fragments, and the charged fragment ions measured by the mass-analyzer and detector, forming the tandem mass spectrum. Peptides tend to fragment along the peptide amino-backbone, revealing the peptide's primary structure, its amino-acid sequence. When the N terminus (left end) fragment retains a proton, the fragment ions are named using the initial letters of the alphabet and the number of amino acids in the fragment. Similarly, when the C terminus (right end) fragment retains a proton, its fragment ions are named using the last letters of the alphabet. The most common fragment ions formed by CID are the *b-ions* and *y-ions*, though *a-ions* can also be observed. Figure 4 shows a peptide fragmentation spectrum of the precursor ion at *m/z* value 898.84 from Fig. 1

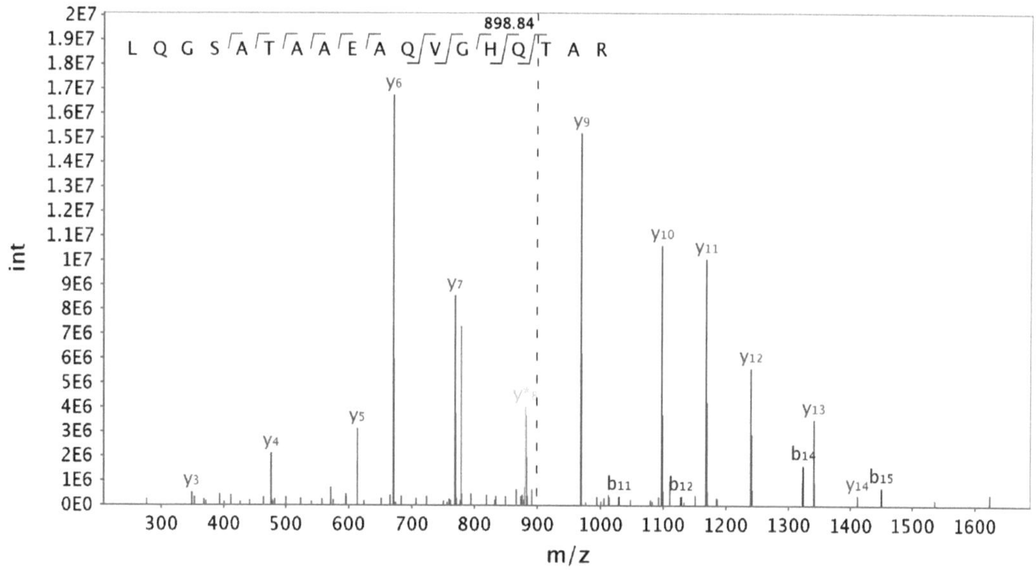

Fig. 4. Tandem mass spectrum (scan #985) of peptide LQGSATAAEAQVGHQTAR, from the charge state +2 precursor ion with *m/z* value 898.84 in scan #984 (Fig. 1), from spectra file raftflow37 of the Peptide Atlas dataset raftflow.

representing the peptide LQGSATAAEAQVGHQTAR in charge state +2. Observed b- and y-ions are marked in the spectrum and on the peptide sequence.

3. Materials

3.1. Tandem Mass-Spectrometry Search Engine

There are many open-source and commercial tandem mass-spectrometry search engines available, although only a few have become widely used. Commercial search engines Mascot (16), from Matrix Science, and SEQUEST (17), from Thermo Fisher are very popular. More recently, a number of free, open-source search engines have found significant adoption, particularly X!Tandem (18) and OMSSA (19).

For small scale or ad hoc analyses, Mascot, X!Tandem, and OMSSA are available for free on the web, at the websites shown in Table 1. Each of these web-based tools allow the user to upload a spectral datafile, select search parameters, execute the search, and browse and interpret the results. Where greater configuration flexibility or search throughput is required, each of these search engines can be installed on users' computers. Each of the available search engines has strengths and weaknesses in terms of its scoring, configuration, file formats, and performance.

3.2. Protein Sequence Database

The protein sequence database provides peptide sequences to be matched against the tandem mass spectra by the search engine. As such, the selected protein sequence database will have a significant impact on the sensitivity, specificity, and speed of the search. Since peptides whose sequence is missing from the protein sequence database will not be matched to spectra, poorly chosen sequence databases will result in spectra going unidentified and their peptides being unobserved. However, larger, more inclusive protein sequence databases take longer to search, and may result in more false-positive identifications and reduced statistical significance.

Table 1
Free web-interfaces to popular tandem mass spectrometry search engines, suitable for small-scale or ad hoc analyses

Search engine	URL
Mascot	http://matrixscience.com
X!Tandem	http://thegpm.org
OMSSA	http://pubchem.ncbi.nlm.nih.gov/omssa

Locally installed search engines generally expect FASTA format protein sequence databases, which can be readily downloaded from the appropriate websites. Installation of protein sequence databases for locally installed search engines may require special configuration and preprocessing of the protein sequence database file, but the analysis flexibility gained is significant. The local installation of specific protein sequence databases is one of the primary reasons to install and run a search engine in-house, as the sequence databases provided by the free web-based search engines are often quite limited.

When available, organism specific sequence databases eliminate false-positives from closely related species, but can leave peptides from common contaminants unidentified. For this reason, keratins, trypsin, and other artifactual protein sequences are sometimes added to organism specific sequence databases, even though they do not inform the biology of the sample.

Where the source of the proteins is a single, well-characterized model-organism, the International Protein Index (IPI) protein sequence databases (20) are a good choice. If the origin of the sample is unknown, or known to be a mixture of organisms, then the Swiss-Prot section of UniProtKB (21) is a good choice. UniProt also provides complete proteome sets for sequenced organisms and tools for selecting and downloading sub-proteomes constrained by protein feature or annotation. NCBI's RefSeq is another good source of protein sequences, and is available in various taxonomic divisions. Organism specific RefSeq sequences can be found in the genomes section of the NCBI FTP site. Web-addresses for these protein sequence databases are provided in Table 2.

The use of NCBI's nr and similar computationally merged protein sequence databases, or protein sequences from poorly annotated genomes is not recommended as redundant peptide sequences and poor protein naming can significantly complicate the interpretation of the results. In some cases, searching ESTs and genome sequences may be appropriate (22), but considerable

Table 2
Sources of protein sequences databases suitable for use with tandem mass spectrometry search engines

Sequence database	URL
IPI	http://www.ebi.ac.uk/IPI
UniProt	http://www.uniprot.org
RefSeq	http://www.ncbi.nlm.nih.gov/RefSeq
RefSeq Genomes	ftp://ftp.ncbi.nih.gov/genomes

post-search analysis must be carried out to make up for the lack of good meta-data and quality control associated with each entry.

3.3. Instrument Characteristics

The technical characteristics of the mass-spectrometer and its configuration for a particular experiment have a huge impact on the tandem mass spectra acquired. As such, these details inform the selection of appropriate search parameters. Some parameters do not change on an experiment to experiment basis, but others do – in each case, the informatics analyst may need to consult with the mass spectrometrist to determine appropriate settings.

It is crucial to establish whether the ionization (see Subheading 2.1) technology used is MALDI or ESI, as MALDI ionization primarily generates charge state +1 peptide ions. Electrospray (ESI) instruments commonly generate peptide ions in charge states +1, +2, and +3, although +4 and +5 charge states are sometimes observed for larger, more basic peptides. In this context, MALDI tandem mass spectra can be assumed to represent charge state +1 precursors, significantly simplifying the analysis.

Appropriate mass-accuracy parameters for precursor and fragment ion m/z values must reflect realistic performance characteristics of the mass spectrometer. These parameters determine when an experimental m/z value is considered to match the peptide or fragment mass computed *in silico*. Fragment ion matches are the foundation of the scoring and evaluation of the quality of a peptide identification assignment. If the fragment mass-accuracy parameter is set too tightly, many valid fragment matches will be missed, reducing peptides' scores. If the fragment mass-accuracy parameter is set too loosely, spurious fragment matches will be observed, inflating all peptides' scores and increasing the number of false-positive identifications.

Precursor mass-accuracy parameters are chosen similarly, depending on the resolution properties of the survey scan. The precursor mass-accuracy parameter is the primary criteria used by the search engines to determine whether or not a peptide sequence will even be scored against a particular spectrum. If set too tight, many valid peptide identifications will be missed. If set too loose, many incorrect peptide sequences will be scored, resulting in slower search times and increased potential for false-positive identifications. Sometimes, an instrument's real-time algorithm for picking precursor ions from the survey scan will select isotope cluster peaks (see Subheading 2.5) other than the monoisotopic peak – in this case, the search engine will require precursor match tolerance of at least 1 Da to match the experimental m/z value with the monoisotopic mass computed from the peptide sequence. As the detrimental effects of loose precursor mass-accuracy parameters are generally pretty mild, a precursor mass tolerance of 2 Da is often used to ensure these peptides are identified.

Note that for some instruments, particularly the LTQ-Orbitrap (Thermo Fisher), the mass-analyzer and its resolution used for survey scans and tandem mass-spectra in any particular experiment is a configuration option set by the mass-spectrometrist – the identity of the instrument is not sufficient to establish mass-accuracy parameters for a particular search.

For older instruments, it is important to understand whether the reported m/z values represent average masses or not (see Subheading 2.6). Current instruments generally have the resolution to measure individual isotope cluster peaks for ions with the relatively low charge states represented here, so experimental m/z values should be matched against monoisotopic masses.

3.4. Tandem Mass Spectra Datafile

The handling and processing of tandem mass spectra datafiles can, unsurprisingly, significantly impact the quality of the peptide identification results. Mass-spectrometers generally store spectral data in binary, vendor (and instrument) specific file formats that can only be read by software provided by the instrument vendor. The mass-spectrometer's "raw" spectra files must be processed and exported to some non-proprietary format for analysis by tandem mass spectrometry search engines. Typically, the vendor software will provide some facility for tandem mass spectrum export, and a number of open-source tools that use the vendor libraries are available as part of the Trans Proteomic Pipeline (TPP) (23) project (see http://tools.proteomecenter.org/wiki/index.php?title=Formats:mzXML) or the ProteoWizard (24) project (see http://proteowizard.sourceforge.net).

Peak-detection or centroiding (see Subheading 2.4) must be carried out if the raw spectra files contain profile spectra, as the tandem mass spectrometry search engines require spectra datafiles of peak lists or centroided spectra. Some tools provide additional spectral processing facilities, which in many cases can improve the quality of the spectra to the analyzed. Common spectral processing options include intensity thresholding, deisotoping, precursor charge state determination or enumeration, and spectrum merging or averaging.

Intensity thresholding removes peaks less than a specific relative intensity, which can reduce spectrum size and eliminate spurious fragment matches. Deisotoping finds isotope clusters and eliminates the non-monoisotopic peaks, reducing spurious fragment matches. Both of these options have the potential to remove valid fragment ions from tandem mass-spectra. The precursor charge states may be determined by examining isotope clusters (see Subheading 2.5) in the survey scans, which are often not exported for the tandem mass-spectrometry search engine. Alternatively, multiple copies of each tandem mass spectrum may be enumerated, each with a different declared charge state, if the software cannot determine the correct charge state. Spectral merging and averaging can be carried out when the same precursor m/z value is selected for fragmentation at multiple nearby time-points in an LC-MS/MS experiment. Merging or averaging these tandem mass-spectra boosts the intensity of common peaks and reduces the intensity of noise, and can make a significant impact in the quality of the peptide identification results.

The open-source tools tend not to provide deisotoping, precursor charge state determination, or spectral averaging facilities, and where the vendor libraries do not provide peak detection routines, they implement crude centroiding algorithms. Nevertheless, until very recently, the vendors did not provide tools for export in convenient open formats and the open-source options were the only option. Due to the dependence on vendor software and libraries, raw spectra conversion must be carried out on the Windows platform.

Commonly used open file formats for tandem mass spectra are indicated by their file extension: .dta which represents a single tandem mass spectrum, with the filename encoding scan number and charge state information; .mgf (Mascot generic format) which represents many tandem mass spectra in a simple, text-based format; and .mzXML, .mzData, and .mzML which represent many tandem mass spectra and their meta-data using XML. Of the XML formats mzXML appeared first, and has been widely adopted; mzData came out of the HUPO Proteomics Standards Initiative; and mzML, which represents an attempt to merge these XML formats, has yet to be widely adopted.

A significant issue with some of these formats is the difficulty in retaining important meta-data associated with each tandem mass-spectrum, particularly the original raw datafile *scan numbers* and LC retention-time in LC-MS/MS experiments.

3.5. Sample Preparation

Information about the manipulation and handling of the protein sample prior to analysis by tandem mass-spectrometry is the final prerequisite for a successful peptide identification analysis.

First, the proteolytic enzyme used to cleave proteins into peptides must be established. Trypsin, which cuts at Arg (R) and Lys (K) unless followed by Pro (P) is most commonly used for cleaving proteins into peptides in the context of proteomics experiments. Second, Cys (C) residues are typically chemically modified, deliberately, to ensure they have a known, predictable mass. Iodoacetamide is the most commonly used reagent for this purpose, subjecting the Cys residues to carbamidomethylation and increasing their mass by 57 Da. Other deliberate chemical labeling of specific residues or peptide or protein termini should also be noted, since these change the expected masses of peptides, too. In particular, many proteomics quantitation workflows use stable-isotope labels, differential chemical modifications on specific amino acids or termini, and these must be considered in setting appropriate peptide identification search parameters.

Where the sample represents a single protein (extracted from a gel band or spot) or is from a specific organism this should also be noted, as this may be useful in selecting appropriate protein sequence databases and ultimately interpreting the results. In addition, specific sample preparation techniques, such as protein separation by 2D gel electrophoresis, can result in an increased likelihood of observing contaminants, such as keratins, in the final result.

4. Methods

4.1. Prepare Spectra Datafile

The vendor and open-source tools available for processing and exporting the tandem mass spectra may not output the spectra in a format supported by your choice of search engine. Once in some open format, however, any number of tools are available for reformatting the data, and if necessary, a custom program can be written for the task. The TPP and ProteoWizard projects (see Subheading 3.4) provide a number of programs for converting between a variety of XML formats and the more basic formats, such as dta, and mgf. The web-based search engines generally permit one spectra file upload per search.

4.2. Specify Search Parameters

Many of the search parameters required by tandem mass spectrometry search engines capture the classic balance between search-time and the potential to miss valid peptide identifications. The informatics analyst must ensure that the parameters selected do not exclude a large number of potential peptide identifications while also keeping running time reasonable. For these parameters, there is always the question of whether or not a more thorough search will yield sufficient additional identifications (spectra, peptides, or proteins) to warrant the additional search-time. While the search-time consequences must always be paid, the benefit is generally impossible to quantify before searching, unless some kind of additional information is available. We make these issues explicit, where they are relevant, in the following steps.

4.2.1. Protein Sequence Database

The selection of the protein sequence database to search represents the classic modeling trade-off between search-time and sensitivity. Larger, more inclusive protein sequence databases will take longer to search, but may identify more peptides. Smaller, more selective protein sequence databases will take less time to search, but important or unexpected peptides may be missed. Where the source of the proteins is a single, well-characterized model-organism, the International Protein Index (IPI) protein sequence databases are a good choice. If the origin of the sample is unknown, or known to be a mixture of organisms, then the Swiss-Prot section of UniProtKB is a good choice. For web-based search engines, some interesting options may be available, depending on the site, but specific exotic options, such as proteins from a specific bacterial genome, may not. See Subheading 3.2 for a discussion of the sequence database options for locally installed search engines.

4.2.2. Instrument Parameters

Having established the instrument characteristics as a prerequisite to the search in Subheading 3.3, all that remains is to map these characteristics to the parameters required by the search engine. Average or monoisotopic masses should be specified as appropriate (see Subheadings 2.6 and 3.3). Mass tolerance parameters are generally specified in Da (Daltons) or ppm (parts per million). The ppm units are used when the mass tolerance is proportional to the measured mass, while Da are used when the mass tolerance is invariant with the measured mass. Low-resolution tandem mass-spectra may require a fragment mass match tolerance setting as large as 0.6 Da, while for higher-resolution fragmentation spectra a setting of 0.1 Da may be appropriate.

Some search engines will require the name of the instrument or a vendor neutral abbreviation of its ionization and mass-analyzer technologies, since these can affect the peptide fragmentation significantly. In a pinch, it is largely sufficient to get the ionization technology (ESI or MALDI, see Subheading 2.1) correct.

4.2.3. Mass Modification Parameters

While the residual, unmodified, mass of amino acids is well established, there is no guarantee that the particular peptide ion observed in the mass spectrometer contains only unmodified amino acids. Some residues, particularly Cys, are chemically modified deliberately (see Subheading 3.5) as part of the sample preparation. In this case, the mass modification is called *fixed* and is applied to every Cys residue in every peptide, before scoring. There is no running-time cost for fixed mass modifications. Incorrectly setting the Cys fixed modification, however, will render Cys containing peptides unidentifiable. The carbamidomethylation of Cys is the most common such modification, and if in doubt, a +57 fixed mass modification on Cys should generally be applied.

So called variable modifications specify additional masses to apply to particular residues. If the oxidized Met variable modification is selected, then every Met residue in the sequence database will be considered first using its residual mass of 131 Da (approximately), and then with mass of about 16 Da more, 147 Da (approximately). Variable modifications can be specified for biological mass modifications, such as for phosphorylation, or common artifactual mass modifications, such as oxidation, on particular residues. The search time will generally increase exponentially in the number of variable modifications so they should only be selected when the variable modifications are expected or are significant for the desired biological conclusion.

4.2.4. Proteolytic Enzyme Parameters

The proteolytic enzyme setting should match the sample preparation conditions established in Subheading 3.5.

Even when the specific proteolysis enzyme and its cleavage motif is known, there is no guarantee that it will cleave at every motif position, or that it will leave non-motif positions alone. As such, the samples' peptides may ultimately have zero, one, or two termini consistent with the enzyme, and the may contain internal motif sites representing a missed cleavage opportunity. By default, search engines will consider only those peptide sequences with both N and C termini consistent with the selected proteolytic enzyme. A semi-specific (or semi-tryptic) search will consider peptide sequence with at least one of the N or C termini consistent with the proteolytic enzyme (trypsin). A non-specific search will consider all peptide sequences, regardless of their N or C termini sequence. The selection of semi-specific proteolysis will typically increase search times by a factor of 20–30, but will often increase the number of identified peptides substantially. A non-specific search is usually only applied in special cases.

The number of internal motif sites a peptide may contain is controlled by a parameter called missed cleavages, which is typically set to a small number, such as 1 or 2, for trypsin.

4.2.5. Precursor Mass Tolerance Parameters

As outlined in Subheading 3.3, appropriate settings for the precursor mass tolerance parameters should be set to ensure that peptide sequences match the precursor mass of their spectra. Due to the selection of non-monoisotopic isotope cluster peaks as precursor ions, the necessary tolerance is often set at 2 Da, regardless of the mass-accuracy characteristics of the precursor measurements. Some search engines will model this behavior explicitly, making it possible to specify a tight precursor mass tolerance and a small number of non-monoisotopic isotope cluster peaks to test, in addition to the monoisotopic mass of the peptide sequence. This second parameter is called $\#^{13}C$ by Mascot, for example, and is generally set to a small number like 1 or 2.

It should be noted that peptide sequences may fail to match the mass of their experimental precursors for a variety of reasons. An incorrect charge state determination for an experimental precursor ion will make it impossible to match against its peptide sequences, while incorrect fixed or missing variable modifications will make it impossible for the *in silico* computation of a peptide mass to match the experimental value.

4.3. Execute Search

Once the spectra are prepared and uploaded and the search parameters are set, the execution of the search is generally straightforward. The informatics analyst should check that the search does not run too quickly or too slowly as a sanity check on the selection of appropriate search parameters.

4.4. Results Interpretation

We provide some general guidelines for results interpretation, which are based on the following basic principles. First, peptide-spectrum match scores merely assess the strength of the fragment evidence and cannot establish the correctness of a match; and second, explicit information about proteins is destroyed by the shotgun proteomics workflow, and must be inferred from the available peptide evidence.

4.4.1. Peptide-Spectrum Match Scores

Each search engine computes a single overall score for each peptide-spectrum match to rank the peptides matched to each spectrum. The score is used to limit the number of retained peptide sequences, per spectrum, to the best few. The range of good and poor values for the peptide-spectrum match scores will vary, per tandem mass spectrum, depending on the precursor ion's charge state, the quality of the spectrum, and the fragmentation characteristics of the (unknown) peptide represented by the precursor. As such, it is impossible to say, for a given spectrum, what the score of the correct peptide should be. However, we can be confident that a peptide-spectrum match score will rank the correct peptide very highly, usually as the best identification, if it is matched with its high-quality fragmentation spectrum by the search engine. However, the inverse is not true – the rank 1 peptide identification associated with each spectrum is not necessarily correct. If the score is sufficiently good, we may conclude the evidence for the peptide-spectrum match is strong. If the score is too weak, we may have to conclude that the correct peptide-spectrum match is unknown, even if the (unknown) correct peptide is ranked 1. Various rule-of-thumb thresholds have been published for selecting the likely correct peptide-spectrum matches in the results from specific search engines, but these have proven considerably less powerful than more formal techniques, such as the statistical significance methods described below.

4.4.2. Peptide-Spectrum Match Characteristics

A peptide identification may also have a variety of qualitative statistics associated with the match between its spectrum and peptide. Commonly derived information include experimental precursor m/z, experimental precursor mass, theoretical precursor mass, presumed charge state, missed cleavages, N- and C-terminal enzyme specificity, and number of matching b- and y- ions, and the peptide rank. All of these match characteristics can be used to assess the quality of the match, though the search engines themselves do not generally factor these characteristics into their scores. These values may also be used to check, or derive, the characteristics of the instrument and sample preparation.

4.4.3. Peptide-Spectrum Match Statistical Significance

In the absence of a way to easily decide whether a specific rank 1 peptide identification should be accepted, various statistical significance techniques have been employed to provide search engines with calibrated primary or derived scores. For a peptide-spectrum match of peptide *P* with spectrum *S*, the per-spectrum *p*-value statistic assesses the ***probability*** that a single "***random***" peptide would score as well as, or better than, peptide *P* when matched against spectrum *S*. The *E*-value statistic assesses the ***expected number*** of "***random***" peptides that would score as well as, or better than, peptide *P* in a search of a "random" sequence database of the same size as the one actually searched. The *E*-value corrects for the increased number of false-positive identifications at a given *p*-value when searching a large sequence database. The scores of random peptides matched with spectrum *S* are used to calibrate the expected range of good scores for spectrum *S*, which hopefully includes that of peptide *P*. Crudely, if the peptide *P*'s score is no better than that of random peptides, the evidence for it being correct is very weak.

E-values make it possible to choose any number significantly less than 1 as a threshold for accepting peptide identifications across the entire dataset. The commonly used threshold of 0.05 indicates that accepted peptide identifications would beat all random peptides in each of 20 researches against a similarly sized database of independently generated "random" sequences.

In practice, each search engine uses a different model of "random" peptides, which may result in an aggressive or conservative estimation of the true *E*-value for a given peptide-spectrum match, and makes the *E*-values essentially non-comparable between search engines. However, well-estimated *E*-values do provide a normalized score for comparisons of peptide-spectrum matches between spectra and peptides, even those of different lengths and spectral properties. *E*-values are always monotonic with the peptide-spectrum match scores for peptides matched to a particular spectrum, and so preserve rank. *E*-values are computed internally by the search engine and output, with the peptide-spectrum match score, in the search engine output, as part of the search results.

Recently, estimates of the false discovery rate (FDR) statistic has become a popular additional or alternative measure of the statistical significance of peptide identification results. The FDR is an estimate of the number of incorrect peptide identifications in any set of selected peptide identifications. Usually, only the rank 1 peptides that pass some score or *E*-value threshold are selected – the FDR statistic estimates the number of incorrect peptide identifications in this set. The FDR statistic can be computed from any score valid for comparing peptide identifications between spectra, so *E*-values should be preferred if available. The FDR statistic can be readily computed using the search engine as a black-box, by applying the search engine both to the sequence database of choice and a decoy sequence database of similar size containing shuffled or reversed protein sequences. Some search engines do this internally too, automatically scoring the intended protein sequences and decoy sequences in one search.

To achieve the best of both worlds, we recommend applying FDR based filtering to *E*-values, filtering at 10% FDR.

4.4.4. Protein Identification and Independence of Peptide Identifications

With the statistically significant peptide spectrum matches identified, it is now possible to characterize the protein content of the sample. Broadly, we look for proteins with multiple independent peptide identifications, as we expect that independent peptide identification errors will not to confirm the same protein.

While a single statistically significant peptide identification with a sufficiently small *E*-value can provide sufficient evidence for a protein, there are also a variety of ways in which a false-positive peptide identification may be statistically significant, but still never the less be incorrect. Sequence homology is one source of such errors. The possibility that two such false-positive identifications occur, and point to the same protein, is unlikely. Furthermore, the desire to boost the number of identified proteins often leads to rather relaxed statistical significance thresholds, increasing the chance that any particular peptide identification is incorrect.

Statistically significant peptide identifications to the same peptide should not increase our confidence that identifications are correct. These repeated identifications are an artifact of the instrument algorithms for determining the precursors to sample in LC-MS/MS workflows. MS/MS spectra of the same precursor ion are often acquired multiple times during its elution envelope, resulting in repeated, related spectra, which tend to have correlated errors. Correlated errors are also likely when multiple peptide ion charge states are seen.

Furthermore, while we usually assume that distinct peptides represent independent peptide identifications, there are occasions when this is not sufficient. Peptides with common N or C terminus are sometimes identified due to precursor charge state enumeration (see Subheading 4.1) or non-specific proteolytic cleavage (see Subheading 4.2.4). These dependent identifications are an extremely common artifact of some spectral processing software, which enumerates identical MS/MS spectra with charges states +2 and +3. In the case of these dependent spectra too, only one should be counted for the purposes of determining matched proteins.

A conservative approach to protein identification is to required the multiple *non-overlapping* statistically significant peptide identifications.

4.4.5. Protein Identification and Peptide Sequence Redundancy

Having established the proteins with a sufficient level of peptide identification support, the next step in protein identification is the elimination of proteins identified solely due to peptide sequences shared with the true proteins of the sample. This is quite common when searching a general sequence database containing protein sequences from related organisms due to sequence homology, but is also becoming more prevalent as protein sequence databases containing sequence variants and protein isoforms become more widely available. Thus, while peptide identification support is *necessary* to conclude the presence of a protein, it is *not sufficient*. We can conclude that one of the proteins supported by shared peptide identification evidence is present, and the difficulty is in determining which one.

When proteins with shared peptide sequences have multiple additional significant peptide identifications to peptide sequences which are not shared, the choice is clear and these should be retained. Proteins whose peptide identifications are a strict subset of some other protein's identifications should be eliminated, as the evidence for their presence in the output can be completely explained by the other protein. Proteins whose peptide identifications are exactly equivalent cannot be distinguished by the evidence in the results and must be treated as equally valid conclusions. The remaining cases require the careful examination of the distinguishing peptide identifications to determine if the evidence is strong enough to support the existence of both protein sequences. Usually this is not the case, and one or the other, or neither should be chosen.

Acknowledgments

The preparation of this manuscript was supported, in part, by CPTI Grant R01 CA126189.

References

1. Aebersold, R. and Mann, M. (2003) Mass spectrometry-based proteomics. *Nature* **422**, 198–207.
2. Deutsch, E. W., Lam, H., and Aebersold, R. (2008) Data analysis and bioinformatics tools for tandem mass spectrometry in proteomics. *Physiological Genomics* **33**, 18–25.
3. Johnson, R., Davis, M., Taylor, J., and Patterson, S. (2005) Informatics for protein identification by mass spectrometry. *Methods* **35**, 223–236.
4. Maccoss, M. (2005) Computational analysis of shotgun proteomics data. *Current Opinion in Chemical Biology* **9**, 88–94.

5. McDonald, W. H. and Yates, J. R. (2003) Shotgun proteomics: integrating technologies to answer biological questions. *Current Opinion in Molecular Therapeutics* **5**, 302–309.
6. Nesvizhskii, A. I. (2007) Mass Spectrometry Data Analysis in Proteomics, volume 367 of *Methods in Molecular Biology*, chapter Protein Identification by Tandem Mass Spectrometry and Sequence Database Searching, 87–119. Humana Press, Totowa, NJ.
7. Sadygov, R. G., Cociorva, D., and Yates, J. R. (2004) Large-scale database searching using tandem mass spectra: looking up the answer in the back of the book. *Nature Methods* **1**, 195–202.
8. Bafna, V. and Edwards, N. (2003) On de novo interpretation of tandem mass spectra for peptide identification. In *RECOMB '03: Proceedings of the Seventh Annual International Conference on Research in Computational Molecular Biology*, 9–18. ACM Press, New York.
9. Chen, T., Kao, M. Y., Tepel, M., Rush, J., and Church, G. M. (2001) A dynamic programming approach to de novo peptide sequencing via tandem mass spectrometry. *Journal of Computational Biology* **8**, 325–337.
10. Frank, A. and Pevzner, P. (2005) Pepnovo: de novo peptide sequencing via probabilistic network modeling. *Analytical Chemistry* 77, 964–973.
11. Taylor, A. and Johnson, R. S. (1997) Sequence database searches via de novo peptide sequencing by tandem mass spectrometry. *Rapid Communications in Mass Spectrometry* **11**, 1067–1075.
12. Mann, M. and Wilm, M. (1994) Error-tolerant identification of peptides in sequence databases by peptide sequence tags. *Analytical Chemistry* **66**, 4390–4399.
13. Tabb, D. L., Ma, Z.-Q., Martin, D. B., Ham, A.-J. L., and Chambers, M. C. (2008) DirecTag: accurate sequence tags from peptide MS/MS through statistical scoring. *Journal of Proteome Research* 7, 3838–3846.
14. Tanner, S., Shu, H., Frank, A., Wang, L. C., Zandi, E., Mumby, M., Pevzner, P. A., and Bafna, V. (2005) Inspect: identification of post-translationally modified peptides from tandem mass spectra. *Analytical Chemistry* 77, 4626–4639.
15. Dass, C. (2001) *Principles and Practice of Biological Mass Spectrometry.* John Wiley & Sons, Inc., New York.
16. Perkins, D. N., Pappin, D. J., Creasy, D. M., and Cottrell, J. S. (1999) Probability-based protein identification by searching sequence databases using mass spectrometry data. *Electrophoresis* **20**, 3551–3567.
17. Eng, J. K., McCormack, A. L., and Yates, J. R. (1994) An approach to correlate tandem mass spectral data of peptides with amino acid sequences in a protein database. *Journal of the American Society of Mass Spectrometry* **5**, 976–989.
18. Craig, R. and Beavis, R. C. (2004) Tandem: matching proteins with tandem mass spectra. *Bioinformatics* **20**, 1466–1467.
19. Geer, L. Y., Markey, S. P., Kowalak, J. A., Wagner, L., Xu, M., Maynard, D. M., Yang, X., Shi, W., and Bryant, S. H. (2004) Open mass spectrometry search algorithm. *Journal of Proteome Research* **3**, 958–964.
20. Kersey, P. J., Duarte, J., Williams, A., Karavidopoulou, Y., Birney, E., and Apweiler, R. (2004) The International Protein Index: an integrated database for proteomics experiments. *Proteomics* **4**, 1985–1988.
21. The Uniprot Consortium (2010) The Universal Protein Resource (UniProt) in 2010. *Nucleic Acids Research* **38**, D142–D148.
22. Edwards, N. J. (2007) Novel peptide identification from tandem mass spectra using ESTs and sequence database compression. *Molecular Systems Biology* **3**, 102.
23. Keller, A., Eng, J., Zhang, N., Li, X.-J. J., and Aebersold, R. (2005) A uniform proteomics MS/MS analysis platform utilizing open XML file formats. *Molecular Systems Biology* **1**, 2005.0017.
24. Kessner, D., Chambers, M., Burke, R., Agus, D., and Mallick, P. (2008) ProteoWizard: open source software for rapid proteomics tools development. *Bioinformatics* **24**, 2534–2536.

Chapter 10

LC-MS Data Analysis for Differential Protein Expression Detection

Rency S. Varghese and Habtom W. Ressom

Abstract

In proteomic studies, liquid chromatography coupled with mass spectrometry (LC-MS) is a common platform to compare the abundance of various peptides that characterize particular proteins in biological samples. Each LC-MS run generates data consisting of thousands of peak intensities for peptides represented by retention time (RT) and mass-to-charge ratio (m/z) values. In label-free differential protein expression studies, multiple LC-MS runs are compared to identify differentially abundant peptides between distinct biological groups. This approach presents a computational challenge because of the following reasons (i) substantial variation in RT across multiple runs due to the LC instrument conditions and the variable complexity of peptide mixtures, (ii) variation in m/z values due to occasional drift in the calibration of the mass spectrometry instrument, and (iii) variation in peak intensities caused by various factors including noise and variability in sample handling and processing. In this chapter, we present computational methods for quantification and comparison of peptides by label-free LC-MS analysis. We discuss data preprocessing methods for alignment and normalization of LC-MS data. Also, we present multivariate statistical methods and pattern recognition methods for detection of differential protein expression from preprocessed LC-MS data.

Key words: Mass spectrometry, LC-MS, Alignment, Normalization, Difference detection

1. Introduction

The introduction of mass spectrometry (MS) as a robust and sensitive technology for protein analysis had a major impact on the analysis of complex proteome samples. In particular, the measurement of peptides obtained from protein digestion by liquid chromatography coupled with mass spectrometry (LC-MS) has paved the road to study a large number of peptides of biological samples in an automated and high-throughput mode.

Although labeling protocols (e.g., ICAT, iTRAQ, ^{18}O-, or ^{15}N-labeling, etc.) remain the core technologies used in

Cathy H. Wu and Chuming Chen (eds.), *Bioinformatics for Comparative Proteomics*, Methods in Molecular Biology, vol. 694, DOI 10.1007/978-1-60761-977-2_10,

LC-MS-based proteomic quantification, increasing efforts have been directed to the label-free approaches. Label-free method is attractive because of cost effectiveness, simpler experimental protocols, fewer measurement artifacts, and limited availability of isotope labeled references (1, 2). The most common label-free method is the spectral count method, where the total number of MS/MS spectra taken on peptides from a given protein in a given LC-MS/MS analysis is used to compare differential abundance between groups of samples (3). This method simply counts the number of spectra identified for a given peptide in different samples and integrates results of all measured peptides for the protein quantified. One of the alternatives to this approach is the comparison of ion intensities, where LC-MS runs are compared to identify differentially abundant ions at specific mass to charge (*m/z*) and retention time points. This approach is based on precursor signal intensity (MS), applicable to data derived from high mass precision spectrometers. The high resolution facilitates extraction of precursor ion signal intensity and thus uncouples quantification from the identification process. It is based on the observation that peak intensity is linearly proportional to the concentration of the ions being detected. A critical challenge in using label-free LC-MS analysis for detection of differential protein expression lies in normalizing and aligning the LC-MS data from various runs to ensure bias-free comparison of the same biological entities across multiple runs. Once the LC-MS data are preprocessed, difference detection can be carried out using multivariate statistical methods and pattern recognition algorithms. Because the number of peaks is typically larger than the number of samples, difference detection raises a problem of multiplicity, where the probability of erroneously declaring significance increases rapidly with the number of tests being performed.

In this chapter, we present computational methods for quantification and comparison of LC-MS runs from multiple samples. We begin with an overview of LC-MS data. We then discuss LC-MS data alignment and normalization methods. This is followed by a description of multivariate statistical methods and pattern recognition algorithms for difference detection from preprocessed LC-MS data. Finally, we provide an overview of existing challenges in label-free LC-MS analysis and future outlook.

2. LC-MS Data

LC-MS experiments generate data that consist of three dimensions (1) the elution time, also called retention (RT) point, (2) the *m/z* value, and (3) the intensity (ion abundance). Figure 1a presents three-dimensional data derived from a typical LC-MS

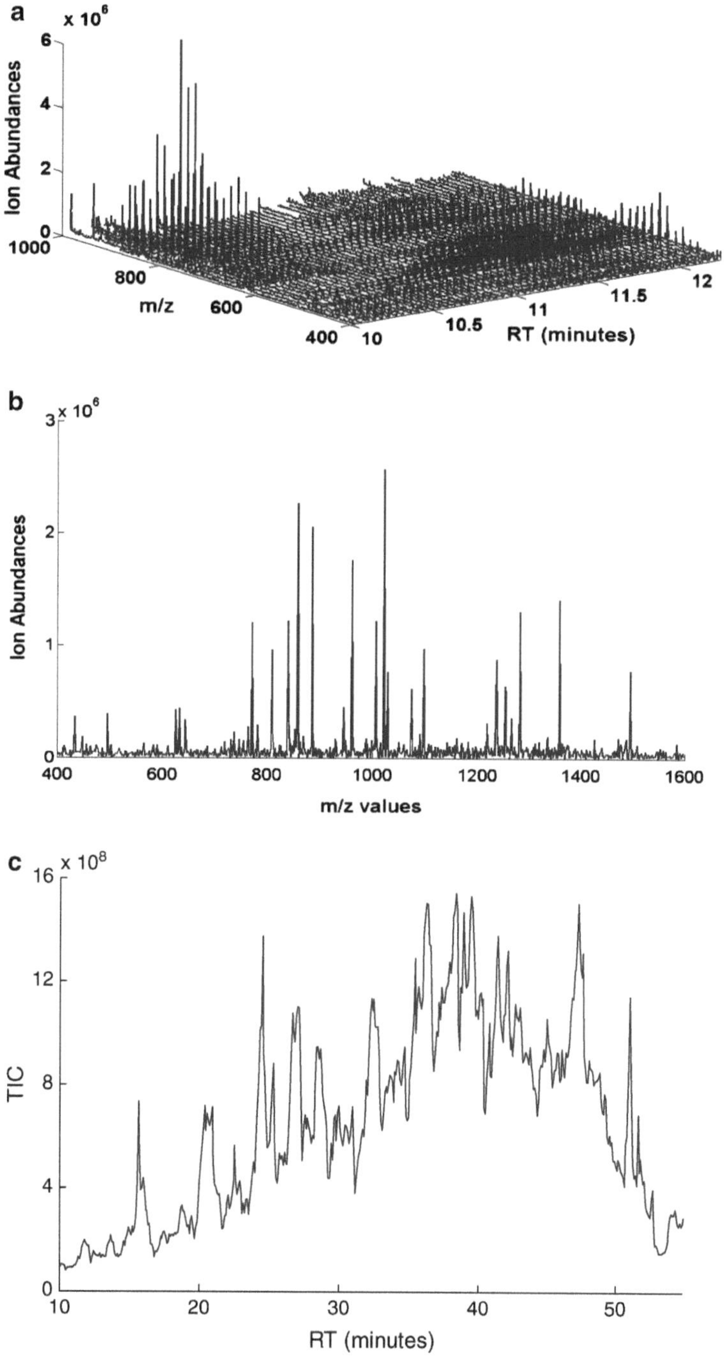

Fig. 1. Data derived from a typical LC-MS experiment. (**a**) Three-dimensional LC-MS data of a sample for RT points between 10 and 12 min and *m/z* values between 400 and 1,000 Da. (**b**) Mass spectrum in the range between 400 and 1,600 Da at RT = 10 min. (**c**) TIC plot of the LC-MS data between 10 and 55 min of RT.

experiment for a single run. As shown in the figure, each LC-MS run generates spectra comprised of thousands of peak intensities for peptides with specific RT and *m/z* values. Figure 1b shows a mass spectrum (ion abundance vs. *m/z*) at a particular RT point

(RT in the figure is 10 min). Figure 1c depicts the total ion chromatogram (TIC) obtained by calculating the sum of the ion abundances across the *m/z* dimension for each RT point. Although RT is a continuous variable, the LC-MS system produces mass spectra at a discrete set of RT points, usually a few seconds apart. It is typical to represent RT points by scan indices, since there is a one-to-one correspondence between RT points and total MS scan numbers.

3. LC-MS Data Preprocessing

Various data preprocessing steps are conducted before LC-MS runs can be compared for differential protein expression. These include deconvolution of multiple charged peaks and isotope clusters (4), outlier screening, binning, baseline correction, smoothing, alignment, and normalization. In the following, we briefly discuss alignment and normalization methods.

3.1. Alignment

Alignment is necessary to correct for chromatographic and mass spectrometric drifts that do not reflect real sample variation. Alignment methods find a common set of features across LC-MS runs to allow quantitative comparison of the same biological entities. Without alignment, the same ion can have different *m/z* or retention time point across multiple runs. Thus, alignment with respect to both *m/z* and retention time is a prerequisite for quantitative comparison of proteins/peptides by LC-MS. Alignment algorithms have traditionally been used on data points and/or feature vectors of fixed dimension (5). Applications of these algorithms for LC-MS data alignment have been reported in the literature (6–16). The most common approaches for aligning LC-MS data are based on the identification of landmarks or structural points (referring to the unique charge species in data) and the use of internal standards, respectively. The landmarks are usually associated with critical or inflection points. Multiple LC-MS runs are then aligned so that the landmarks are synchronized. In this framework, the most widely used algorithm is dynamic time warping (DTW) that performs the alignment in time axis by stretching or shrinking the time series data. Another common method is correlation optimized warping (COW), which computes a piecewise linear transformation by dividing the time series into segments and then performing a linear warp within each segment to optimize overlap while constraining segment boundaries. The parameters for the best linear transformation are determined by maximizing the sum of correlation coefficients or covariance between data segments in pairs of samples. Most of the existing algorithms including DTW and COW are either limited to a consensus combination of pair-wise alignment or use a reference (template) for alignment. This limitation leads to suboptimal results compared to global alignment techniques.

Alignment methods that rely on optimization of global fitting function provide an alternative solution to address the above challenges without requiring landmarks or internal standards. For example, a recently introduced method called continuous profile model (CPM) has been applied for alignment and normalization of continuous time-series data and for detection of differences in multiple LC-MS data (6, 17). Similarly, we developed a probabilistic mixture regression model (PMRM) for global alignment of LC-MS data (18, 19). We approach the problem of LC-MS data alignment with an ultimate goal of detecting differences among groups of LC-MS runs. This is accomplished by estimating a model that has the following two functions (1) modeling and correcting the variation within each class and (2) identifying systematic changes across classes. Specifically, we use a mixture model that incorporates LC-MS runs clustered into K components. For each of these groups, a prototypical LC-MS intensity profile is estimated by nonlinear regression with spline basis functions.

A particular advantage of PMRM is its ability to model non-Gaussian multimodal density functions using simpler component density functions that can be defined on nonvector data such as LC-MS data. Moreover, the framework lends itself to an expectation-maximization (EM) algorithm with the following features (i) the explicit use of transformation priors for modeling the ion abundance (peak intensity) variability in both RT and m/z dimensions of the data, (ii) the use of a probabilistic metric that allows estimation of the distance among multiple LC-MS data instead of computing pair-wise distances, and (iii) the ability to extend the method for alignment and normalization of LC-MS data involving multiple groups. We demonstrated that analysis of LC-MS data via PMRM has the potential to address critical concerns such as unequal intervals across multiple runs and misalignment both in time and measurement space (18, 19). We assume that the observed dataset D representing multiple groups is generated with the following three features (i) an individual is randomly drawn from a population of M objects (i.e., the dataset D); (ii) the individual is assigned to the kth group with probability α_k, where $\sum_{k=1}^{K} \alpha_k = 1$. These are the prior weights corresponding to all K groups, where $K << M$; and (iii) for an individual i, there is a density function $p_k(\mathbf{y}_k \mid \theta_k)$ that generates the observed functional data $\mathbf{y}_i$.

The observed density on the $\mathbf{y}_i$'s is a mixture model, i.e., a convex combination of component models $p(\mathbf{y}_i \mid \theta) = \sum_{k=1}^{K} \alpha_k p_k(\mathbf{y}_i \mid \theta_k)$. Thus, we estimate the most likely values for the parameters θ_k and α_k using the assumed functional densities $p_k(\cdot)$ on the observed data $\mathbf{y}_i$'s. This is accomplished by using the EM algorithm, which is a general procedure for finding the maximum-likelihood estimators of the parameters from the mixture models (20–22). Thus, this probabilistic-based framework allows us to find the best group alignment in time and measurement spaces from the observed dataset D.

The goal is to pull out the mixture of components from the full joint density, using the observed dataset D as a guide, so that the underlying group behavior can be inferred. A standard approach to deal with the hidden data is to utilize the EM algorithm for consistent estimation. The estimation algorithm is implemented by taking advantage of the connection between smoothing B-splines (at the design points) and mixed regression models. Splines are recommended for data fitting whenever there is no particular reason for using a single polynomial or other elementary functions. Spline functions have the following useful properties: smooth and flexible, easy to evaluate along with their derivatives and integrals, and easy to generalize to higher dimensions.

3.2. Normalization

Normalization is one of the important preprocessing tasks in LC-MS-based studies. Because of lack of reliable methods, internal standards spiked in biological samples are typically used for normalization. For example, the mzMine toolbox uses multiple internal standard compounds injected to samples to calculate a set of normalization factors, one for each standard compound based on either searching for a standard compound peak closest to the peak or using weighted contribution of each standard compound (23). However, as the authors themselves noted that this method suffers from the ad hoc assignments of internal standards for each component based on a subset of relevant chemical properties (24). Also, in the context of the need for universally applicable analytical tools and that internal standards vary depending on the instrument used and samples under study, it is desired to develop normalization methods that do not rely on internal standards.

As summarized by Karpievitch et al. (25), several normalization methods that do not use internal standards have been used, including global scaling, lowess (26), quantile normalization (27), and ANOVA models (28). A global scaling method shifts all ion intensities of a sample by a constant amount, so that all samples have the same mean, median, or total ion current. However, this approach cannot capture complex bias trends like those commonly seen in LC-MS data. When the sources of bias are known exactly, ANOVA models can effectively estimate and remove systematic biases (29). Nevertheless, it is generally not possible to identify all of the relevant sources of bias to sufficiently address them with ANOVA models. To address this, Karpievitch et al. recommend EigenMS that uses singular value decomposition to capture and remove biases from LC-MS peak intensity measurements.

4. Difference Detection

Difference detection deals with the identification of peaks that represent differentially abundant ions with specific RT and *m/z* values. Various unsupervised and supervised methods have been proposed for peak selection from LC-MS data. For example, principal component analysis (PCA) transforms the data to a new coordinate system such that the variables in the new data space (known as scores or principal components) are orthogonal and are sorted in the decreasing order of their variances. The peaks that contribute to the top factors are identified by using the eigenvalue plot (30). A similar approach can used in a supervised way [e.g., partial least squares (PLS)], where the training samples with known phenotypes are used to calculate the factors. The weight plot obtained from this PLS analysis provides as a tool to select useful peaks (30, 31).

Another commonly used supervised approach applies statistical analyses such as *t*-test, which recognizes differentially abundant peaks between biological groups involving multiple subjects. However, thousands of peaks need to be tested against the null hypothesis of no difference. This raises a problem of multiplicity, where the probability of erroneously declaring significance increases rapidly with the number of tests being performed. To address this, false discovery rate (FDR) procedures are typically used. For example, FDR procedure by Benjamini and Hochberg (32) controls the proportion of errors among rejected tests and provides a less conservative approach than traditional approaches that control the family-wise error rate.

Another concern is the use of an appropriate test statistic when performing the required hypothesis tests. Various statistical methods are proposed to address this concern. For example, the shrinkage *t*-statistics of Opgen-Rhein and Strimmer (33) derives *t*-statistics on the basis of a model-free shrinkage estimator of the variances. In order to compute the *p*-values without making parametric assumptions, the null distribution of the test statistics can be simulated by permutation of the sample labels. For example, when comparing two groups, the class labels are randomly reassigned and the shrinkage *t*-statistics are recomputed. This procedure is repeated many times to obtain an approximation to the null distribution for data with structure similar to the one on hand. The corresponding *p*-value is obtained by evaluating the probability of observing a test score at least as extreme as the observed one in this simulated null distribution. The permutation test is repeated for each peak one at a time resulting in thousands of *p*-values, to which the false discovery rate control procedure is applied.

The selected peaks are typically used as inputs to a pattern classification algorithm such as random forest (RF) and support vector machine (SVM). The RF method can be utilized for sample classification and identification of peaks that contribute strongly to classification. It is an ensemble of unpruned classification or regression trees, induced from bootstrap samples of the training data, using random feature selection in the tree induction process. It is a classification method based on "growing" an ensemble of decision tree classifiers. In order to classify a new sample, the input is analyzed using each of the classification trees in the forest. Each tree gives a classification, "voting" for that class. The forest chooses the classification having the most votes (over all the trees in the forest). A measure of the importance of classification variables is also calculated by considering the difference between the results from original and randomly permuted versions of the data set. Prediction is made by aggregating (majority vote for classification or averaging for regression) the predictions of the ensemble. RF generally exhibits a substantial performance improvement over the single tree classifier such as classification and regression tree. It has been successfully applied in proteomics profiling studies to construct a classifier and discover peak intensities most likely responsible for the separation between classes (34, 35).

The SVM recursive feature elimination (SVM-RFE) algorithm recursively classifies samples with SVM and selects peaks according to their SVM weights (36). Benefiting from the good performance of SVMs in high-dimensional gene expression data, SVM-RFE is often considered as one of the best feature selection algorithms in the literature. Also, stochastic global optimization methods such as genetic algorithms, simulated annealing, and swarm intelligence methods have been used to systematically select features from a high-dimensional search space without the need for an exhaustive search. A hybrid of SVM and ant colony optimization (ACO) was also developed to select a panel of peaks (37).

To evaluate the generalization capability of the peaks and the SVM classifier determined by the training data set, the SVM classifier should be tested using a blind validation set, i.e., a test set that is set aside during the process of data preprocessing, peak selection, and building the SVM classifier. An important weakness of many pattern recognition algorithms is that they are not based on a probabilistic model. There is no probability level or confidence interval associated with predictions derived from using them to classify a new set of data. The confidence that an analyst can have in the accuracy of the results produced by a given classifier is based purely on its historical accuracy—how well it has predicted the desired response in other similar circumstances. Thus, after learning is completed, a pattern recognition paradigm is evaluated for its performance through previously unseen testing data set (also known as a blind validation set). The purpose of this testing is to prove the adequacy or to detect the inadequacy of the selected peaks or the classifier built. Inadequate performance could be attributed to insufficient or redundant features, inappropriate selection of model structure for the classifier, too few or too many model parameters, insufficient training, overtraining, error in the program code, or complexity of the underlying system such as presence of highly nonlinear relationships, noise, and systematic bias. The aim of evaluating a classifier is to ensure that it serves as a general model. A general model is one whose input-output relationships (derived from the training data set) apply equally well to new sets of data (previously unseen test data) from the same problem not included in the training set. Thus, the goal of a pattern recognition algorithm is the generalization to new data of the relationships learned on the training set (38).

Various methods have been used to test the generalization capability of a classifier. These include the *k*-fold cross-validation, bootstrapping, and hold-out methods. In *k*-fold cross-validation, the data is divided into *k* subsets of (approximately) equal size. The model is then trained several times, each time leaving out one of the subsets from training, but using only the omitted subset to compute the classification error. If *k* equals the sample size, this is called "leave-one-out" cross-validation. In the leave-one-out method, one sample is selected as a validation sample and feature selection and classifier building are performed using the remaining data set. The resulting model is tested on the validation sample. The process is repeated until all samples appear in the validation set. In the hold-out method, only a single subset (also known as validation set) is used to estimate the generalization error. Thus, the hold-out method does not involve crossing. In bootstrapping, a subsample is randomly selected from the full training data set with replacement.

Common bootstrapping methods include bagging and boosting. Bagging can be used with many classification methods and regression methods to reduce the variance associated with prediction, and thereby improve the prediction process. In bagging, many bootstrap samples are drawn from the available data, some prediction method is applied to each bootstrap sample, and then the results are combined by voting. Boosting can be used to improve the accuracy of classification. Unlike bagging, the samples used at each step are not all drawn in the same way from the same population, but rather the incorrectly predicted cases from a given step are given increased weight during the next step. Hence, boosting uses a weighted average of results obtained from applying a prediction method to various samples.

5. Challenges and Future Outlook

As large volume and high dimensional data are being generated by the rapidly expanding use of LC-MS technologies, the number of reported applications of proteomic pattern recognition algorithms is expected to increase. To reduce false-positive discoveries, significant development on bioinformatics and robust validation methods will be required. However, with increasing demand comes the need for further improvements that can make implementation of these algorithms for high dimensional LC-MS data analysis more efficient. Key improvements include (i) careful study design to minimize the effect of factors that may introduce bias to the data; (ii) enhanced computational power to handle the high dimensionality and large volume data; (iii) improved high-throughput technologies with less background noise and technical variability; (iv) enhanced quality control and protocol development/implementation; (v) improved data preprocessing methods to minimize the impact of background noise, sample degradation, and variability in sample preparation and instrument settings; (vi) improved visualization tools to assess data quality and interpret results; (vii) adequate data storage and retrieval systems; and (viii) advances in multivariate statistical methods and patter recognition algorithms to enhance their speed and make them more accessible to the user.

Careful study design is needed to make sure that a protocol is in place that enables appropriate randomization and replication to avoid bias in sample collection and sample preparation (39). As the future of mass spectrometry and proteomics unfolds, it will produce improved understanding of the data and the underlying biology. Additional developments in label-free protein quantification technologies will help to meet the demands of both proteomics and clinical applications.

Acknowledgments

This work was supported in part by the National Science Foundation Grant IIS-0812246 awarded to HWR.

References

1. Lill, J. (2003) Proteomic tools for quantitation by mass spectrometry. *Mass Spectrom Rev* **22**, 182–194.
2. Goodlett, D. R. and Yi, E. C. (2003) Stable isotopic labeling and mass spectrometry as a means to determine differences in protein expression. *TrAC Trends Anal Chem* **22**, 282–290.
3. Old, W. M., Meyer-Arendt, K., Aveline-Wolf, L., Pierce, K. G., Mendoza, A., Sevinsky,

J. R., Resing, K. A., and Ahn, N. G. (2005) Comparison of label-free methods for quantifying human proteins by shotgun proteomics. *Mol Cell Proteomics* **4**, 1487–1502.

4. Zhongqi, Z., Shenheng, G., and Marshall, A. G. (1997) Enhancement of the effective resolution of mass spectra of high-mass biomolecules by maximum entropy-based deconvolution to eliminate the isotopic natural abundance distribution. *J Am Soc Mass Spectrom* **8**, 659–670.
5. Ramsay, J. O. and Silverman, B. W. (2002) *Applied functional data analysi : methods and case studies.* Springer, New York.
6. Listgarten, J., Neal, R. M., Roweis, S. T., Wong, P., and Emili, A. (2007) Difference detection in LC-MS data for protein biomarker discovery. *Bioinformatics* **23**, e198–e204.
7. Wang, P., Tang, H., Fitzgibbon, M. P., McIntosh, M., Coram, M., Zhang, H., Yi, E., and Aebersold, R. (2007) A statistical method for chromatographic alignment of LC-MS data. *Biostatistics* **8**, 357–367.
8. Wiener, M. C., Sachs, J. R., Deyanova, E. G., and Yates, N. A. (2004) Differential mass spectrometry: a label-free LC-MS method for finding significant differences in complex peptide and protein mixtures. *Anal Chem* **76**, 6085–6096.
9. Radulovic, D., Jelveh, S., Ryu, S., Hamilton, T. G., Foss, E., Mao, Y., and Emili, A. (2004) Informatics platform for global proteomic profiling and biomarker discovery using liquid chromatography-tandem mass spectrometry. *Mol Cell Proteomics* **3**, 984–997.
10. Sadygov, R. G., Maroto, F. M., and Huhmer, A. F. (2006) ChromAlign: a two-step algorithmic procedure for time alignment of three-dimensional LC-MS chromatographic surfaces. *Anal Chem* **78**, 8207–8217.
11. Prakash, A., Mallick, P., Whiteaker, J., Zhang, H., Paulovich, A., Flory, M., Lee, H., Aebersold, R., and Schwikowski, B. (2006) Signal maps for mass spectrometry-based comparative proteomics. *Mol Cell Proteomics* **5**, 423–432.
12. Jaitly, N., Monroe, M. E., Petyuk, V. A., Clauss, T. R., Adkins, J. N., and Smith, R. D. (2006) Robust algorithm for alignment of liquid chromatography-mass spectrometry analyses in an accurate mass and time tag data analysis pipeline. *Anal Chem* **78**, 7397–7409.
13. America, A. H., Cordewener, J. H., van Geffen, M. H., Lommen, A., Vissers, J. P., Bino, R. J., and Hall, R. D. (2006) Alignment and statistical difference analysis of complex peptide data sets generated by multidimensional LC-MS. *Proteomics* **6**, 641–653.
14. Pierce, K. M., Wood, L. F., Wright, B. W., and Synovec, R. E. (2005) A comprehensive two-dimensional retention time alignment algorithm to enhance chemometric analysis of comprehensive two-dimensional separation data. *Anal Chem* **77**, 7735–7743.
15. Horvatovich, P., Govorukhina, N. I., Reijmers, T. H., van der Zee, A. G. J., Suits, F., and Bischoff, R. P. H. (2007) Chip-LC-MS for label-free profiling of human serum. *Electrophoresis* **28**, 4493–4505.
16. Mueller, L. N., Rinner, O., Schmidt, A., Letarte, S., Bodenmiller, B., Brusniak, M. Y., Vitek, O., Aebersold, R., and Muller, M. (2007) SuperHirn – a novel tool for high resolution LC-MS-based peptide/protein profiling. *Proteomics* **7**, 3470–3480.
17. Listgarten, J., Neal, R. M., Roweis, S. T., and Emily, A. (2005) Multiple alignment of continuous time series. *Neural Inf Process Syst* **17**, 817–824.
18. Befekadu, G. K., Tadesse, M. G., Hathout, Y., and Ressom, H. W. (2008) Multiclass alignment of LC-MS data using probabilistic-based mixture regression models. *Proceedings of the 30th Annual International Conference of the IEEE Engineering in Medicine and Biology Society*, Vancouver, BC, 4094–4097.
19. Ressom, H. W., Befekadu, G. K., and Tadesse, M. G. (2009) Analysis of LC-MS data using probabilistic-based mixture regression models. *at – Automatisierungstechnik* **57**, 453–465.
20. Dempster, A. P., Laird, N. M., and Rubin, D. B. (1977) Maximum likelihood from incomplete data via the EM algorithm. *J R Stat Soc Series B (Methodol)* **39**, 1–38.
21. Jordan, M. I. and Jacobs, R. A. (1994) Hierarchical mixtures of experts and the EM algorithm. *Neural Comput* **6**, 181–214.
22. Redner, R. A. and Walker, H. F. (1984) Mixture densities, maximum likelihood and the EM algorithm. *SIAM Rev* **26**, 195–239.
23. Katajamaa, M. and Oresic, M. (2005) Processing methods for differential analysis of LC/MS profile data. *BMC Bioinformatics* **6**, 179.
24. Sysi-Aho, M., Katajamaa, M., Yetukuri, L., and Oresic, M. (2007) Normalization method for metabolomics data using optimal selection of multiple internal standards. *BMC Bioinformatics* **8**, 93.
25. Karpievitch, Y. V., Taverner, T., Adkins, J. N., Callister, S. J., Anderson, G. A., Smith, R. D., and Dabney, A. R. (2009) Normalization of peak intensities in bottom-up MS-based proteomics using singular value decomposition. *Bioinformatics* **25**, 2573–2580.
26. Yang, Y. H., Dudoit, S., Luu, P., Lin, D. M., Peng, V., Ngai, J., and Speed, T. P. (2002)

Normalization for cDNA microarray data: a robust composite method addressing single and multiple slide systematic variation. *Nucleic Acids Res* **30**, e15.
27. Bolstad, B. M., Irizarry, R. A., Astrand, M., and Speed, T. P. (2003) A comparison of normalization methods for high density oligonucleotide array data based on variance and bias. *Bioinformatics* **19**, 185–193.
28. Kerr, M. K., Martin, M., and Churchill, G. A. (2000) Analysis of variance for gene expression microarray data. *J Comput Biol* 7, 819–837.
29. Hill, E. G., Schwacke, J. H., Comte-Walters, S., Slate, E. H., Oberg, A. L., Eckel-Passow, J. E., Therneau, T. M., and Schey, K. L. (2008) A statistical model for iTRAQ data analysis. *J Proteome Res* 7, 3091–3101.
30. Purohit, P. V. and Rocke, D. M. (2003) Discriminant models for high-throughput proteomics mass spectrometer data. *Proteomics* **3**, 1699–1703.
31. Chen, C., Gonzalez, F. J., and Idle, J. R. (2007) LC-MS-based metabolomics in drug metabolism. *Drug Metab Rev* **39**, 581–597.
32. Benjamini, Y. and Hochberg, Y. (1995) Controlling the false discovery rate: a practical and powerful approach to multiple testing. *J R Stat Soc Series B* **57**, 289–300.
33. Opgen-Rhein, R. and Strimmer, K. (2007) Accurate ranking of differentially expressed genes by a distribution-free shrinkage approach. *Stat Appl Genet Mol Biol* **6**, Article9.
34. Datta, S. (2008) Classification of breast cancer versus normal samples from mass spectrometry profiles using linear discriminant analysis of important features selected by random forest. *Stat Appl Genet Mol Biol* 7, Article7.
35. Wu, B., Abbott, T., Fishman, D., McMurray, W., Mor, G., Stone, K., Ward, D., Williams, K., and Zhao, H. (2003) Comparison of statistical methods for classification of ovarian cancer using mass spectrometry data. *Bioinformatics* **19**, 1636–1643.
36. Guyon, I., Weston, J., Barnhill, S., and Vapnik, V. (2002) Gene Selection for cancer classification using support vector machines. *Mach Learn* **46**, 389–422.
37. Ressom, H. W., Varghese, R. S., Drake, S. K., Hortin, G. L., Abdel-Hamid, M., Loffredo, C. A., and Goldman, R. (2007) Peak selection from MALDI-TOF mass spectra using ant colony optimization. *Bioinformatics* **23**, 619–626.
38. Wang, Z., Wang, Y., Xuan, J., Dong, Y., Bakay, M., Feng, Y., Clarke, R., and Hoffman, E. P. (2006) Optimized multilayer perceptrons for molecular classification and diagnosis using genomic data. *Bioinformatics* **22**, 755–761.
39. Zhang, Z. and Chan, D. W. (2005) Cancer proteomics: in pursuit of “true” biomarker discovery. *Cancer Epidemiol Biomarkers Prev* **14**, 2283–2286.

Chapter 11

Protein Identification by Spectral Networks Analysis

Nuno Bandeira

Abstract

While advances in tandem mass spectrometry (MS/MS) steadily increase the rate of generation of MS/MS spectra, standard algorithmic approaches for peptide identification recently seemed to be reaching the limit on the amount of information that could be extracted from MS/MS spectra. However, a closer look reveals that a common limiting procedure is to analyze each spectrum in isolation, even though high throughput mass spectrometry regularly generates many spectra from related peptides. By capitalizing on this redundancy we show that, similarly to the alignment of protein sequences, unidentified MS/MS spectra can also be aligned for the identification of modified and unmodified variants of the same peptide. Moreover, this alignment procedure can be iterated for the accurate grouping of multiple modification variants of the same peptides. Furthermore, the combination of shotgun proteomics with the alignment of spectra from overlapping peptides led to the development of Shotgun Protein Sequencing – similarly to the assembly of DNA reads into whole genomic sequences, we show that assembly of MS/MS spectra enables the highest ever de novo sequencing accuracy, while recovering nearly complete protein sequences. We further show that shotgun protein sequencing has the potential to overcome the limitations of current protein sequencing approaches and thus catalyze the otherwise impractical applications of proteomics methodologies in studies of unknown proteins.

Key words: Tandem mass spectrometry, MS/MS, Alignment, Assembly, Spectral networks, Shotgun protein sequencing, Algorithms

1. Introduction

Tandem mass spectrometry (MS/MS) is nowadays the technology of choice for the identification of proteins and posttranslational modifications (1). Fast-paced technological developments have delivered high-throughput analysis of thousands of proteins in a mere couple of hours at unprecedented levels of mass resolution and accuracy (2). However, the major computational approaches to the automated identification of the millions of MS/MS spectra generated on a daily basis still interpret every single MS/MS spectrum in isolation like the original techniques for *de novo*

Cathy H. Wu and Chuming Chen (eds.), *Bioinformatics for Comparative Proteomics*, Methods in Molecular Biology, vol. 694,
DOI 10.1007/978-1-60761-977-2_11,

sequencing introduced by Klaus Biemann's group in the 1960s (3) and *database searching* first proposed in the early 1990s (4, 5). In database searching, each MS/MS spectrum is compared against a given database of known peptides and significant matches are selected for protein identification. Elaborate scoring functions have been derived to provide statistical significance to observed identifications and help make this the approach of choice for the analysis of model organisms (6, 7). However, database search is only applicable when the proteins sequences are obtained in advance through other experimental procedures such as DNA sequencing or Edman degradation. Conversely, de novo sequencing becomes the mass spectrometric approach of choice for studies of unknown proteins. Nevertheless, fully automated de novo analysis has remained an elusive goal because of difficulties in sequencing accuracy – the best algorithms for individual ion trap MS/MS spectra still predict one incorrect amino acid out of every five predictions (8). In this chapter, we propose to approach the MS/MS identification problem from a different perspective – first combine uninterpreted MS/MS spectra from overlapping peptides and only then determine the consensus identifications (of sequences *and* modifications) for *sets* of aligned MS/MS spectra.

2. Data Acquisition Protocols

Most experimental protocols use enzymatic digestion to generate smaller peptides that are then analyzed by mass spectrometry to identify proteins in the sample. Trypsin digestion is often used because its strong cleavage specificity tends to be reproducible and facilitates the analysis of complex samples by generating only a few different peptides per protein. Alternatively, less specific enzymes or combinations of enzymes may be used to generate extensive protein coverage (9, 10). As illustrated in Fig. 1a, b, these procedures tend to generate many overlapping peptides covering the same protein regions. Although the specificity of trypsin digestion leads to many spectra covering the same protein regions, nonspecific digestion tends to generate spectra covering large portions of the protein sequences.

From a computer science perspective, a protein or peptide sequence can be thought of as a string over a weighted alphabet of 20 amino acids, with the mass of each amino acid given by $m(a)$; the parent mass of a peptide $\rho = a_1,\ldots, a_n$ is defined as $m(\rho) = \sum_{i=1}^{n} m(a_i)$. Additionally, the i-th prefix (*suffix*) mass of a peptide, referred to as $b_i(y_i)$, is simply the summed mass of its prefix (suffix) string with i amino acids. Mass spectrometry instruments measure $\frac{mass}{charge}$ ratios of ionized molecules, or simply

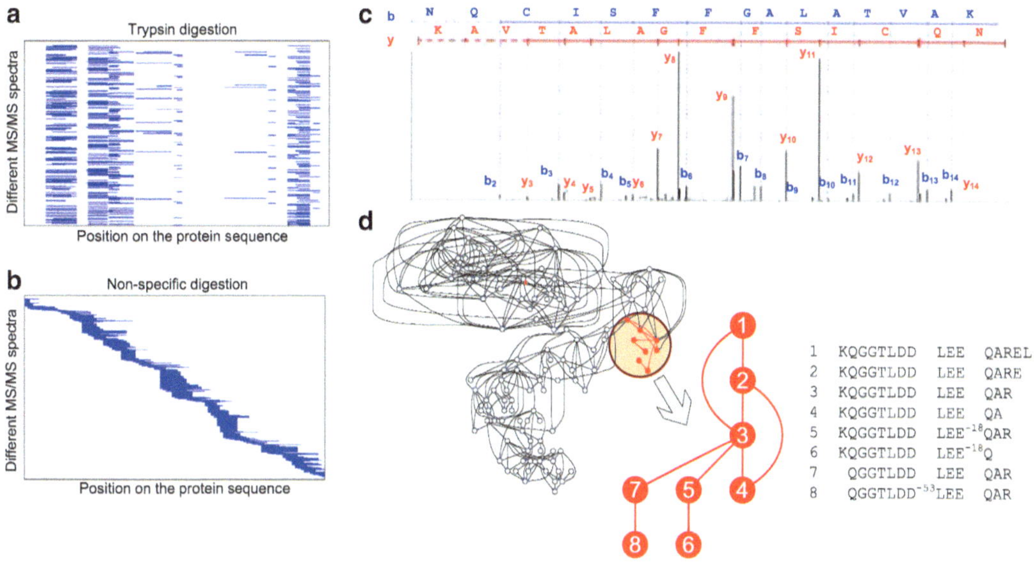

Fig. 1. Spectral coverage of overlapping peptides resulting from enzymatic digestion of a target protein; *horizontal axes* represent peptide location on the protein and *vertical axes* separate different MS/MS spectra: (**a**) Spectral coverage resulting from trypsin digestion; (**b**) Spectral coverage resulting from non-specific enzymatic digestion or digestion with multiple enzymes of different specificities. (**c**) MS/MS spectrum for peptide NQCISFFGALATVAK; **b**-ions (prefix masses) are shown in *blue*, *y*-ions (suffix masses) are shown in *red*. Note that the *b*/*y* peak assignments are not known in advance but can only be determined for identified spectra. (**d**) Spectral network formed by a set of 117 IKKβ spectra (11); each node corresponds to a different spectrum and nodes are connected by an edge if the corresponding spectra were paired by spectral alignment. A subcomponent of the spectral network is shown in *red* along with the corresponding peptides. For example, the edge between nodes 1 and 3 indicates that the spectrum for peptide 1 was significantly aligned to the spectrum from peptide 3.

measure mass if we make the simplifying assumption that all fragments have charge one.[1] Conceptually, when applied to the analysis of peptides, these instruments proceed through the following three stages (1, 12):

1. The first MS stage snapshots the parent masses of the peptides passing through the instrument (MS).
2. A parent mass is selected and the many copies of (usually) the same peptide are dissociatedinto fragments by a collision-induced random process. Peptides tend to break only once and between consecutive amino acids, often generating complementary pairs of detectable fragment masses: one corresponding to a prefix mass (*b*-ion) and another corresponding to a suffix mass (*y*-ion).
3. The second MS stage determines the masses of the peptide fragments (MS/MS).

[1] We remark that the term precursor mass is commonly used to denote the term $\frac{M+18+Z}{Z}$, where M is a peptide's parent mass and Z its parent charge.

Because many copies of the same peptide are initially present in the sample the same masses are detected several times with different masses having different relative abundances. As such, a tandem mass spectrum or MS/MS spectrum S of an unspecified peptide ρ with parent mass $m(S) = m(\rho)$ is a list of fragment masses, each with an assigned intensity proportional to the relative abundance of the corresponding fragment mass. Figure 1c shows an experimental MS/MS spectrum for the peptide NQCISFFGALATVAK (acquired on a Thermo LTQ ion trap mass spectrometer).

3. Spectral Networks

Samples of digested proteins often contain multiple overlapping peptides, i.e., different peptides covering the same region of a protein sequence. The simplest example is the acquisition of multiple spectra from the same peptide (sometimes detected and merged using spectral clustering techniques (13–15)). However, these samples also commonly contain spectra from similar but different peptides such as prefix peptides (e.g., PEPTI/PEPTIDES), suffix peptides (e.g., TIDES/PEPTIDES) or partially-overlapping peptides (e.g., PEPTIDES/TIDESHIGH). MacCoss et al. (9) were the first to realize the potential of overlapping peptides for the identification of posttranslationally modified proteins and have recently demonstrated the increased throughput of modified digestion schemes on the identification of proteins from complex mixtures (16). Rich peptide ladders, such as illustrated in Fig. 1b, are also routinely generated for hydrogen-exchange DXMS studies (10). Also, even samples digested with trypsin typically have many peptides that differ from each other by a deletion of terminal amino acids (semi-tryptic peptides). In addition, the existing experimental protocols already unintentionally generate many chemical modifications (sodium, potassium, Fe(III), etc.) and it has been shown that existing MS/MS datasets often contain modified versions for many peptides (17–20).

If the peptide sequences were known in advance, determining their overlap would be a straightforward application of the standard sequence alignment algorithms (21). Conversely, spectral alignment is defined as the alignment of matching peaks between spectra from overlapping peptides (22, 23). This concept is illustrated in Fig. 2a with the matching *b*-ions highlighted in blue. The surprising outcome of spectral alignment is that even though one does not know the peptide sequences in advance, it turns out that the sequence information encoded in the masses of the *b*/*y*-ions suffices to detect pairs of MS/MS spectra from overlapping peptides.

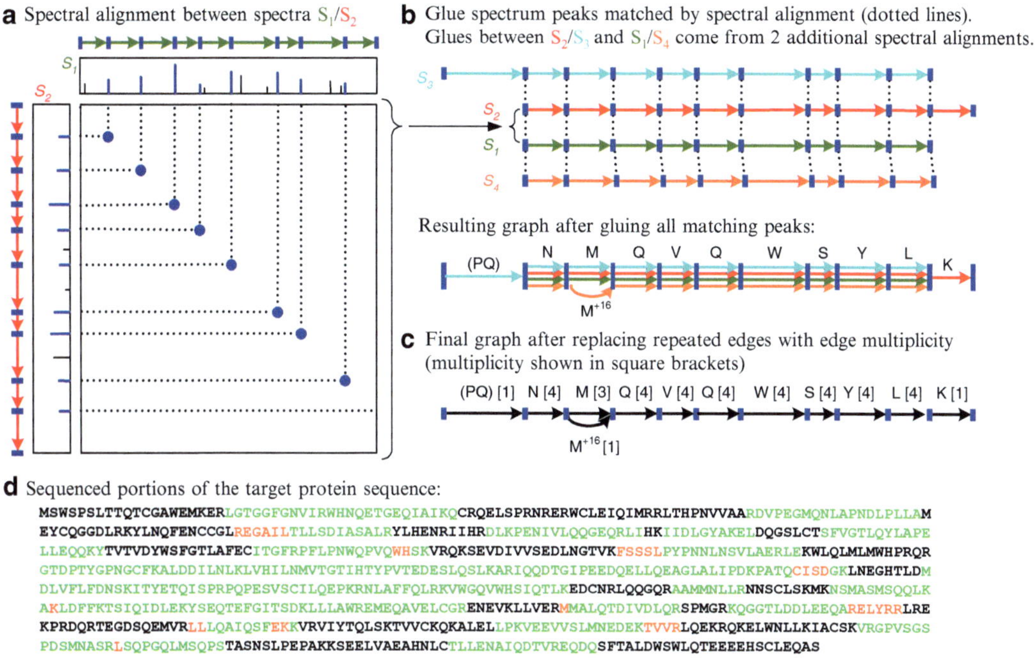

Fig. 2. Shotgun Protein Sequencing (SPS) via assembly of tandem mass spectra; (**a**) Spectral alignment between spectrum S_1 (from peptide NMQVQWSYL) and spectrum S_2 (from peptide NMQVQW-SYLK) reveals the common sequence information in both spectra. Next to each spectrum is a graph representation of the corresponding peptide sequence with consecutive *b*-ions represented as nodes connected by *arrow edges*. (**b**) Matching peaks in spectral alignments become pairwise gluing instructions between every pair of aligned spectra. Additional spectra S_3 (from PQNMQVQWSYL) and S_4 (from NM^{+16}QVQWSYL) respectively illustrate assembly of additional types of spectral alignment: partially overlapping peptides and modified/unmodified variants of the same peptide; (**c**) repeated edges are replaced by single edges with weight proportional to their multiplicity and the consensus sequence for all assembled spectra is found by the heaviest path in this graph; (**d**) Recovered portions of a target protein in the sample. Correct amino acid predictions are shown in *green* (93%) and incorrect in orange (7%).

In principle, the score of the spectral alignment between two given spectra could simply be defined as the maximum number of matched ions over all possible offsets of one spectrum in relation to the other. Although this would work to a limited extent, we have found that taking into account ion intensities and correlated occurrences of multiple ion types leads to a much more accurate separation between true spectral pairs (spectra from overlapping peptides) and false spectral pairs (spurious matches between spectra from unrelated peptides). In fact, it turns out that the reliability of spectral alignment allows one to discern the high-scoring true spectral pairs from the many millions of possible spectral pairs in high-throughput proteomics experiments (11, 23). Moreover, because each spectrum may align to several other spectra, the set of detected spectral pairs defines a spectral network where each node corresponds to a different spectrum and nodes are connected by an edge if the corresponding spectra were found to be significantly aligned. This concept is illustrated in Fig. 1d with a particular network found on a set of IKKβ spectra (11). Note that because most spectra usually come from noncontiguous protein regions, the consequent outcome of this approach is not a single spectral network but rather multiple spectral networks, one for each set of spectra from overlapping peptides.

4. Shotgun Protein Sequencing

The limited availability of sequenced genomes and multiple mechanisms of protein variation often refute the common assumption that all proteins of interest are known and present in a database. Well-known mechanisms of protein diversity include variable recombination and somatic hyper-mutation of immunoglobulin genes (24). The vital importance of some of these novel proteins is directly reflected in the success of monoclonal antibody drugs such as Rituxan™, Herceptin™, and Avastin™ (25, 26), all derived from proteins that are not directly inscribed in any genome. Similarly, multiple commercial drugs have been developed from proteins obtained from species whose genomes are not known. In particular, peptides and proteins isolated from venom have provided essential clues for drug design (27, 28) – examples include drugs for controlling blood coagulation (29–31) and drugs for breast (32, 33) and ovarian (34) cancer treatment. Even so, the genomes of the venomous snakes, scorpions, and snails are unlikely to become available anytime soon.

Despite this vital importance of novel proteins, the mainstream method for protein sequencing is still initiated by restrictive and low-throughput Edman degradation (35, 36) – a task made difficult by protein purification procedures, posttranslational modifications and blocked protein N-termini. In the mid-1980s, Klaus Biemann's group (37) had already recognized the potential of tandem mass spectrometry for protein sequencing and manually sequenced a complete protein from rabbit bone marrow. In 2006, this approach was resurrected by Genentech researchers who were able to sequence antibodies by a combination of MS/MS and Edman degradation (38). But while these approaches relied on the separate interpretation of each MS/MS spectrum, the pattern of overlapping peptides illustrated in Fig. 11.1b leads to particularly exciting possibilities for computational analysis – as in the assembly of genomic sequences from DNA reads, it now becomes feasible to assemble uninterpreted MS/MS spectra into protein sequences (15, 39).

The assembly of spectra from overlapping peptides can be likened to a simple allegory – imagine you have a jewelry box containing many copies of a particular model of bead necklaces. In this allegory, all necklaces are made from the same type of bead and thread but different necklace models are characterized by designer-specified varying thread distances between consecutive beads. Thus, any given necklace model is completely defined by a sequence of consecutive inter-bead distances. But what if, after collecting many copies of your favorite necklace model, you 1 day find that someone cut each necklace multiple times at randomly chosen bead positions? In this context, the necklace-model recovery problem is that of rediscovering the original necklace model given only the leftover pieces in the jewelry box. Although mass spectrometry adds a fair amount of complexity to this problem, this allegory captures the essence of the spectral assembly problem where amino acid masses correspond to inter-bead distances and beads represent the amide bonds between consecutive amino acids.

The shotgun protein sequencing (SPS) approach to de novo sequencing is a three-stage approach to the assembly of MS/MS spectra into amino acid sequences: (a) find pairs of spectra from overlapping peptides using spectral alignment, (b) assemble the aligned spectra, and (c) determine a consensus amino acid sequence for each set of assembled spectra. As illustrated in Fig. 2, this approach is not unlike (a) finding necklace pieces with matching interbead distances, (b) gluing the matching beads, and (c) determining the necklace model from the recovered distances between glued beads.

By capitalizing on the correlated ion occurrences in all assembled spectra, shotgun protein sequencing leads to significant improvements in de novo sequencing accuracy and, on average, only makes one mistake out of every ten amino acid predictions, even on low-accuracy ion trap MS/MS spectra. Using this approach, we were able to resequence large portions of multiple proteins in pure venom extract from western diamondback rattlesnake (39). In addition, compelling evidence was found for novel *Crotalus atrox* peptides featuring strong homology to venom peptides from other species (Table 1).

5. Spectral Networks from Spectra of Modified Peptides

In traditional DNA sequence alignment, it often happens that query sequences differ from the reference sequences by the insertion or deletion of one or more nucleotides (21). Although the insertion/deletion of amino acids is also usually allowed when aligning protein sequences, an additional factor needs to be considered when aligning peptides from experimental samples – the occurrence of posttranslational modifications. Recently, Tsur et al. (19) and Savitski et al. (41) argued that the phenomenon of modifications is much more widespread than previously thought

Table 1
Homologous contig sequences obtained with shotgun protein sequencing on venom proteins extracted from Western Diamondback Rattlesnakes (39) (*Crotalus atrox*). On the de novo sequences, *parentheses* indicate sequences where the order of the amino acids was not determined; *square brackets* indicate indistinguishable amino acid masses (on ion trap spectra). On the homologous sequences (identified using blastp (NCBI) and SPIDER (40)), the segments identical to the de novo reconstructions are shown underlined. It turned out that all homologies were either matched to a different snake species or can be explained by single nucleotide polymorphisms of the previously known *Crotalus atrox* sequences, which were also detected in the same sample

De novo sequence	Homologous sequences	Species
L(TP)GSQCAD(GV) CCDQCRF[Q,K]	LTPGSQCADGVCCDQCRFT	Agkistrodon contortrix laticinctus
	LRPGSQCAEGMCCDQCRFM	Crotalus durissus durissus
	LRPGAQCADGLCCDQCRFI	Crotalus atrox
KVLNEDEQTRD(PK)	KVLNEDEQTRDPK	Trimeresurus jerdonii
	KVPNEDEQTRNPK	Crotalus atrox
(LTNCSPK)(TD)IYSYSWKR	LTNCSPKTDIYSYSWKR	Crotalus viridis viridis
Y(MF)(YL)DFLCTDPSEKC	YMFYLDFLCTDPSEK	Crotalus viridis viridis
(IVS)WGGDI(CA)Q(PH) EPGVY(TK)	IVSWGGDICAQPHEPGHYTK	Agkistrodon acutus
	IVSWGGDPCAQPREPGVYTK	Trimeresurus stejnegeri
	IVSWGGDICAQPREPEPYTK	Crotalus durissus durissus

and advocated blind database search for the identification of these modifications. Alignment-based blind database search was first described in detail by Pevzner et al. (42) and essentially allows for efficient database search while allowing for one modification of any mass. The application of this approach recently resulted in the most comprehensive set of posttranslational modifications ever identified in aged human lenses (20).

From a sequence alignment perspective, a modification could be modeled by following the modified residue with a special character for each type of modification. Thus, the alignment of a modified peptide PEPT*IDE with its unmodified counterpart PEPTIDE would result in a single difference caused by the insertion of the modification "*." Although MS/MS spectra represent peptides as a sequence of peaks, computing the spectral alignment between spectra from modified and unmodified variants of the same peptide is substantially similar to the sequence alignment problem. This correspondence can be illustrated by representing each spectrum as sequence of 1/0 symbols respectively corresponding to "peak"/"no-peak" events at each mass value. Thus, for any integer mass m, let $s(m)$ be a sequence of $m-1$ zeros followed by a single one. For example, if an imaginary peptide of mass 12 was composed by amino acids XYZ (with masses 3, 4, 5, respectively) then its theoretical spectrum would contain peaks at masses 3, 7, 12 and the corresponding 0/1-sequence representation would be $s(3)s(4)s(5) = 001000100001$. In this framework, any sequence of masses (such as a peptide or a modified peptide) can be expressed as a sequence of 0/1 symbols and pairs of sequences can then be aligned using standard sequence alignment algorithms (21). As such, a modification of mass m' corresponds to the insertion of m' additional zeros right before the sequence for the modified residue (i.e., the mass of the residue becomes larger). Conversely, if the modification causes a loss of m'' Daltons (mass units) from the modified residue then the corresponding effect is the deletion of m'' zeros from the sequence for the modified residue. Although spectral alignment algorithms (11, 15, 42) do not explicitly convert spectra to sequences of zeros and ones, this model illustrates the essential concepts behind the approach. Figure 3a illustrates the spectral alignment between MS/MS spectra from the peptides TETMA and TET^{+80}MA.

When first analyzing a sample possibly containing modified peptides one does not know a priori that residues or peptides will be modified. Thus, spectral alignment considers every possible spectral pair and every possible location for the mass difference (e.g., modification mass) between the aligned spectra. By requiring a significant match between the aligned spectrum peaks (11) but placing no restrictions on which modifications to consider, this approach can be used to discover novel or unexpected modifications. In fact, when applied to a set of spectra from cataractous lenses proteins from a 93-year-old patient, spectral networks were able to rediscover the modifications identified by database search methods and additionally discovered several novel modification events (11, 19).

The identification of peptides containing multiple modifications via database search is a challenging problem because of the combinatorial explosion in the number of possible modification variants for all the peptides in a database (19). Not only can the large number of possible peptide variants make this approach much slower, but the increased number of peptide candidates for any given spectrum significantly increases the risk of incorrect identifications. However, samples containing peptides with two or more modifications often also contain variants of the same peptide with only one or no modification. In these cases, we have found that spectral alignment is able to group these related spectra

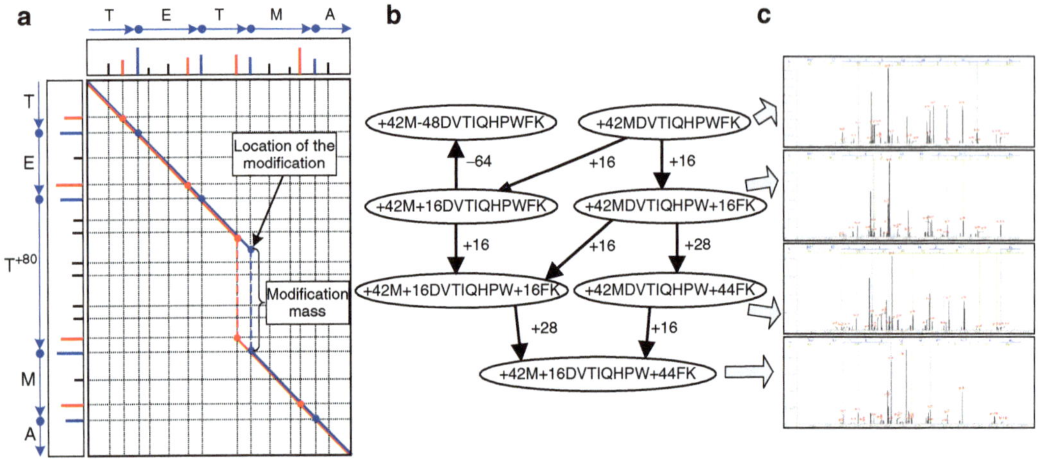

Fig. 3. Identification of posttranslational modifications through spectral networks; (**a**) spectral alignment between modified and unmodified variants of the peptide TETMA (*b*-ions shown in *blue*, *y*-ions in *red*, *blue/red lines* track consecutively matched *b*/*y*-ions); (**b**) grouped modification states of the peptide MDVTIQHPWFK from a sample of cataractous lenses; (**c**) Highly correlated MS/MS spectra from the indicated peptide variants.

from multiple modification variants of the same peptide into small spectral networks. Figure 3b illustrates the spectral network for a particular peptide in a sample of cataractous lenses proteins.

By grouping together spectra from multiple variants of the same peptide, spectral networks additionally contribute to the reliable identification of highly modified peptides. Although database searching is restricted to matching ion masses between theoretical and observed spectra, spectral networks further capitalize on the correlated cooccurrences of ions at corresponding masses and with similar peak intensities (Fig. 3c). In general terms, it becomes easier to identify a highly modified peptide if one additionally observes highly-similar spectra from the intermediate modification states. Thus, spectral alignment not only allows one to discover unexpected modifications (instead of only identifying expected modifications) but additionally defines an alternative way to reliably identify highly modified peptides.

6. Comparative Shotgun Protein Sequencing

Monoclonal antibodies have been exploited as indispensable reagents for biomedical research and as diagnostic and therapeutic agents (43, 44); Fig. 4 illustrates the recombinant nature of immunoglobulins. The specificity and effector functions of antibodies are highly dependent on the amino acid sequence and the presence (or absence) of specific modifications (24). Although DNA sequencing is routinely used in the initial characterization

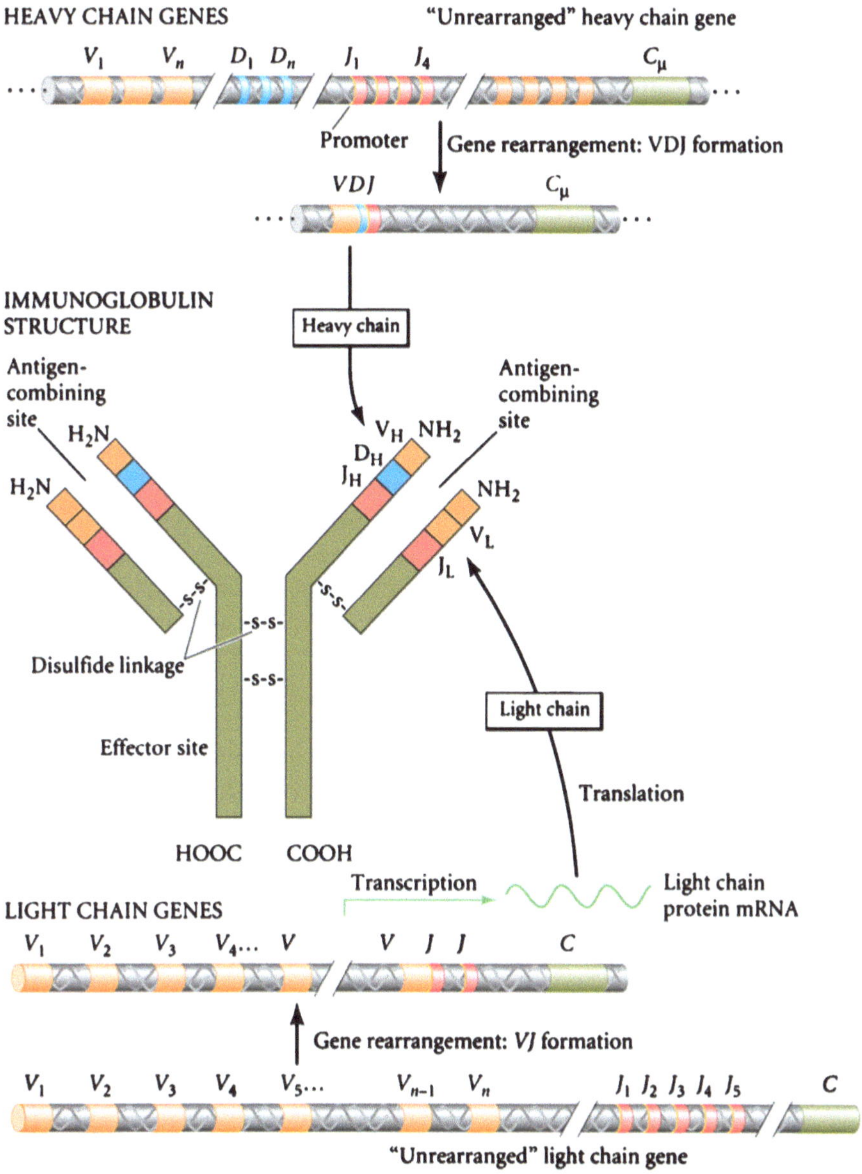

Fig. 4. *Center:* Structure of a typical immunoglobulin (antibody) protein. Two identical heavy chains and two identical light chains are connected by disulfide linkages. The antigen-binding site is composed of the variable regions of the heavy and light chains, whereas the effector site of the antibody is determined by the amino acid sequence of the heavy chain constant region. *Bottom:* Rearrangement of the light chain genes during B lymphocyte differentiation. While the developing B cell is still maturing in the bone marrow, one of the 300 or more V gene segments combines with one of the 5 J gene segments and moves closer to the constant (C) gene segment. *Top:* Rearrangement of the heavy chain genes. A heavy chain gene contains three segments (V, D, and J) that come together to form the variable region, as well as a constant region.

of monoclonal antibodies, subsequent mutations and modifications are typically recognized by analysis at the protein level. It is therefore critical to sequence the antibodies in order to monitor the integrity of the molecule, to troubleshoot performance in preclinical assays, to regenerate cDNA by reverse engineering, and to perform quality control. In addition, protein-level rearrangements (such as observed on IG4 antibodies) can only be revealed by protein level analysis.

Comparative SPS (CSPS) complements SPS by using homologous sequences from known proteins (e.g., known antibodies) as templates to assemble unknown proteins. The key computational challenge is not unlike "comparative fragment assembly" in classical DNA sequencing when a known genome (e.g., human) is used as a template for assembling another genome (e.g., macaque). CSPS first constructs a set of homologous proteins by matching SPS contigs against the protein database and further scores each protein by the overall alignment score of all contigs matched to this protein. All proteins with scores above the threshold are selected and the theoretical spectra of these proteins are constructed. For our purposes, the theoretical spectrum of a protein is the set of all possible *b*-ions representing an "idealized" top-down spectrum of the protein. The resulting "long" theoretical spectra of the selected proteins are further assembled with real spectra/contigs using Shotgun Protein Sequencing. The theoretical protein spectra serve as the "glue" connecting SPS contigs that map to at least one common mass on the same theoretical spectrum (Clustal W alignments are used to map multiple homologous proteins to the same reference protein). Sets of contigs matched to the same protein but without common masses on the protein spectrum are still ordered but not glued into the same CSPS contig. After application of SPS, a consensus sequence is again derived using only the mass differences determined from the overlapped spectra (i.e., homology glues contigs but does not directly influence the resulting protein sequence).

CSPS was first demonstrated on two monoclonal antibodies (45) that had been raised against the B- and T-cell Lymphocyte Attenuator molecule (BTLA): a first-generation antibody (aBTLA) and a mutated version of the original species (mt-aBTLA). Antibodies were raised in mice against human BTLA and were selected for their ability to attenuate T cell responses in vitro to protect against graft versus host disease. The antibodies were separately digested with Lys-C, Glu-C, Asp-N, chymotrypsin, pepsin and trypsin and the resulting peptide mixtures were analyzed with LTQ-FTMS and LTQ-Orbitrap instruments.

The CSPS assembly of aBTLA heavy chain contigs is illustrated in Fig. 5. To validate our approach, all resulting CSPS contigs were compared with the aBTLA sequence obtained by manual Edman degradation sequencing (and MS/MS database search in the constant regions).

SPS resulted in 63 contigs covering 95% of the aBTLA heavy chain (not counting contigs from proteases and contaminants); grouped by CSPS into three long contiguous regions (CSPS

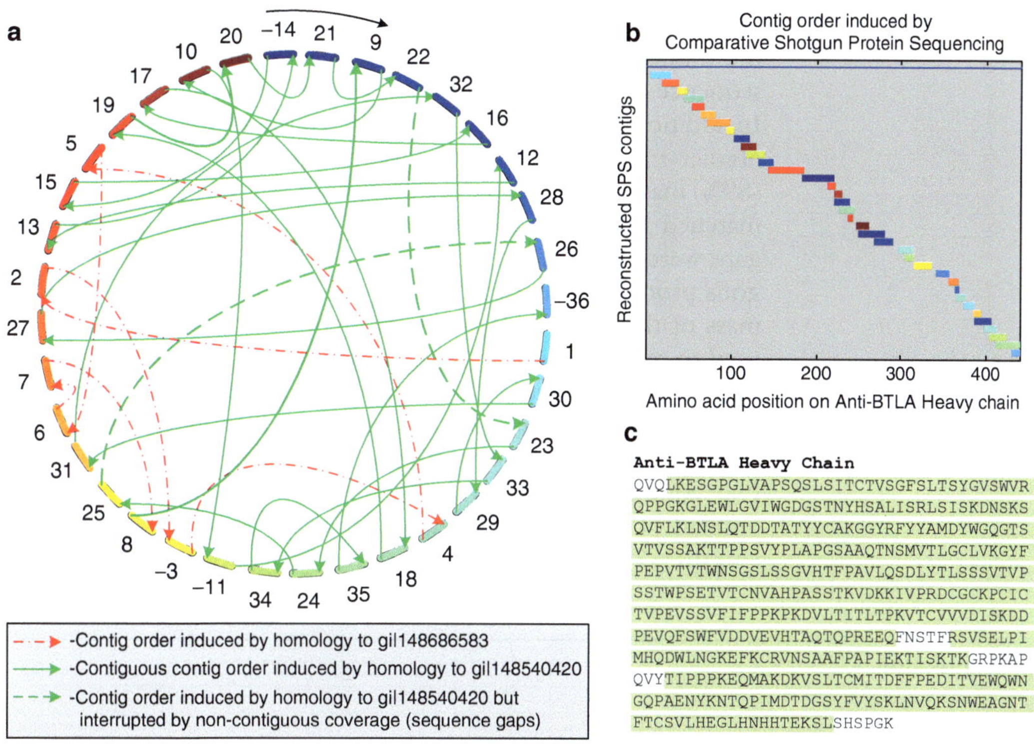

Fig. 5. Comparative protein sequencing. The heavy chain contigs matched to two different proteins (gi|148540420 and gi|148686583) homologous to different regions of the aBTLA heavy chain – 9 SPS-contigs matched gi|148540420, 47 SPS-contigs matched gi|148686583 and 8 contigs matched both (*see* Fig. 2). The protein regions matched by the latter were confirmed by a corroborating CLUSTALW alignment of gi|148540420/gi|148686583. (**a**) Homology-derived order of 36 aBTLA heavy chain protein contigs. Each protein contig is represented as a *colored dash* on the circumference; the *color gradient* and *black arrow* at the *top* indicate the arbitrary contig order (because the contig order is unknown beforehand). *Colored arrows* reveal the contiguous contig order induced by the homologous proteins and the *dashed colored arrows* indicate protein contigs in the correct order but separated by sequence gaps. The recovered contig order is indicated by *indices* next to each contig; negative indices −*j* indicate that the *j*-th contig resulted in a reversed sequence (i.e., inferred from a sequence of *y*-ions rather than *b*-ions). One can derive the contig order by starting at contig 1 (*blue*), that is connected to contig 2 (*red*), that is connected to contig 3 (*yellow*), etc. resulting in the reconstructed aBTLA heavy chain. (**b**) Linear rendering of the homology-induced contig order illustrated in (**a**). (**c**) The complete aBTLA heavy chain sequence recovered by our approach; highlighted sections were covered by protein contigs (95% coverage) and the missing amino acids were obtained from the homologous protein sequences.

contigs) of lengths 288, 40, and 92 aa using two homologous proteins. Comparison of the CSPS contigs with the Edman degradation data revealed that the three sequence gaps not covered by CSPS contigs had no coverage by MS/MS spectra. Thus, these gaps were caused by particularities of the sequence that hinder MS/MS analysis rather than by shortcomings of CSPS algorithm. For example, the [(N)STFRSV(S)] gap contains the NXT motif indicative of glycosylation. Indeed Asn297 is typically glycosylated and this impedes the identification of these fragments. In addition to this area the first three N-terminal amino are missing because the N-terminal peptides were either too short (<6 aa) or too long (>18 aa) for MS/MS identifications. CSPS results on the mutant BTLA antibody (mt-aBTLA) were similar with 97% sequence coverage with three contigs of lengths 292, 40, and 97. In addition, this sequence clearly illustrates the ability of CSPS to predict multiple mutations and modifications – 25 out of 28 (89%) mass offsets from the closest homologous protein correctly matched the target sequence. It turned out that the sequence gaps were identical to the corresponding regions in the homologous proteins. When combined with the resulting match to the mass of the intact protein, these identical homologies could be used to connect the long contigs into a contiguous sequence (this step should be taken with caution because multiple mutations may result in compensatory offsets of total mass zero). Even without this final step, sequencing the aBTLA light chain resulted in two contigs (34 and 179 aa) covering 97% of the sequence. Similarly, the mt-aBTLA light chain resulted in a single contig of length 217 covering 99% of the target sequence.

7. Discussion

Spectra from overlapping peptides or modification-variants of the same peptide deliver a wealth of correlated sequence information that can be explored with a new generation of algorithms based on spectral networks. In a departure from standard procedures, having spectra from modified/unmodified variants of the same peptide allows one to directly discover the modifications in the sample rather than having to guess in advance the list of modifications to search for. Spectra from multiple modification-variants can be combined into spectral networks and correlated ion masses and intensities used to increase the confidence in the identification of highly modified peptides.

Tandem mass spectra are inherently noisy and mass-spectrometrists have long been trying to reduce the noise and achieve reliable de novo interpretations by advancing both instrumentation and experimental protocols. In particular, Zubarev and colleagues (46, 47) have demonstrated the power of combining CID and ECD spectra. However, this technique as well as the approach described in Frank et al. (48) require either special instrumentation or highly accurate Fourier transform mass-spectrometry. While one can also reduce the complexity of spectrum identification by using stable isotope labeling (49), the impact of this approach (for peptide identification) has been restricted, in part, by the cost of the isotope and the high mass resolution required. Alternative end-labeling chemical modification approaches have disadvantages such as low yield, complicated reaction conditions, and unpredictable changes in ionization and fragmentation. As a result, the impact of these important techniques is mainly in protein quantification rather than identification (49). The key difference between spectral networks analysis and labeling techniques is that, instead of trying to introduce a specific modification in a controlled fashion, spectral networks take advantage of the multiple modifications naturally present in the sample. This approach allows one to decode modifications (without knowing in advance what they are) and thus provides a computational (rather than instrumentation-based) solution to the problem of MS/MS spectra identification.

From a protein sequencing perspective, the extensive sequence coverage achievable with nonspecific proteolytic digestion enables the assembly of spectra from overlapping peptides into long *protein contigs*. Moreover, by capitalizing on the correlated sequence information in sets of assembled spectra, the shotgun protein sequencing approach is able to significantly increase the de novo sequencing accuracy even on low mass-accuracy ion trap MS/MS spectra. In general, using mass spectrometry for shotgun protein sequencing results in certain limitations that are without counterpart in the DNA sequencing realm. In particular, the sampling frequency of the amino acids across a protein sequence is not uniform and is dictated by local sequence context and thus the coverage of a protein by its peptides is biased by the specificity and distribution of cleavage sites of the proteases employed. Also, certain combinations of amino acids have identical elemental compositions that are indistinguishable by mass and may leave ambiguity in the draft (or even finished) sequences depending on the completeness of fragmentation in the MS/MS spectra ($I = L = 113$, $GG = N = 114$, $GA = Q = 128$). Others have the same nominal mass, but not elemental composition, and are distinguishable only in MS/MS from high resolution instruments ($Q = K = 128$ and $W = DA = VS = 186$). High-resolution mass spectrometers, such as Thermo's LTQ-Orbitrap, may seamlessly elevate Shotgun Protein Sequencing to a whole new level of productivity. In principle, higher mass accuracy should be directly translatable into higher sequencing accuracy and much more sensitive detection of overlaps between spectra with poor *b*/*y*-ion ladders.

Nonetheless, even with a standard experimental setup and using only a relatively small MS/MS dataset from a modest resolution mass spectrometer, shotgun protein sequencing very rapidly generated much more information about western diamondback rattlesnake venom proteins than some of the most laborious Edman degradation/cloning studies (50). Moreover, these contigs can be easily produced with minimal experimental and computational effort while Edman degradation projects often take months to complete. Furthermore, the recovered protein contigs may be readily aligned and ordered by comparative protein sequencing that, akin to comparative DNA sequencing, utilizes previously determined protein sequences from evolutionarily close species. CSPS opens up many possibilities for sequence discovery in the biotechnology industry compared to traditional methods. Replacing Edman degradation with CSPS significantly increases the resulting coverage from the same amounts of material (95–99% sequence coverage vs. ≅10% for Edman sequencing), greatly speeds up the analytical protocol and allows one to automatically discover posttranslational modifications. Thus CSPS opens a possibility to correlate unexpected modifications with changes in antibody efficiency while simultaneously tracking mutations. Also, CSPS is already faster than the cDNA sequencing route commonly used in many laboratories.

References

1. Aebersold, R. and Mann, M. (2003) Mass spectrometry-based proteomics. *Nature*, **422**, 198–207.
2. Yates, J. R. (2004) Mass spectrometry as an emerging tool for systems biology. *Biotechniques*, **36**, 917–919.
3. Biemann, K., Cone, C., Webster, B., and Arsenault, G. (1966) Determination of the amino acid sequence in oligopeptides by computer interpretation of their high-resolution mass spectra. *J Am Chem Soc*, **88**, 5598–5606.
4. Henzel, W. J., Billeci, T. M., Stults, J. T., Wong, S. C., Grimley, C., and Watanabe, C. (1993) Identifying proteins from two-dimensional gels by molecular mass searching of peptide fragments in protein sequence databases. *Proc Natl Acad Sci USA*, **90**, 5011–5015.
5. Yates, J., Eng, J., and McCormack, A. (1995) Mining genomes: correlating tandem mass spectra of modified and unmodified peptides to sequences in nucleotide databases. *Anal Chem*, **67**, 3202–3210.
6. Keller, A., Nesvizhskii, A., Kolker, E., and Aebersold, R. (2002) Empirical statistical model to estimate the accuracy of peptide identifications made by MS/MS and database search. *Anal Chem*, **74**, 5383–5392.
7. Nesvizhskii, A. I. and Aebersold, R. (2005) Interpretation of shotgun proteomic data: the protein inference problem. *Mol Cell Proteomics*, **4**, 1419–1440.
8. Fischer, B., Roth, V., Roos, F., Grossmann, J., Baginsky, S., Widmayer, P., Gruissem, W., and Buhmann, J. M. (2005) Novohmm: a hidden Markov model for de novo peptide sequencing. *Anal Chem*, **77**, 7265–7273.
9. MacCoss, M., et al. (2002) Shotgun identification of protein modifications from protein complexes and lens tissue. *Proc Natl Acad Sci USA*, **99**, 7900–7905.
10. Englander, J., Del Mar, C., Li, W., Englander, S., Kim, J., Stranz, D., Hamuro, Y., and Woods, V. (2003) Protein structure change studied by hydrogen-deuterium exchange, functional labeling, and mass spectrometry. *Proc Natl Acad Sci USA*, **100**, 7057–7062.
11. Bandeira, N., Tsur, D., Frank, A., and Pevzner, P. (2007) Protein identification via spectral networks analysis. *Proc Natl Acad Sci USA*, **104**, 6140–6145.
12. Siuzdak, G. (2003) *Mass Spectrometry in Biotechnology*. MCC Press, San Diego.
13. Tabb, D., MacCoss, M., Wu, C., Anderson, S., and Yates, JR. (2003) Similarity among tandem mass spectra from proteomic experiments: detection, significance, and utility. *Anal Chem*, **75**, 2470–2477.
14. Beer, I., Barnea, E., Ziv, T., and Admon, A. (2004) Improving large-scale proteomics by clustering of mass spectrometry data. *Proteomics*, **4**, 950–960.
15. Bandeira, N., Tang, H., Bafna, V., and Pevzner, P. (2004) Shotgun protein sequencing by tandem mass spectra assembly. *Anal Chem*, **76**, 7221–7233.
16. Klammer, A. A. and MacCoss, M. J. (2006) Effects of modified digestion schemes on the identification of proteins from complex mixtures. *J Proteome Res*, **5**, 695–700.
17. Hunyadi-Gulyas, E. and Medzihradszky, K. (2004) Factors that contribute to the complexity of protein digests. *DDT Targets*, **3**, 3–10.
18. Tanner, S., Shu, H., Frank, A., Wang, L., Zandi, E., Mumby, M., Pevzner, P., and Bafna, V. (2005) InsPecT: identification of posttranslationally modified peptides from tandem mass spectra. *Anal Chem*, **77**, 4626–4639.

19. Tsur, D., Tanner, S., Zandi, E., Bafna, V., and Pevzner, P. A. (2005) Identification of post-translational modifications by blind search of mass spectra. *Nat Biotechnol*, **23**, 1562–1567.
20. Wilmarth, P. A., Tanner, S., Dasari, S., Nagalla, S. R., Riviere, M. A., Bafna, V., Pevzner, P. A., and David, L. L. (2006) Age-related changes in human crystallins determined from comparative analysis of post-translational modifications in young and aged lens: does deamidation contribute to crystallin insolubility? *J Proteome Res*, **5**, 2554–2566.
21. Smith, T. F. and Waterman, M. S. (1981) Identification of common molecular subsequences. *J Mol Biol*, **147**(1), 195–197.
22. Pevzner, P., Dancík, V., and Tang, C. (2000) Mutation-tolerant protein identification by mass spectrometry. *J Comput Biol*, **7**, 777–787
23. Bandeira, N., Tsur, D., Frank, A., and Pevzner, P. (2006) A New Approach to Protein Identification. Apostolico, A., Guerra, C., Istrail, S., Pevzner, P. A., and Waterman, M. (eds.), *Proceeding of the Tenth Annual 21 International Conference in Research in Computational Molecular Biology (RECOMB 2006)*, vol. 3909 of Lecture Notes in Computer Science, pp. 363–378, Springer, Germany.
24. Gearhart, P. J. (2002) Immunology: the roots of antibody diversity. *Nature*, **419**, 29–31.
25. Wiles, M. and Andreassen, P. (2006) Monoclonals – the billion dollar molecules of the future. *Drug Discov World*, Fall 2006, 17–23.
26. Haurum, J. S. (2006) Recombinant polyclonal antibodies: the next generation of antibody therapeutics? *Drug Discov Today*, **11**, 655–660.
27. Lewis, R. J. and Garcia, M. L. (2003) Therapeutic potential of venom peptides. *Nat Rev Drug Discov*, **2**, 790–802.
28. Pimenta, A. M. and De Lima, M. E. (2005) Small peptides, big world: biotechnological potential in neglected bioactive peptides from arthropod venoms. *J Pept Sci*, **11**, 670–676.
29. Joseph, J. and Kini, R. (2004) Snake venom prothrombin activators similar to blood coagulation factor Xa. *Curr Drug Targets Cardiovasc Haematol Disord*, **4**, 397–416.
30. Swenson, S., Toombs, C., Pena, L., Johansson, J., and Markland, F.(2004) Alpha-fibrinogenases. *Curr Drug Targets Cardiovasc Haematol Disord*, **4**, 417–435.
31. Kini, R., Rao, V., and Joseph, J. (2001) Procoagulant proteins from snake venoms. *Haemostasis*, **31**, 218–224.
32. Swenson, S., Costa, F., Minea, R., Sherwin, R., Ernst, W., Fujii, G., Yang, D., and Markland, F. (2004) Intravenous liposomal delivery of the snake venom disintegrin contortrostatin limits breast cancer progression. *Mol Cancer Ther*, **3**, 499–511.
33. Pal, S. K., Gomes, A., Dasgupta, S. C., and Gomes, A. (2002) Snake venom as therapeutic agents: from toxin to drug development. *Indian J Exp Biol*, **40**, 1353–1358.
34. Markland, F., Shieh, K., Zhou, Q., Golubkov, V., Sherwin, R., Richters, V., and Sposto, R. (2001) A novel snake venom disintegrin that inhibits human ovarian cancer dissemination and angiogenesis in an orthotopic nude mouse model. *Haemostasis*, **31**, 183–191.
35. Zugasti-Cruz, A., Maillo, M., López-Vera, E., Falcón, A., Heimer de la Cotera, E. P., Olivera, B. M., and Aguilar, M. B. (2006) Amino acid sequence and biological activity of a gamma-conotoxin-like peptide from the worm-hunting snail Conus austini. *Peptides*, **27**, 506–511.
36. Ogawa, Y., Yanoshita, R., Kuch, U., Samejima, Y., and Mebs, D. (2004) Complete amino acid sequence and phylogenetic analysis of a long-chain neurotoxin from the venom of the African banded water cobra, *Boulengerina annulata*. *Toxicon*, **43**, 855–858.
37. Johnson, R. and Biemann, K. (1987) The primary structure of thioredoxin from *Chromatium vinosum* determined by high-performance tandem mass spectrometry. *Biochemistry*, **26**, 1209–1214.
38. Pham, V., Henzel, W. J., Arnott, D., Hymowitz, S., Sandoval, W. N., Truong, B. T., Lowman, H., and Lill, J. R. (2006) De novo proteomic sequencing of a monoclonal antibody raised against ox40 ligand. *Anal Biochem*, **352**, 77–86.
39. Bandeira, N., Clauser, K., and Pevzner, P. (2007) Shotgun protein sequencing: assembly of tandem mass spectra from mixtures of modified proteins. *Mol Cell Proteomics*, **6**, 1123–1134.
40. Han, Y., Ma, B., and Zhang, K. (2005) Spider: software for protein identification from sequence tags with de novo sequencing error. *J Bioinform Comput Biol*, **3**, 697–716.
41. Savitski, M. M., Nielsen, M. L., and Zubarev, R. A. (2006) Modificomb, a new proteomic tool for mapping substoichiometric post-translational modifications, finding novel types of modifications, and fingerprinting complex protein mixtures. *Mol Cell Proteomics*, **5**, 935–948.
42. Pevzner, P., Mulyukov, Z., Dancik, V., and Tang, C. (2001) Efficiency of database search

for identification of mutated and modified proteins via mass spectrometry. *Genome Res*, **11**, 290–299.

43. Ferrara, N., Hillan, K. J., Gerber, H. P., and Novotny, W. (2004) Discovery and development of bevacizumab, an anti-vegf antibody for treating cancer. *Nat Rev Drug Discov*, **3**, 391–400.
44. Reichert, J. M. and Valge-Archer, V. E. (2007) Development trends for monoclonal antibody cancer therapeutics. *Nat Rev Drug Discov*, **6**, 349–356.
45. Bandeira, N., Pham, V., Pevzner, P., Arnott, D., and Lill, J.R. (2008) Automated de novo protein sequencing of monoclonal antibodies. *Nat Biotechnol*, **26**, 1336–1338.
46. Savitski, M. M., Nielsen, M. L., and Zubarev, R. A. (2005) New data base-independent, sequence tag-based scoring of peptide ms/ms data validates mowse scores, recovers below threshold data, singles out modified peptides, and assesses the quality of ms/ms techniques. *Mol Cell Proteomics*, **4**, 1180–1188.
47. Savitski, M. M., Nielsen, M. L., Kjeldsen, F., and Zubarev, R. A. (2005) Proteomics-grade de novo sequencing approach. *J Proteome Res*, **4**, 2348–2354.
48. Frank, A. M., Savitski, M. M., Nielsen, M. L., Zubarev, R. A., and Pevzner, P. A. (2007) De novo peptide sequencing and identification with precision mass spectrometry. *J Proteome Res*, **6**, 114–123.
49. Shevchenko, A., Chernushevich, I., Ens, W., Standing, K. G., Thomson, B., Wilm, M., and Mann, M. (1997) Rapid "de novo" peptide sequencing by a combination of nanoelectrospray, isotopic labeling and a quadrupole/time-of-flight mass spectrometer. *Rapid Commun Mass Spectrom*, **11**, 1015–1024.
50. Zhou, Q., Smith, J. B., and Grossman, M. H. (1995) Molecular cloning and expression of catrocollastatin, a snake-venom protein from *Crotalus atrox* (western diamondback rattlesnake) which inhibits platelet adhesion to collagen. *Biochem J*, **307(Pt 2)**, 411–417.

Chapter 12

Software Pipeline and Data Analysis for MS/MS Proteomics: The Trans-Proteomic Pipeline

Andrew Keller and David Shteynberg

Abstract

The LC-MS/MS shotgun proteomics workflow is widely used to identify and quantify sample peptides and proteins. The technique, however, presents a number of challenges for large-scale use, including the diverse raw data file formats output by mass spectrometers, the large false positive rate among peptide assignments to MS/MS spectra, and the loss of connectivity between identified peptides and the sample proteins that gave rise to them. Here we describe the Trans-Proteomic Pipeline, a freely available open source software suite that provides uniform analysis of LC-MS/MS data from raw data to quantified sample proteins. In a straightforward manner, users can extract MS/MS information from raw data of many instrument formats, submit them to search engines for peptide identification, validate the results to remove false hits, combine together results of multiple search engines, infer sample proteins that gave rise to the identified peptides, and perform quantitation at the peptide and protein levels.

Key words: Shotgun proteomics, Freeware, Machine learning, Protein inference

1. Introduction

The LC-MS/MS shotgun proteomics workflow is widely used to identify and quantify sample peptides and proteins (1). The technique, however, presents a number of challenges for large-scale use, including the diverse raw data file formats output by mass spectrometers, the large false positive rate among peptide assignments to MS/MS spectra, and the loss of connectivity between identified peptides and the sample proteins that gave rise to them. Many tools addressing these challenges are available from different groups, but must be assembled together in ways that are usually not convenient. There are only a few suites of tools that aim to provide a single environment for performing all or most steps in

Cathy H. Wu and Chuming Chen (eds.), *Bioinformatics for Comparative Proteomics*, Methods in Molecular Biology, vol. 694, DOI 10.1007/978-1-60761-977-2_12, © Springer Science+Business Media, LLC 2011

the shotgun workflow. These include the OpenMS Proteomics Pipeline (2), MaxQuant (3), and the oldest and most comprehensive, Trans-Proteomic Pipeline (4).

The Trans-Proteomic Pipeline (TPP) is a fully open-source suite of software tools that facilitates and standardizes the analysis of LC-MS/MS data. It includes components for MS data representation, MS data visualization, peptide identification and validation, protein inference, and quantitation. The pipeline is integrated together behind a web-based GUI interface and is compatible with Windows, Linux, and MacOS platforms. In a straightforward uniform manner, users can extract MS/MS information from raw data of many instrument formats, submit them to search engines for peptide identification, validate the results to remove false hits, infer sample proteins that gave rise to the identified peptides, and perform quantitation at the peptide and protein levels. Importantly, statistical methods are used to provide predicted errors associated with peptide and protein identifications, and quantitation abundance ratios. Visualization tools enable users to explore both MS^1 and MS^2 data to evaluate analysis results. Both CAD (collision-activated dissociation) and ETD (electron transfer dissociation) types of MS/MS data are supported by the TPP.

2. Materials

The TPP is freely available for multiple platforms. Downloading and installation instructions for Windows are found at (5), and for Linux and MacOS, at (6). A demo describing how to use the software is available at (7). Support for the software is provided via the spctools-discuss email discussion list at Google Groups. Users can browse the list archive, or join the list (currently with over 960 members) to pose questions to members of the TPP community. In addition, twice each year the Institute for Systems Biology offers a 5-day course on proteomics data analysis and use of the TPP (8). All programs of the TPP are open source and available at (9) for use consistent with their licenses. Users familiar with C++ can modify the source code to customize the analysis to their specific needs.

3. Methods

The TPP web powered user interface, Petunia, facilitates use of the software from processing of raw LC-MS/MS files all the way to protein level quantitation. In addition, it accesses several utilities for creating search databases and viewing data. All tools in the

TPP can also be run on a command line terminal. Learning the command line interface for the tools will empower the user to activate newly introduced features and fixes before they become integrated into Petunia (see Note 1).

At the home page of Petunia, users must first select the search engine for the current analysis. The Analysis Pipeline then displays a series of steps that can be run in order, as shown in Fig. 1. Users commence analysis by converting the raw LC-MS/MS data files to either mzXML (10) or mzML (11) using the converter utilities in Petunia. The TPP transparently supports both formats.

3.1. Peptide Identifications: Searching Sequence and Spectral Libraries

The first major algorithmic step in moving from raw MS/MS data to identified peptides and proteins is the database search, whereby peptides are assigned to spectra. Some search engines use a sequence database of known protein or nucleotide sequences

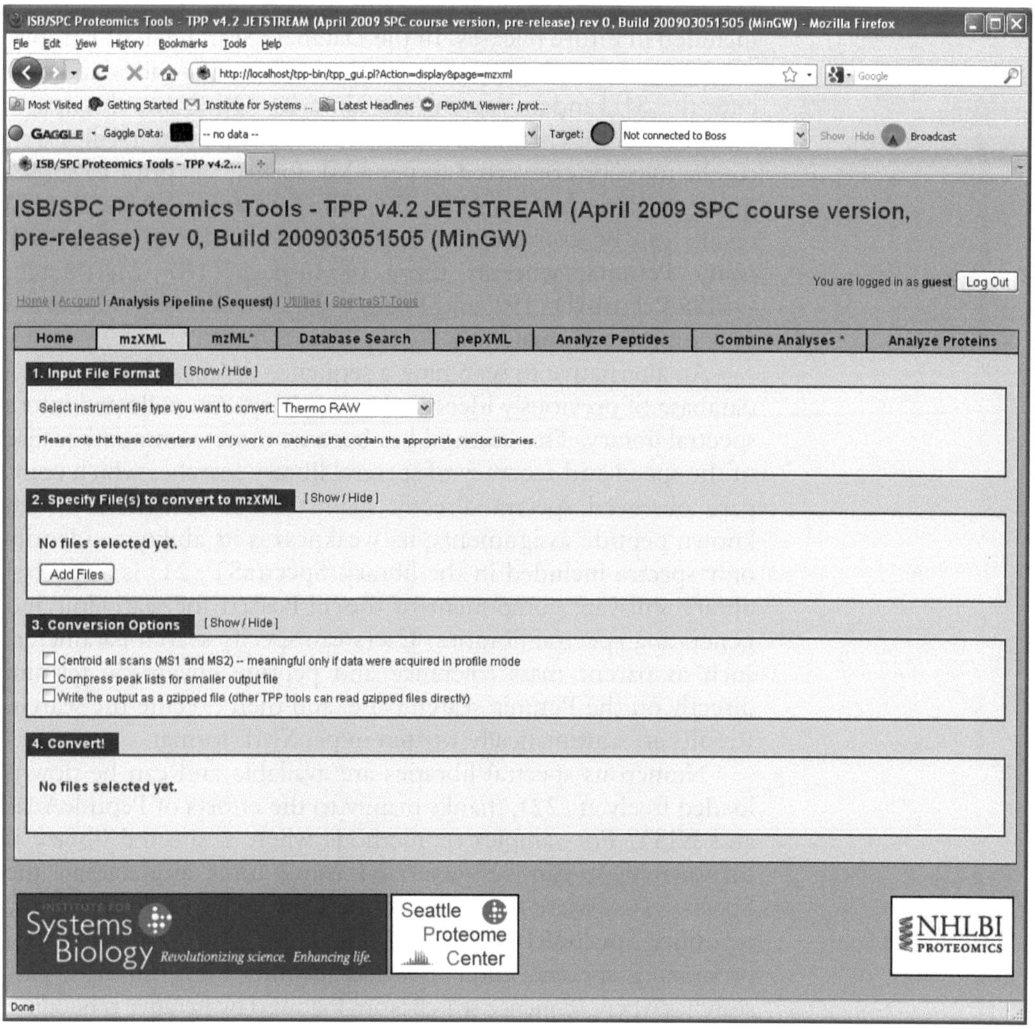

Fig. 1. Petunia web-based GUI Interface for use of the TPP tools.

to extract peptides and generate theoretical spectra for scoring against observed spectra. Users must specify search conditions including parent mass tolerance, peptide enzymatic termini, amino acid modifications, and database. Careful attention must be devoted to the selection of the database. It should include the sequence of the proteins that are expected to be in the sample, and may also include a set of decoy sequences known to be absent in the sample, distinguished by a name with fixed prefix. Their identifications are expected to be a representation of true negatives (12). Decoy sequences provide both a set of known incorrect results that can aid in distinguishing correct from incorrect results, and a measure of error (false discovery rate based on decoys). These aspects are discussed further in Subheading 3.2 below.

X! Tandem is an open-source sequence search engine distributed in the TPP with the *K*-score module optimized for use with other TPP components (13). Additional open-source search engines such as OMSSA (14) and MyriMatch (15) will also be included in future releases. In the Database Search tab of Petunia, users can set search parameters in the built-in text editor and execute the X! Tandem search (see Note 2). Other search engines must be run independently and their results imported. All search results must be converted to pepXML format (4) prior to subsequent analysis in the TPP. SEQUEST (16) and Mascot (17) results can be converted to pepXML and analyzed in the TPP using Petunia, whereas those of Inspect (18), MyriMatch, OMSSA, ProbID (19), and Phenyx (20) are currently supported only on the command line.

An alternative to searching a sequence database is to search a database of previously identified MS/MS spectra collected into a spectral library. This method has become more common because of the speed and accuracy of spectral library searches which compare observed spectra directly against those in a library with known peptide assignments; its weakness is its ability to identify only spectra included in the library. SpectraST (21) is a spectral library software component of the TPP used for searching and generating spectral libraries. Users can specify search parameters such as parent mass tolerance and peptide enzymatic termini directly on the Petunia search page, and then execute the search. Results are automatically written in pepXML format.

Numerous spectral libraries are available and can be downloaded freely at (22), thanks mainly to the efforts of PeptideAtlas and NIST. For samples or methods where a spectral library is unavailable, it can be generated using tools available in the SpectraST software (see Note 3). Perhaps the most powerful way to utilize SpectraST in the TPP (assuming one starts without a preexisting spectral library) would include doing multiple pass searches where sequence searches are followed by iterative rebuilding of the spectral library. Each subsequent search would start with a spectral library search to quickly identify previously seen spectra (effectively constituting a data-reduction step), followed by searching of the high-quality unidentified data against a sequence database with one or more search engines and conditions. Each new round of sequence searching would be followed by a comprehensive expansion of the spectral library with newly identified spectra, as shown in Fig. 2. In theory, this process could continue until all peptide ion fragment spectra are identified.

3.2. Validation of Search Results: PeptideProphet

PeptideProphet (23) is a critical component of the TPP, used to automatically identify the often small fraction of peptide assignments to MS/MS spectra in a data set that are correct. It employs the Expectation-Maximization (24) algorithm to compute probabilities that each top-ranking result is correct based on search engine scores and peptide properties, such as the number of termini consistent with enzymatic cleavage. It does so by partitioning the observed distributions of those features into inferred correct and incorrect distributions, which then contribute to the computed probability that any result is correct. Because it learns from each data set how to distinguish correct from incorrect results, it is able to derive in a robust manner accurate probabilities truly reflective of the confidence that each result is correct. This means that 90 and 50% of all results in a data set assigned probabilities of 0.9 and 0.5, respectively, are expected to be correct.

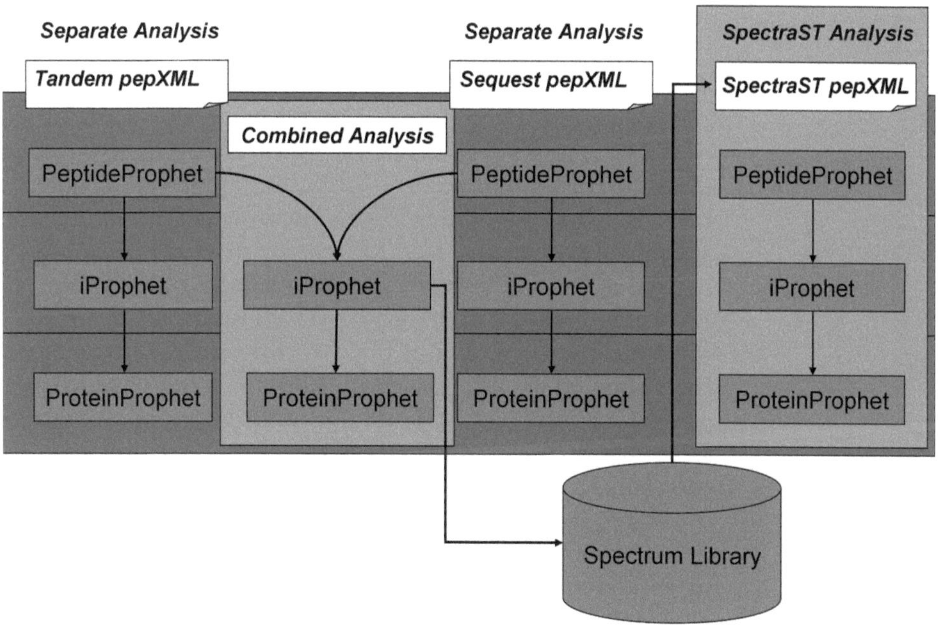

Fig. 2. Workflow utilizing sequence and spectral library searches. After each new sequence database search, results are combined together, validated, and used to expand the spectral library which in turn is searched to identify duplicate spectra. Protein inference and validation is performed using the ProteinProphet tool, and can be done either on the single standard search engine analysis, the combined analysis of several search engines, or on the results of a spectrum library search done with SpectraST.

In general, analyzing together as many results as possible is advised because this helps the program better learn distributions of search scores and peptide properties among the correct and incorrect results. However, it is important to include only search results with similar expected distributions: Results generated from similar samples and mass spectrometers, and searched with similar search conditions. In addition, the software requires the presence of some incorrect results in the data set in order to learn how to distinguish them from correct results. It is therefore advisable to include in the analysis search results for all MS/MS spectra submitted to the search engine, without any filtering of low scoring data. Finally, searches employing databases with decoy sequences are generally recommended because they provide PeptideProphet with a set of known incorrect results thereby facilitating its analysis.

PeptideProphet models the results of each parent charge independently. Some parent charges may have too few results to enable an analysis. In this case, PeptideProphet roughly estimates whether a result is likely correct or incorrect based on learned distributions of results of an adjacent parent charge. In addition, when a single MS/MS spectrum is searched both as a doubly and triply charged parent ion, the probabilities of both results are adjusted to ensure that the sum of their probabilities does not exceed unity.

3.2.1. Running PeptideProphet

PeptideProphet can be run from the Analyze Peptides tab of Petunia (see Note 4). Most user options specify peptide properties to be used, in addition to search scores, to compute the probabilities that results are correct. By default, NTT (number of tolerable termini consistent with enzymatic cleavage) and NMC (number of missed enzymatic cleavages) are used. Low resolution mass difference information (between parent and assigned peptide) is used unless the "Use accurate mass binning" option, appropriate when the MS/MS parent mass is determined on a high resolution instrument, is selected. Select the "Use Hydrophobicity/RT information" option to compare the observed parent retention times with those predicted based on their assigned peptide sequences. Additional options such as "Use pI information" and "Use N-glyc motif information" should be selected when relevant, such as Free Flow Electrophoresis (FFE) (25) and N-glycocapture (26) sample data, respectively.

When search results were generated using a database with decoy protein sequences (distinguishable from non-decoy entries by an identifying name prefix), the user can specify that the program take advantage of decoy hits as known incorrect results by selecting the "Use decoy hits to pin down..." option and entering the decoy protein name prefix. With this selection, the decoy hits will contribute only to the distributions of search score and peptide properties among incorrect results. It requires that sufficient numbers of decoy hits of each parent charge are present, and is most useful for improving the model when there are few correct results in the sample and the standard unsupervised method is unable to identify those from the bulk of incorrect results. Additionally, the "Use Non-parametric..." option can be selected to employ a smoothed bar graph modeling of discriminant score distributions. This has the advantage of not requiring the learned distributions of search score among correct and incorrect results to adhere to pre-designated parameterized distributions. For search engines where the parametric model is not very robust or does not exist (e.g., SpectraST, Inspect, MyriMatch, OMSSA, Phenyx) PeptideProphet must be run in this mode.

3.2.2. Viewing and Assessing Results

After running PeptideProphet, a probability is assigned to each search result in the data set indicating its likelihood of being correct. These can be viewed in the pepXML viewer in the probability column. Note that by default, all results with a probability less than 0.05 are discarded after analysis to make the pepXML file sizes more manageable. If you would like to retain all results, you can set the minimum probability value to 0 before running PeptideProphet.

It is important to view the distributions of search score and peptide properties learned by the program among correct and incorrect results of each parent charge. This model summary information, visible on clicking on any probability link, can be used to be sure the program did an adequate job and as a diagnostic. For example for results generated from a tryptic sample searched with a semi-tryptic or unconstrained database, an easy validation test of the analysis is the expected large fraction of inferred correct results that are fully tryptic, and the expected majority of inferred incorrect results that are *not* fully tryptic. In a similar manner, for searches with a parent mass tolerance wider than the measurement precision of the mass spectrometer, the inferred correct results are expected to have a smaller range of mass differences than the inferred incorrect results. When PeptideProphet was run with the accurate mass binning option for high resolution data, users can view a graphic display of the learned mass difference model distributions, as shown in Fig. 3.

The distributions learned by PeptideProphet among *correct* results can also be used diagnostically to gain insight into properties of the sample and MS/MS spectra. For example, the learned NTT and NMC distributions can be used to assess trypsinization quality, the learned mass difference information, to assess the accuracy and precision of the mass spectrometer, and the learned retention time information, to assess the reverse phase chromatography. When relevant, the learned pI information can be used to assess the FFE characteristics, and the learned N-glyc motif information, the efficiency of N-glycocapture.

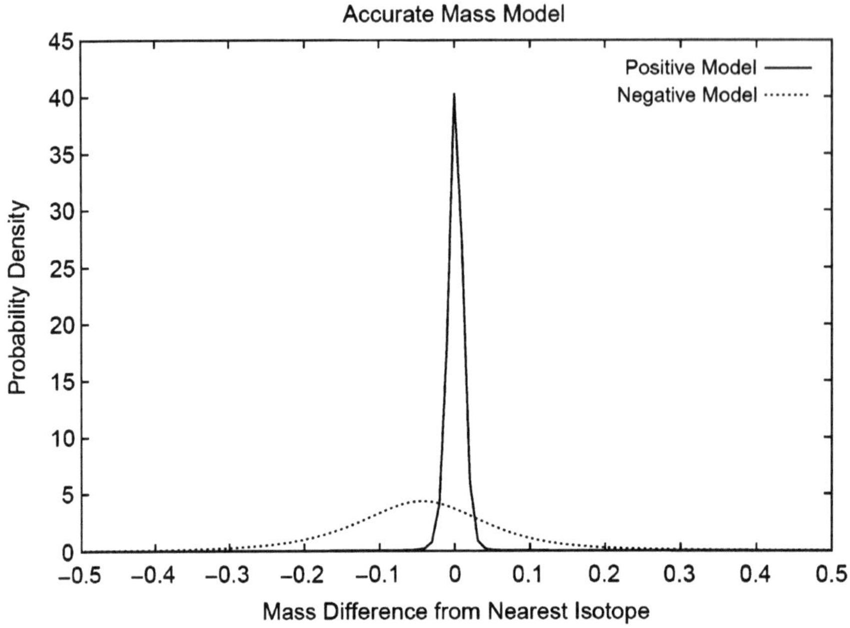

Fig. 3. Mass difference distributions among correct (positive) and incorrect (negative) X! Tandem LTQ-FT search results learned by PeptideProphet.

When the database search was made with a forward/decoy database and the decoy sequences were *not* revealed to PeptideProphet (to be used as known incorrect results), the decoy hits can be used to independently assess the accuracy of computed probabilities. To generate a plot of decoy versus PeptideProphet based false discovery rates, go to the Utilities "Decoy Peptide Validation" tab, select the pepXML file to analyze, and enter the decoy protein name prefix. A series of minimum PeptideProphet probability thresholds is then applied to the data and the resulting PeptideProphet (based on the computed probabilities) and decoy (based on the fraction of decoy results) error rates are plotted, as shown in Fig. 4. The accuracy of computed probabilities can be assessed by the correspondence of PeptideProphet and decoy false discovery rates, particularly in the important range close to 0.

The predicted error rate is an objective measure that can be used to compare two analyses, for example search results using different search conditions or different search engines. A table of minimum probability thresholds and corresponding predicted error rates is displayed in the model summary of each analysis. To compare two analyses, select and apply the minimum probability thresholds in each analysis that correspond to a fixed desired predicted error rate. This is a useful means to determine the search conditions and search engine that confer the greatest number of obtainable correct results.

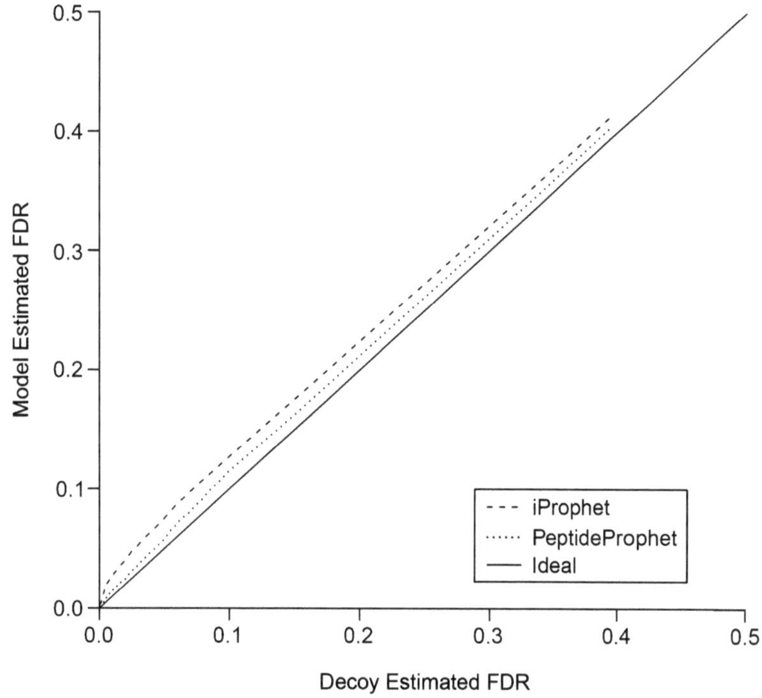

Fig. 4. Comparison of model-estimated vs. decoy-estimated false discovery rates for PeptideProphet and iProphet probabilities computed for a Yeast SILAC Orbitrap data set searched with X! Tandem.

If learned distributions are not consistent with expectations, the data should be re-analyzed using different user options. For example, one can change the optional peptide properties (de-selecting one that was not consistent with expectations in the previous analysis). If decoys are present, one can switch between using parameterized and non-parameterized discriminant score distributions. If results of specific charge states in particular gave a poor analysis, one can specify to exclude them from analysis. If results assigned short peptides are a problem, one can increase the specified minimum peptide length.

3.2.3. Export to Tab-Delimited File

From the pepXML viewer, select the columns to display in the Pick Columns tab, then in the Other Actions tab, select the "Export Spreadsheet" button. This creates a tab-delimited XLS file.

3.2.4. Adapting Analysis to New Search Engines

PeptideProphet can generally be adapted to any additional search engine with the help of a training data set of search results of known validity, such as those generated using publicly available MS/MS data sets (27). The training data set is first used to derive a discriminant score that combines together in a linear combination multiple search scores, when available, into a single score that has a high power to distinguish correct from incorrect results.

The data set is further used to identify parameterized distributions (e.g. Gaussian, Gamma, etc.) that accurately model the observed discriminant score distributions among correct and incorrect training data set results. This information can then be easily added to the PeptideProphet source code for inclusion in the TPP run on the command line.

3.3. Additional Peptide-Level Validation: iProphet

The iProphet (28) tool of the TPP provides additional validation beyond PeptideProphet, employing statistical models that can apply to multiple searches and experiments. This analysis can increases the discriminating power of PeptideProphet probabilities, and allows for the integration of results of multiple search engines on the same data. For example, whether the same peptide was assigned to a spectrum with multiple search engines can be used to increase confidence in the validity of that assignment. Currently, iProphet implements five statistical models not considered in PeptideProphet. For each, it computes a new score whose distributions among correct and incorrect results are learned from the data and used to recalculate the probabilities that results are correct.

3.3.1. Number of Sibling Searches

Number of sibling searches (NSS) is a statistic that is based on the output of multiple search engines for the same set of spectra, processed through PeptideProphet. For each search result, an NSS value is computed by summing the probabilities of other search engine results of the same spectrum that agree on the peptide sequence and subtracting the probabilities of search engine results of the same spectrum that disagree on the peptide sequence. This model should be used only for searches with similar search parameters.

3.3.2. Number of Replicate Spectra

Number of replicate spectra (NRS) is a statistic that represents multiple identifications of the same peptide ion in one experiment. This statistic attempts to model the fact that in a typical dataset multiple observations (of high probability) should increase confidence of an identification, whereas high-signal-to-noise spectra that are misidentified will often be misidentified with the same incorrect peptide ion each time, but with marginal or low probability. It is computed as the sum of probabilities of all other spectra of the same charge assigned to the same peptide minus the number of such spectra. This method of computing NRS attempts to preserve the probabilities of "lucky" peptide ions identified only once with a high probability, but nonetheless may favor the identification of abundant peptides that give rise to multiple MS/MS spectra.

3.3.3. Number of Sibling Experiments

Number of sibling experiments (NSE) is a statistic that represents multiple identifications of the same peptide ion across different experiments. It is up to the user to define the boundaries between experiments. This statistic is used to model the fact that correct identifications are likely to be observed in multiple experiments.

3.3.4. Number of Sibling Ions

Number of sibling ions (NSI) is a statistic used to count occurrences of the same peptide sequence and corresponding modifications that are identified in different charge states. This NSI model is based on the notion that identifications of correct peptide sequences are often seen in multiple charge states, whereas identifications of the same incorrect peptide sequence are unlikely to be observed by multiple charge states.

3.3.5. Number of Sibling Modifications

Number of sibling modifications (NSM) is a statistic used to count occurrences of the same peptide sequence in different modified states. The NSM model is based on the fact that correct identification of peptide sequences are often seen in two or more modified states when a search with variable modifications is performed, whereas identifications to the same incorrect peptide sequence with different modifications are unlikely to be observed.

3.3.6. Running iProphet

iProphet is a powerful tool that is very simple to run. Unless one is testing the performance of the various models iProphet applies, there is no need to set any iProphet parameters. The easiest way to run the tool is on a single analysis (one experiment, one search engine), which can be done in the Analyze Peptides tab of Petunia (as of TPP version 4.3.). In this mode only three of the five models will get used: NRS, NSI, and NSM. To utilize all of the models a more complex analysis can be setup for combining multiple searches and experiments in the Combine Analysis tab of Petunia (see Note 5).

3.4. Inference of Sample Proteins: ProteinProphet

In the shotgun workflow, the connectivity between peptides and proteins is lost during sample processing, so must be computationally restored. The ProteinProphet (29) program of the TPP infers the most likely proteins in the sample based on the observed peptides assigned to MS/MS spectra and their computed probabilities of being correct. Users have the option to run the analysis immediately following PeptideProphet or iProphet, or independently as a second step anytime afterward. Running ProteinProphet independently enables users to include validated results of multiple search engines and search conditions of the same data set. The program combines together peptide evidence corresponding to each protein to compute a probability that the protein was present in the sample. Importantly, it addresses two critical issues for protein inference based on MS/MS spectra: Peptides corresponding to "single-hit" proteins are less likely to be correct than those corresponding to "multi-hit" proteins, and many peptides are present in more than a single database protein entry (isoforms, homologous proteins, etc.).

The non-random grouping of correct and incorrect peptide identifications among corresponding proteins leads to an amplification of the error rate from peptide to protein level. For example, among 100 peptides with a computed probability of 0.5, the 50 that are correct assignments likely correspond to a small subset of correct proteins, whereas the 50 that are incorrect likely all correspond to single-hit proteins, those to which no other peptides correspond. As a result, the error rate at the protein level would likely be much higher than 0.5. To counteract this amplification and ensure that computed protein probabilities are accurate, peptide probabilities are first adjusted to take into account whether or not their corresponding protein is single-hit or multi-hit. This information is incorporated into an NSP (number of sibling peptides) score computed for each search result, reflecting the estimated number of other correct peptides in the data set corresponding to the same protein. Distributions of NSP among correct and incorrect results are learned from the data and used to adjust the probabilities that the results are correct. Probabilities of peptides corresponding to single-hit proteins are penalized, and those of peptides corresponding to multi-hit proteins, boosted. Because the distributions are learned from each data set, NSP adjustments for low coverage data sets (those with few MS/MS spectra corresponding to each sample protein) will be negligible, and those for high coverage data sets, significant. The NSP adjustments to PeptideProphet probabilities are made as an initial step before peptide evidence is combined together into probabilities that proteins are present in the sample. This ensures accurate peptide probabilities following protein grouping, and thus accurate computed protein probabilities.

A large fraction of peptides are present in more than a single database protein entry. This is particularly true in the case of higher eukaryotes which have related protein family members, alternative splice forms, and partial sequences. ProteinProphet apportions shared peptides among all their corresponding proteins in proportion to the estimated likelihood that the proteins were present in the sample. The apportionment weights in turn are incorporated into each peptide's contribution when calculating the likelihood of a protein being present in the sample. Peptide apportionment weights and protein probabilities are updated iteratively until convergence is achieved. The result is the minimal list of proteins that are sufficient to explain the observed peptides. It is important to keep in mind that the list may not be fully inclusive because it cannot be ruled out that some additional proteins with common peptides are present in the sample.

3.4.1. Running ProteinProphet

ProteinProphet can be run in the Analyze Peptides tab for a single LC-MS/MS run, or from the Analyze Proteins tab for one or more runs (see Note 6). Results are stored in protXML format (4). Learned NSP distributions are used to penalize peptides corresponding to single-hit proteins an appropriate amount. However, short proteins are likely to be single-hit whether or not they are correct identifications. Correct peptides corresponding to short proteins are likely to be penalized as a result. This can be avoided with the "Normalize NSP using protein length" option, where the NSP score is computed as the number of siblings divided by protein length. Peptides corresponding to short proteins would thus be assigned higher NSP values and penalized to a lesser degree. The tradeoff is that incorrect peptides corresponding to short proteins will be similarly treated.

The "Check peptide's total weight…" option affects whether or not probabilities for proteins that belong to a protein group (with many shared peptides) are computed using each peptide's weight contribution to the protein alone or to all members of the protein group. This feature allows protein probabilities for proteins to be computed under the assumption that the given protein is the *only* protein in the protein group present, and allows comparing which proteins in a large protein group (where at least one is correct) are more likely.

The iProphet option (see Note 6) has two effects. First, it tells ProteinProphet to read the iProphet probabilities in the pepXML file, and second it forces ProteinProphet to use only the top probability for each unique unmodified peptide sequence rather than for each unique parent charge and peptide sequence combination. This is recommended because the iProphet probabilities already account for repeated observations of the same unique peptide sequence in the different charges and modified states. This option helps to reduce the ProteinProphet false positive rate when analyzing very large data sets consisting of many LC-MS/MS runs. Use the "Import XPRESS protein ratios" and "Import ASAPRatio protein ratios and pvalues" options to compute and display XPRESS and ASAPRatio protein ratios, respectively. Libra protein ratios are automatically computed when available at the peptide level.

3.4.2. Viewing Results

Results of ProteinProphet can be viewed as HTML to show identified proteins with their assigned probabilities of being present in the sample, along with all of their corresponding peptides. Each peptide has a weight, the fraction ranging from 0 to 1 of its apportionment to the protein. Clicking on the link to any weight retrieves all protein entries that contain the peptide. This is useful when exploring whether proteins attributed to shared peptides are valid identifications. Unambiguous peptides found in no other protein are indicated with an asterisked weight of 1. Each peptide also has a computed NSP value reflecting whether or not the protein is single-hit or multi-hit. Clicking on the link to any NSP value displays the learned NSP distributions among correct and incorrect results, and the probability adjustment factor derived for ranges of NSP values. This is useful for assessing the learned fraction of correct peptides corresponding to single-hit proteins.

When quantitation was performed at the protein level, protein ratios are displayed with links to their contributing peptide quantitation information.

When the search was made with a forward/decoy database and the decoy sequences were not announced to PeptideProphet (to be used as known incorrect results), the decoy proteins can be used to independently assess the accuracy of computed probabilities. To generate a plot of decoy vs. ProteinProphet based false discovery rates, go to the Utilities "Decoy Protein Validation" tab, select the protXML file to analyze, and enter the decoy protein name prefix. A series of minimum ProteinProphet probability thresholds is then applied to the data and the resulting ProteinProphet (based on the computed probabilities) and decoy (based on the fraction of decoy results) error rates are plotted. The accuracy of computed probabilities can be assessed by the correspondence of ProteinProphet and decoy false discovery rates, particularly in the important range close to 0.

When using the Firefox browser for viewing ProteinProphet results, it is possible to make use of the Firegoose plugin (30) for automatically exporting filtered protein lists available in the ProteinProphet viewer to various sources of additional information. Perhaps the most useful is the TPP's own Protein Information and Property Explorer (PIPE) (31) which can look up all of the different names of the proteins/genes identified in the sample and examine these in the context of different databases, as shown in Fig. 5. For instance, using the PIPE it is possible to identify the KEGG pathways that include identified proteins or look for enrichment of specific GO terms in the set. More information on the PIPE and direct access to the tool are available at (32). ProteinProphet results can also be exported to tab delimited XLS format for subsequent analysis, and imported to SBEAMS-Proteomics database (33).

3.5. Quantitation: XPRESS, ASAPRatio, Libra

The TPP supports both isotopic and isobaric labeling strategies for determining relative quantitation levels of peptides and proteins. Programs XPRESS (34) and ASAPRatio (35) compute abundance ratios based on the extracted ion currents of heavy and light labeled peptide pairs generated with techniques such as ICAT (36) and SILAC (37). The program Libra (38) performs peptide quantitation based on relative peak intensities of MS/MS reporter peaks used in isobaric techniques such as iTRAQ and TMT. Programs are executed at the peptide level in the Analyze Peptides tab of Petunia, and at the protein level, in the Analyze Proteins tab.

3.5.1. Isotopic Labeling Quantitation Analysis

In the isotopic labeling strategy, two samples are labeled with heavy and light adducts differing from one another only by their isotopic composition. In the ICAT technique, protein samples are harvested and reacted with light or heavy labels, then combined

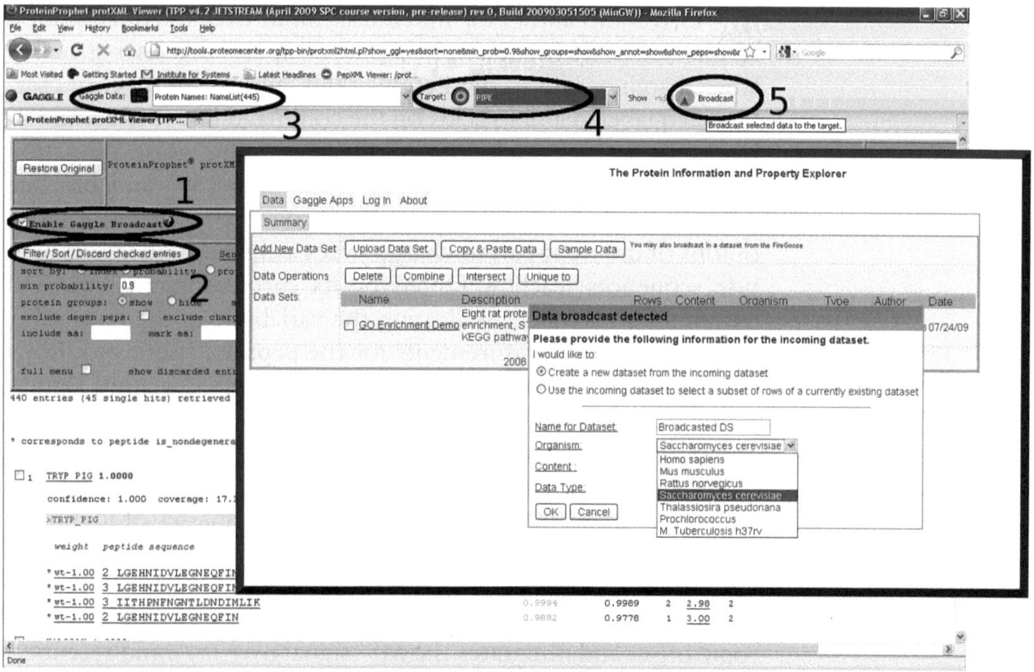

Fig. 5. Steps to enable export of ProteinProphet results to the PIPE and other resources. (1) Check "Enable Gaggle Broadcast" on the ProteinProphet page; (2) click the "Filter/Sort/..." button. When the page is finished loading, the "Gaggle Data" drop-down list will be populated in the Firegoose toolbar; (3) Select "Protein Names: NameList" on the "Gaggle Data" dropdown list; (4) select "PIPE" (or other broadcast destination) from the Target dropdown list; (5) click "Broadcast" to open the PIPE tab with display options.

together, trypsinized, and selected for ICAT containing peptides for injection into a mass spectrometer. In the SILAC technique, tissue culture cell samples are grown with either isotopically heavy or light amino acids (often arginine or lysine), combined together, purified, trypsinized, and injected into a mass spectrometer. In both cases, relative quantitation of the two samples can be determined from the LC-MS/MS data as the ratio of the light and heavy parent ion extracted ion currents. This requires correctly assigning a peptide to an MS/MS spectrum associated with the light and/or heavy parent ion, properly pairing together the heavy and light parent ion partners, and integrating the parent ion peak volumes. Pairing parent ions must take into account whether or not the heavy and light parents co-elute, and if not, what time difference offset and order to expect them.

The TPP has two programs to compute abundance ratios from isotopically labeled heavy and light samples, XPRESS and ASAPRatio. Users are free to use one of both depending on preference. XPRESS was originally developed for ICAT labeling. However, the program now also supports up to three user-specified labeled amino acids, as well as metabolic ^{15}N labeling which involves all 20 amino acids. Users can also select the peak time width (in number of scans) to account for differences in chromatography, and the mass tolerance for finding parent ion partners.

ASAPRatio allows for much user input over the quantitation results for each ion pair. It supports up to five user-specified labeled amino acids in the Petunia GUI, and any number of user-specified labeled amino acids on the command line (including ^{15}N labeling). It determines the peak areas using noise reduction methods such as Savitzky–Golay smoothing (39) and baseline subtraction. It furthermore combines together abundance ratios of ions of different parent charge states assigned to identical peptide sequences, each weighted by its chromatogram area, and computes error values indicating the variability in the ratio value among different measurements for the peptide. It does so, however, only after removing outliers, those ratios deviating significantly from the majority. It can also handle cases where the light and heavy labeled samples are not combined prior to mass spectrometry, but are represented in the data set as separate runs.

ASAPRatio computes the monoisotopic masses of identified peptides independently from the search engine and uses these to identify isotopic envelopes corresponding to identified peptides in the data. It opens a user selected mass range around each isotopic offset and integrates the MS^1 signal over m/z and retention time range. The user can specify different mass ranges depending on the instrument type (see Note 7).

Users specify the labels and whether heavy and light parent ions co-elute. They also must specify the mass tolerance for matching parent ion partners. Once run, hands-on functionality allows users to look at the ion chromatographs to verify proper quantitation. Importantly, users can modify the quantitation by specifying peak start and stop times, modes of background signal removal, and contributing parent ion charges, as shown in Fig. 6.

Abundance ratios of peptides can be used to compute ratios at the protein level. For example, ProteinProphet groups together all peptides attributed to a particular protein. In general, the log ratios of corresponding peptides are combined together, each weighted by the inverse of its error. The result is a computed protein ratio and accompanying error indicating the variability in the log ratio value of different corresponding peptides. It does so, however, only after removing outliers, those peptide ratios deviating significantly from the majority. This handles cases in which individual peptides have distinct ratio values because of differential post-translational modification of the peptide in the two samples. p-values of protein abundance ratios are computed by modeling the null hypothesis of unchanged peptide ratios as a Gaussian distribution from the observed log ratio data. Protein ratios are then normalized to the mean peptide ratio, and p-values are computed for each individual ratio based on the z-score of its log ratio value and its error, using the null hypothesis. These p-values reflect the likelihood of observing a protein abundance ratio because of chance alone. This assumes that the majority of peptides in the data set will not be expressed at significantly different levels in the two samples.

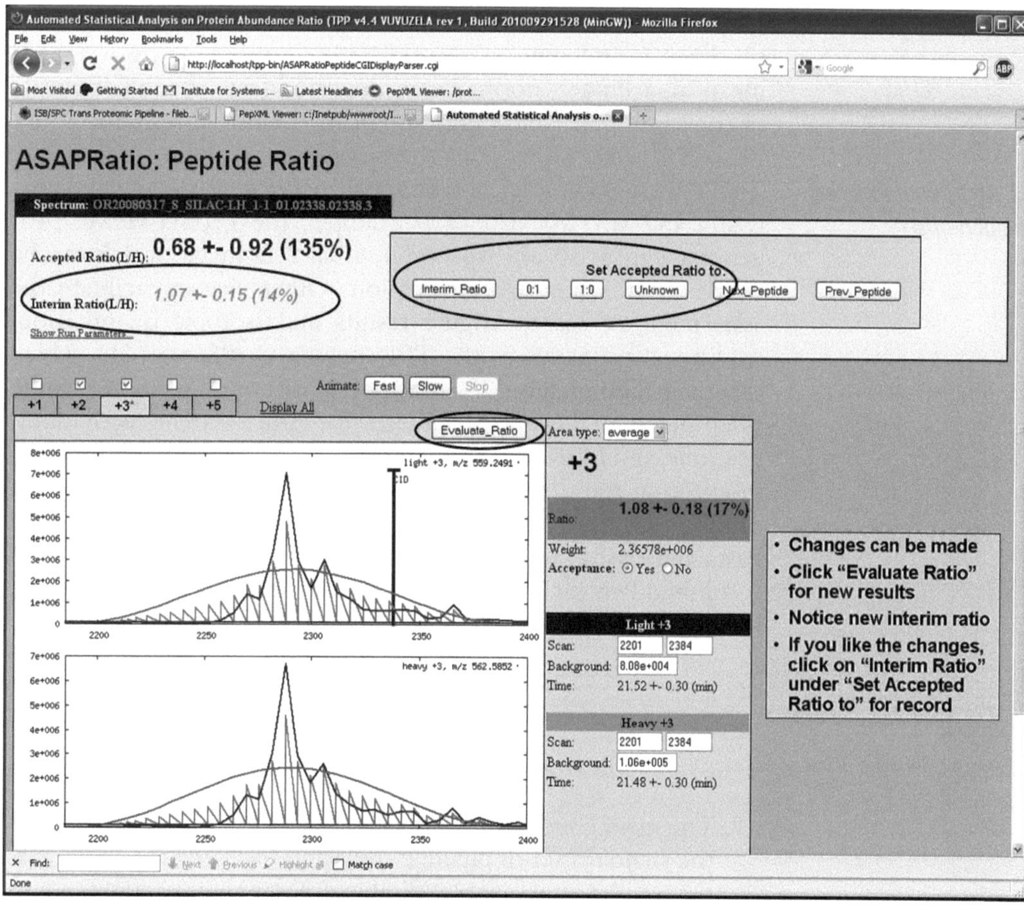

Fig. 6. Interface for exploring ASAPRatio peptide quantitation.

3.5.2. Isobaric Tag Quantitation Analysis

In the isobaric labeling strategy using iTRAQ or TMT, up to eight samples are reacted with labels that have identical mass, yet on fragmentation of the parent ion, give rise to distinct signature peaks near 100 m/z. Samples are combined together, purified, trypsinized, and injected into a mass spectrometer. Relative quantitation information is extracted from each MS/MS spectrum based on the integrated areas of its signature peaks attributed to the various samples.

Libra software computes relative abundances given specified signature fragments for the various labeled samples. Users specify options in a Libra condition file prior to running the software. This includes description of the signature ions for each sample, and for each the fraction of peak intensity encountered as −2, −1, +1, and +2 Da offset peaks (including overflow into adjacent signature peaks). This enables the software to properly deconvolute the signature peaks into the contributions of the various samples. A dropdown enables automatic specification of values for iTRAQ

4 and 8 channel, and TMT 6 channel labeling. In addition, users specify mass tolerance, centroiding method (intensity weighted mean vs. average), normalization method (one channel intensity or the sum of all), and minimum peak intensity threshold.

3.6. Additional Pipeline Components

The TPP contains a set of additional tools for visualizing and analyzing LC-MS/MS data. For example, the Pep3D viewer (40) generates from all MS^1 spectra a visual 2-D image which can be overlaid with MS/MS information. Other viewers facilitate the assessment of search engine results and isotopic quantitation. Additional programs in the TPP include QualScore (41), which identifies high quality MS/MS spectra without confident peptide assignments so they can be targeted for future searches with modifications, and MaRiMba (42), a tool for selecting multiple reaction monitoring transitions for targeted proteomics. The Utilities tab of Petunia hosts programs to create decoy databases from existing Fasta databases, and perform peptide level decoy validation of computed PeptideProphet and iProphet probabilities, and protein level decoy validation of computed ProteinProphet probabilities.

4. Notes

1. The main command line tool in the TPP is xinteract, it drives the standard set of pipeline tools and has a myriad of options that can be set to enable/disable different user parameters and features. Running xinteract on the command line without any options produces a long and detailed usage statement. This tool executes an analysis using a single search engine/single experiment analysis. Programs InterProphetParser and ProteinProphetParser can be run to combine multiple search engine/experiment analyses together.
2. To perform a database search, go to the Database Search tab and specify the desired mzXML file to analyze and the database to search. Search parameters are specified differently for each search engine. Most search engines including X! Tandem require a separate parameter file. A fresh TPP installation includes suggested parameter files for X! Tandem (ISB_input_kscore.xml and tandem_params.xml). When executing a new search, the provided tandem_params.xml file should be copied to the search directory and edited to reflect the parameters to be used.
3. SpectraST can build a spectral library using identifications from a sequence database search, given a pepXML file that has been processed with PeptideProphet or iProphet, with the following command:

```
spectrast -cNlibname -cP0.9 interact.pep.xml.
```

This will import all identifications with a minimum probability of 0.9 from the pepXML file interact.pep.xml, and produce a library named libname.splib (with the corresponding libname.sptxt, libname.spidx, and libname.pepidx files).

4. The general workflow is to validate search results (PeptideProphet), perform quantitation (XPRESS, ASAPRatio, Libra) if specified, combine multiple validated search results together for additional peptide-level validation (iProphet), and finally perform protein inference (ProteinProphet) and protein level quantitation if specified. In the Analyze Peptides tab users can specify all these steps at once, or merely the first validation step. Subsequent steps can be performed independently from the Combine Analyses and Analyze Proteins tabs.
5. In the Combine Analyses tab of Petunia, one or more pepXML files containing PeptideProphet probabilities can be processed. If multiple experiments are being combined, each should be labeled with a different Experiment Label option on the Analyze Peptides tab in Petunia (TPP version 4.3 and above). No options have to be selected unless there is a specific need to disable any of the iProphet models.
6. In the Analyze Proteins tab of Petunia, select all pepXML files (validated search results) to use as input for the inference step. The iProphet option (as of TPP version 4.3.0) must be selected to use iProphet peptide probabilities and to use the top probability for each unique unmodified peptide sequence identified. This can also be achieved on the command line as:

```
ProteinProphet interact.iproph.pep.xml inter-
act.iproph.prot.xml IPROPHET.
```

7. It is especially important to note that for high-mass-accuracy instruments, when modifications are specified in the search as average masses, they must be redefined for ASAPRatio as monoisotopic masses for all modifications (even those that are not part of the label such as Oxidized Methionine, Iodoacetamide, etc.). For high mass accuracy instruments a value of 0.05 Da for the mass tolerance in ASAPRatio should work nicely, for an LTQ instrument the default setting will suffice.

Acknowledgments

We would like to thank Eric Deutsch and Luis Mendoza for valuable discussions.

References

1. Aebersold, R. and Mann, M. (2003) Mass spectrometry-based proteomics. *Nature* **422**, 198–207.
2. Kohlbacher, O., Reinert, K., Gropl, C., Lange, E., Pfeifer, N., Schulz-Trieglaff, O., and Sturm, M. (2007) TOPP-the OpenMS proteomics pipeline. *Bioinformatics* **23**, e191–e197.
3. Cox, J. and Mann, M. (2008) MaxQuant enables high peptide identification rates, individualized p.p.b.-range mass accuracies and proteome-wide protein quantification. *Nat. Biotechnol.* **26**, 1367–1372.
4. Keller, A., Eng, J., Zhang, N., Li, X.J., and Aebersold, R. (2005) A uniform proteomics ms/ms analysis platform utilizing open xml file formats. *Mol. Syst. Biol.* **1**, 2005.0017.
5. TPP Windows Installation Guide. http://tools.proteomecenter.org/wiki/index.php?title=Windows_Installation_Guide.
6. TPP Source code Installation Guide for Linux. http://tools.proteomecenter.org/wiki/index.php?title=Software:TPP#Source_code_Installation_.28For_Linux_systems.29.
7. TPP demo. http://tools.proteomecenter.org/wiki/index.php?title=TPP_Demo2009.
8. TPP training course. http://www.systemsbiology.org/Resources_and_Development/Current_Course_Offerings.
9. Sashimi site. http://sourceforge.net/projects/sashimi/.
10. Pedrioli, P.G., Eng, J.K., Hubley, R., Vogelzang, M., Deutsch, E.W., Raught, B., Pratt, B., Nilsson, E., Angeletti, R.H., Apweiler, R., Cheung, K., Costello, C.E., Hermjakob, H., Huang, S., Julian, R.K., Kapp, E., McComb, M.E., Oliver, S.G., Omenn, G., Paton, N.W., Simpson, R., Smith, R., Taylor, C.F., Zhu, W., and Aebersold, R. (2004) A common open representation of mass spectrometry data and its application to proteomics research. *Nat. Biotechnol.* **22**, 1459–1466.
11. Deutsch, E. (2008) mzML: a single, unifying data format for mass spectrometer output. *Proteomics* **8**, 2776–2777.
12. Elias, J.E. and Gygi, S.P. (2007) Target-decoy search strategy for increased confidence in large-scale protein identifications by mass spectrometry. *Nat. Methods* **4**, 207–214.
13. MacLean, B., Eng, J.K., Beavis, R.C., and McIntosh, M. (2006) General framework for developing and evaluating database scoring algorithms using the TANDEM search egnine. *Bioinformatics* **22**, 2830–2832.
14. Geer, L.Y., Markey, S.P., Kowalak, J.A., Wagner, L., Xu, M., Maynard, D.M., Yang, X., Shi, W., and Bryant, S.H. (2004) Open mass spectrometry search algorithm. *J. Proteome Res.* **3**, 958–964.
15. Tabb, D.L., Fernando, C.G., and Chambers, M.C. (2007) MyriMatch: Highly accurate tandem mass spectral peptide identification by multivariate hypergeometric analysis. *J. Proteome Res.* **6**, 654–661.
16. Eng, J., McCormack, A.L., and Yates, J.R. (1994) An approach to correlate tandem mass spectral data of peptides with amino acid sequences in a protein databases. *J. Am. Soc. Mass Spectrom.* **5**, 976–989.
17. Perkins, D.N., Pappin, D.J., Creasy, D.M., and Cottrell, J.S. (1999) Probability-based protein identification by searching sequence databases using mass spectrometry data. *Electrophoresis* **20**, 3551–3567.
18. Tanner, S., Shu, H., Frank, A., Wang, L., Zandi, E., Mumby, M., Pevzner, P.A., and Bafna, V. (2005) Inspect: Fast and accurate identification of post-translationally modified peptides from tandem mass spectra. *Anal. Chem.* 77, 4626–4639.
19. Zhang, N., Aerbersold, R., and Schwikowski, B. (2002) ProbID: A probabilistic algorithm to identify peptides through sequence database searching using tandem mass spectral data. *Proteomics* **10**, 1406–1412.
20. Colinge, J., Masselot, A., Cusin, I., Mahé, E., Niknejad, A., Argoud-Puy, G., Reffas, S., Bederr, N., Gleizes, A., Rey, P.A., and Bougueleret, L. (2004) High-performance peptide identification by tandem mass spectrometry allows reliable automatic data processing in proteomics. *Proteomics* **4**, 1977–1984.
21. Lam, H., Deutsch, E.W., Eddes, J.S., Eng, J.K., King, N., Stein, S.E., and Aebersold, R. (2007) Development and validation of a spectral library searching method for peptide identification from MS/MS. *Proteomics* 7, 655–667.
22. Spectral libraries. http://www.peptideatlas.org/speclib/.
23. Keller, A., Nesvizhskii, A., Kolker, E., and Aebersold, R. (2002) Empirical statistical model to estimate the accuracy of peptide identifications made by MS/MS and database search. *Anal. Chem.* **74**, 5383–5392.
24. Dempster, A., Laird, N., and Rubin, D.B. (1977) Maximum likelihood from incomplete data via the EM algorithm. *J. R. Stat. Soc.* **B39**, 1–38.

25. Malmstrom, J., Lee, H., Nesvizhskii, A., Shteynberg, D., Mohanty, S., Brunner, E., Ye, M., Weber, G., Eckerskorn, C., and Aebersold, R. (2006) Optimized peptide separation and identification for mass spectrometry based proteomics via free-flow electrophoresis. *J. Proteome Res.* **5**, 2241–2249.
26. Zhang, H., Yi, E.C., Li, X., Mallick, P., Spratt, K., Masselon, C.D., Camp, D.G., Smith, R.D., Kemp, C.J., and Aebersold, R. (2004) High throughput quantitative analysis of serum proteins using glycopeptide capture and liquid chromatography mass spectrometry. *Mol. Cell Proteomics* **4**, 144–155.
27. Keller, A., Purvine, S., Nesvizhskii, A., Stoliar, S., Goodlett, D., and Kolker, E. (2002). Experimental protein mixture for validating tandem mass spectral analysis. *OMICS* **6**, 207–212.
28. Shteynberg, D., Deutsch, E.W., Lam, H., Eng, J.K., Sun, Z., Tasman, N., Mendoza, L., Moritz, R., Aebersold, R., and Nesvizhskii, A. Post-processing and validation of tandem mass spectrometry datasets improved by iProphet, in preparation.
29. Nesvizhskii, A., Keller, A., Kolker, E., and Aebersold, R. (2003). A statistical model for identifying proteins by tandem mass spectrometry. *Anal. Chem.* **75**, 4646–4658.
30. Firegoose Installation Guide. http://gaggle.systemsbiology.org/docs/geese/firegoose/install/.
31. Ramos, H., Shannon, P., and Aebersold, R. (2008) The Protein Information and Property Explorer: an easy-to-use, rich-client web application for the management and functional analysis of proteomic data. *Bioinformatics* **24(18)**, 2110–2111.
32. Protein Information and Property Explorer. http://pipe.systemsbiology.net/.
33. Marzolf, B., Deutsch, E.W., Moss, P., Campbell, D., Johnson, M.H., and Galitski, T. (2006) SBEAMS-Microarray: database software supporting genomic expression analyses for systems biology. *BMC Bioinformatics* **7**, 286.
34. Han, D.K., Eng, J., Zhou, H., and Aebersold, R. (2003) Quantitative profiling of differentiation-induced microsomal proteins using isotope-coded affinity tags and mass spectrometry. *Nat. Biotechnol.* **19**, 946–951.
35. Li, X.J., Zhang, H., Ranish, J.A., and Aebersold, R. (2003) Automated statistical analysis of protein abundance ratios from data generated by stable-isotope dilution and tandem mass spectrometry. *Anal.Chem.* **75**, 6648–6657.
36. Gygi, S.P., Rist, B., Gerber, S.A., Turecek, F., Gelb, M.H., and Aebersold, R. (1999) Quantitative analysis of complex protein mixtures using isotope-coded affinity tags. *Nat. Biotechnol.* **17**, 994–999.
37. Ong, S.E. and Mann, M. (2007) Stable isotope labeling by amino acids in cell culture for quantitative proteomics. *Methods Mol. Biol.* **359**, 37–52.
38. Pedrioli, P.G., Raught, B., Zhang, X.D., Rogers, R., Aitchison, J., Matunis, M., and Aebersold, R. (2006) Automated identification of SUMOylation sites using mass spectrometry and SUMmOn pattern recognition software. *Nat. Methods* **3**, 533–539.
39. Savitzky, A. and Golay, M.J.E. (1964). Smoothing and differentiation of data by simplified least squares procedures. *Anal. Chem.* **36**, 1627–1639.
40. Li, X.J., Pedrioli, P.G., Eng, J., Martin, D., Yi, E.C., Lee, H., and Aebersold, R. (2004) A tool to visualize and evaluate data obtained by liquid chromatography-electrospray ionization-mass spectrometry. *Anal. Chem.* **76**, 3856–3860.
41. Nesvizhskii, A.I., Vogelzang, M., and Aebersold, R. (2004) Measuring MS/MS spectrum quality using a robust multivariate classifier. In Proc. 52nd ASMS Conf. Mass Spectrom., Nashville, TN.
42. Sherwood, C., Eastham, A., Peterson, A., Eng, J.K., Shteynberg, D., Mendoza, L., Deutsch, E., Risler, J., Lee, L.W., Tasman, N., Aebersold, R., Lam, H., and Martin, D.B. (2009) MaRiMba: a software application for spectral library-based MRM transition list assembly. *J. Proteome Res.* **8(10)**, 4396–4405.

Chapter 13

Analysis of High-Throughput ELISA Microarray Data

Amanda M. White, Don S. Daly, and Richard C. Zangar

Abstract

Our research group develops analytical methods and software for the high-throughput analysis of quantitative enzyme-linked immunosorbent assay (ELISA) microarrays. ELISA microarrays differ from DNA microarrays in several fundamental aspects and most algorithms for analysis of DNA microarray data are not applicable to ELISA microarrays. In this review, we provide an overview of the steps involved in ELISA microarray data analysis and how the statistically sound algorithms we have developed provide an integrated software suite to address the needs of each data-processing step. The algorithms discussed are available in a set of open-source software tools (http://www.pnl.gov/statistics/ProMAT).

Key words: ELISA, Microarray, Standard curve, Bioinformatics, Calibration, ProMAT, ELISA-BASE

1. Introduction

Our research is focused on the early detection of breast cancer based on changes in circulating proteins. It is widely recognized that breast cancer is a heterogeneous disease, and that it is unlikely that a single protein biomarker will be able to detect all forms of this disease. Although early detection of this disease is likely to decrease mortality and morbidity, it is unlikely that a small (i.e., early) tumor will significantly alter levels of abundant proteins in the blood. Therefore, early detection of breast cancer based on proteins in blood will likely require the analysis of a panel of low-abundance proteins. For this reason, we have been developing sandwich enzyme-linked immunosorbent assay (ELISA) microarrays for biomarker analysis. ELISAs are an exceptionally sensitive and specific method for measuring the concentrations of trace proteins in complex biological fluids, such as blood serum or plasma. Indeed, the ELISA appears to be the only established analytical approach that is routinely used to measure low-abundance

Cathy H. Wu and Chuming Chen (eds.), *Bioinformatics for Comparative Proteomics*, Methods in Molecular Biology, vol. 694, DOI 10.1007/978-1-60761-977-2_13, © Springer Science+Business Media, LLC 2011

proteins in complex biological solutions. ELISA microarrays have other advantages: they consume only trace amounts of each sample, and they are suited for high-throughput (e.g., hundreds or thousands of samples) analysis. As such, ELISA microarrays are an ideal tool for the evaluation of panels of biomarkers in large numbers of samples. This capability is important, as it seems likely that these types of large-scale studies are needed to determine if a biomarker panel is truly useful for the detection of a rare disease such as cancer.

Critical to the ELISA microarray technology platform to reach its high-throughput potential is the ability to effectively manage and process the extensive amounts of data generated in a large study. We have developed an integrated suite of software specifically for this purpose. This software addresses three important needs of a large-scale ELISA microarray study: calibration, standard curve estimation and sample concentration prediction. A key feature of this software is the diagnostic images generated to aid the user in the evaluation of both the data quality and the original analysis of the data. In this article, we review the features and use of each program in this software suite.

2. Materials

2.1. Data

Data from an ELISA microarray experiment consists of a set of spot intensity values, along with the corresponding spot and sample information such as spot position, chip, slide, sample type, etc. These data are a combination of researcher input and the output from a microarray image analysis tool. The necessary data support calibration, standard curve estimation and sample concentration prediction. Calibration data is obtained from specific calibration assays (i.e., calibration spots). To generate data for the standard curves, a set of chips are processed with a serial dilution of a standard mixture of purified proteins (i.e., antigens) at known concentrations. Other chips are treated with samples to predict the concentrations of target proteins from the standard curves.

Generating suitable data for the analyses described below requires good experimental design (i.e., replication, randomization and blocking) before the experiment is performed, including

- Replicates of each assay (i.e., "spots") per chip (we typically use four replicate spots).
- Replicate chips for each sample and standard dilution (we typically use at least three replicates).
- Chip calibration spots, if normalization is to be performed.
- Randomized assignment of standards and samples to chips.

2.2. Software

1. ProMAT and ProMAT Calibrator: (http://www.pnl.gov/statistics/ProMAT) ProMAT (1) and ProMAT Calibrator (2) tools are free and open-source ELISA microarray tools we have developed in Java (http://java.sun.com) and R (http://www.r-project.org), which is an open-source statistical programming language.
2. ELISA-BASE: (http://www.pnl.gov/statistics/ProMAT/ELISA-BASE.stm) ELISA-BASE (3) is an ELISA microarray database tool which extends the BioArray Software Environment (BASE) (4) system, and includes ProMAT in addition to the ability to track metadata associated with study design, reagents and data processing. ELISA-BASE is also free and open-source.

3. Methods

3.1. Data Analysis Overview

There are three steps in analyzing ELISA microarray data. First, the data typically are calibrated for analytical biases between chips introduced by processing. Second, standard curves are estimated with the calibrated data, and finally the protein concentrations are predicted using the standard curves. Our tools are specifically designed to address these steps (Fig. 1). ProMAT Calibrator adjusts the data using one or more calibration assays chosen by the user. ProMAT estimates standard curves and can predict protein concentrations if sample data are provided. These two programs are written in Java and R (http://www.r-project.org) and can be installed on the user's computer for routine use.

The third tool, ELISA-BASE, is an extension of the BioArray Software Environment (BASE) (4), which is a web-enabled database tool for tracking and analysis of DNA microarray data. ELISA-BASE allows the user to track ELISA microarray reagents, experiment protocols and data processing steps, and also includes ProMAT (and will soon include ProMAT Calibrator).

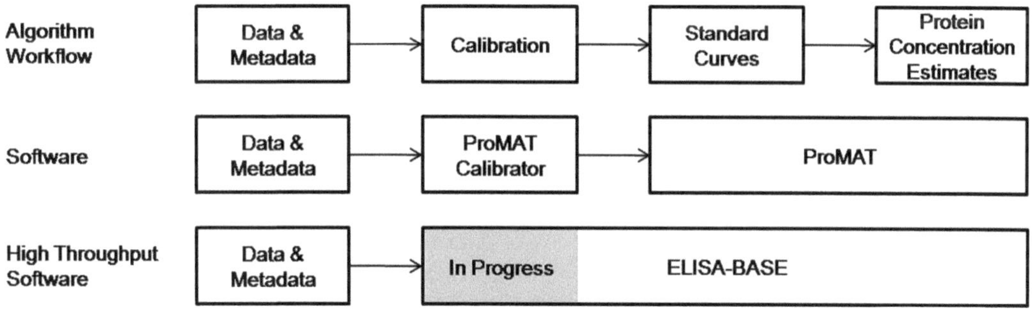

Fig. 1. Overview of the data analysis process for ELISA microarrays.

3.2. Preparing Data for Analysis

ProMAT, ProMAT Calibrator and ELISA-BASE all import data from comma-delimited (.csv) text files containing spot intensity data, which are created by most microarray image analysis tools. See Note 1 for instructions on creating a CSV file, if needed. Along with spot intensity measurements, the tools require information about the array design (i.e., spot characteristics and positioning), the experiment design and array incubation and processing procedures. This information should be assembled in the following files:

1. Slide layout file: defines assay pattern on individual chip and the maximum concentration of each antigen standard.
2. Experiment information file: defines which standards or samples are printed on each slide (and the positions of subarrays, if applicable) and what the dilution of each standard or sample.

   Make sure that none of the following characters:

   , \ / : * ? “ < > |

are used in file names or in any user-defined fields such as antigen name or slide identifier in the slide layout and experiment information files, as this will cause an error.

3.2.1. Slide Layout

ProMAT and ProMAT Calibrator need information about the slide layout to associate an antigen with each spot intensity. This slide layout file contains several columns of data, where each column has a column name on the first line. (This file is not needed for ELISA-BASE, see Subheading 3.5.2.) The essential columns are:

1. Antigen name or spot name. This column can be given any name and the user will enter the column name in the ProMAT or ProMAT Calibrator window in the *Spot ID Column* field.
2. The maximum concentration for the standard data for this antigen (no units). This column must be called *max.concentration*. The maximum concentration column lists the antigen concentration in the standard mixture before dilution.
3. Spot position columns for matching the slide layout metadata to the file containing the signal intensity data for the individual spots. For example, if the data files have columns *Spot Row* and *Spot Column*, and these are the columns that uniquely determine the position within a chip of a specific capture antibody, then these column names should appear in your slide layout file and with *the exact same name as in the intensity data files.* You may have as few or as many columns as are necessary to specify the antigen to data mapping.

Any spot intensity data that are not matched to a line in the slide layout file will be ignored. This characteristic can be advantageous, because the slide layout file may also be used to filter the data, such that ProMAT only extracts data from a subset of the ELISA tests. An example slide layout file is provided with ProMAT and ProMAT Calibrator.

3.2.2. Experiment Information

The three tools also need information about the incubation and processing of each array, which is stored in a comma-delimited text file. Separate experiment information files must be prepared for the samples and the standards. The sample experiment information file contains information about the chip location (i.e., slide and array) and the dilution of each sample. The standards experiment information file contains information on the chip locations and relative concentrations of the antigen mix. For ELISA-BASE, both the standards and samples information may be in one file. The experiment information file(s) must contain the following columns:

1. Name of the intensity data file name. This column must be named *file.name*.
2. A slide identifier, which may be numbers and/or characters. This column must be named *slide.number*.
3. A sample identifier, which may be either numbers and/or characters. This column must be called *sample.id*.
4. The dilution factor of the standard or sample prior to incubation. This must be less than or equal to 1 (i.e., 0.25 is a fourfold dilution). This column must be called *dilution*.
5. If one intensity data file contains multiple samples, then the experiment information file must include columns sufficient to uniquely specify each sample. For example, if each array was treated with a different sample and the intensity data files contain columns called *Array Row* and *Array Column*, then the experiment information file should contain columns with the same names. In this case, each array would correspond to a single row in the experiment information file. *These columns must be labeled exactly as they appear in the data files.*
6. If multiple scans were taken of each slide (e.g., image replicates, or images under different scanner settings) additional columns may be added for that information. These columns may be given any name, and in the analysis tools the user should enter those column names in the "Imager settings column names" parameter.
7. If using ELISA-BASE, a column called *detection* should be included, with the ID of the detection antibody mixture.

The experiment information files determine which data are used in the analysis, so any data files and/or any arrays not listed in either the standards or the samples experiment information will not be used. An example experiment information file is provided with ProMAT and ProMAT Calibrator.

3.3. Data Calibration

3.3.1. Overview

In DNA microarray data analysis, data normalization algorithms are well defined. ELISA microarray assaying differs in crucial ways that make DNA normalization algorithms inappropriate.

Most importantly, DNA microarrays typically assay thousands of genes, most of which are not expected to be differentially expressed between samples, whereas ELISA microarrays use a small number of targeted assays, many of which may vary across study groups. To address this problem we have developed a statistically sound calibration protocol and algorithm using calibrant assays (2 and unpublished results). Using this approach requires an appropriate calibrant sandwich ELISA in each chip (see Note 2 for choosing an appropriate calibrant assay).

The first step is to determine if calibration is likely to be useful, which means determining whether there are inherent biases in the data because of experimental processing steps that calibration may reduce. Every step of the experiment, from sample preparation to slide printing to slide imaging and spot intensity estimation, may introduce variance or biases into the intensity measurements. Assuming the dataset contains calibrant spots for each chip (i.e., spots whose intensities are expected to be nominally constant across all chips) the next step is to evaluate whether there are chip-level biases for which calibration may be effective. A helpful action at this point is to plot the calibrant spot intensities against the processing step(s) for which bias is a concern. For example, Fig. 2 shows simulated data with an induced spot print order effect, as well as a difference between the two columns of chips on a slide. Overall, there is a downward trend in spot intensity versus print order, and the intensities in the left slide columns (in blue) show less variance than those in the right columns (in red). A diagnostic plot like this, generated for the calibration spots

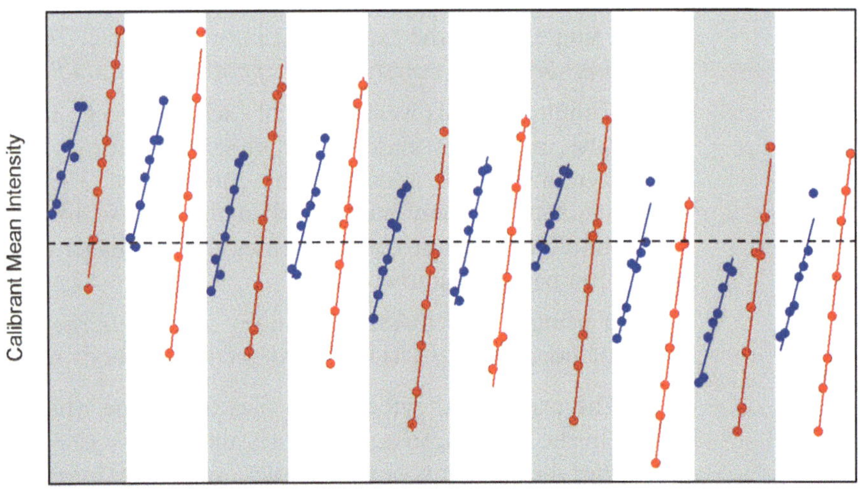

Fig. 2. Simulated data of mean calibrant intensity for each chip in print order for ten slides each with two columns and eight rows of chips. *Blue* and *red spots* denote eight chips in the *left* and *right columns* of a slide, respectively. The *blue* and *red lines mark* the fits of linear models to the affected data. The *plot* shows an overall downward trend with print order, and the chips in the *left columns* have smaller variances than those in the *right column.*

versus any suspected processing variable, can quickly show if data biases are present and if data calibration may be beneficial.

3.3.2. Using ProMAT Calibrator

When ProMAT Calibrator is started, the window in Fig. 3 is displayed. The parameters are:

- *Analysis name*: (optional) whatever information is entered here will be used as the first part of the name for all output files. If left blank, output files will be assigned common names that do not differentiate between studies.
- *Standards directory*: directory containing the data files for the standard assays.
- *Standards slide layout*: the slide layout file for the standards data described in Subheading 3.2.
- *Standards experiment info*: the experiment information file for the standards data described in Subheading 3.2.
- *Samples directory*: directory containing the data files for the sample assays.
- *Samples slide layout*: the slide layout file for the sample data described in Subheading 3.2.
- *Samples experiment info*: the experiment information file for the sample data described in Subheading 3.2.

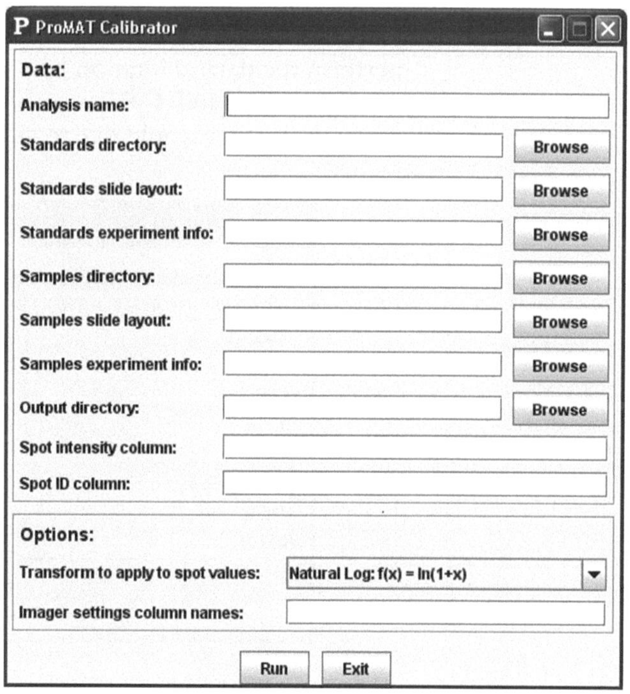

Fig. 3. ProMAT Calibrator window.

- *Output directory*: the directory in which the output files will be created.
- *Spot intensity column*: the name of the column in the spot intensity data files listing spot intensities (see Note 3).
- *Spot ID column*: the name of the column in the slide layout file providing the spot assay name.
- *Transform to apply to spot values*: option to log transform the data prior to calibration. See Note 4 for an explanation of why and when you might want to log transform.
  - Natural Log $f(x) = \ln(1 + x)$
  - Identity $f(x) = x$
- *Imager settings column names*: if the experiment information file includes imager settings (e.g. multiple laser or PMT settings) or image replicate IDs, then put those column names here, separated by commas (e.g., Laser, PMT, Image Replicate). Data from different imager settings will be analyzed separately, such that sample data will only be compared to standard data analyzed in the same manner. Also, if the same slides are repeatedly scanned, perhaps using different scanner settings, this function allows all of the data to be processed in a single run.

Click *Run* to start the analysis. ProMAT Calibrator will load the data and then present the user with the assay names so the user can identify the calibrant(s) (Fig. 4). Choose one or morc (use the Ctrl key to select multiple values) then click Okay to perform the data calibration based on the selected assays.

When ProMAT Calibrator is finished, a message will appear asking if the user would like to view the results. If Yes is selected,

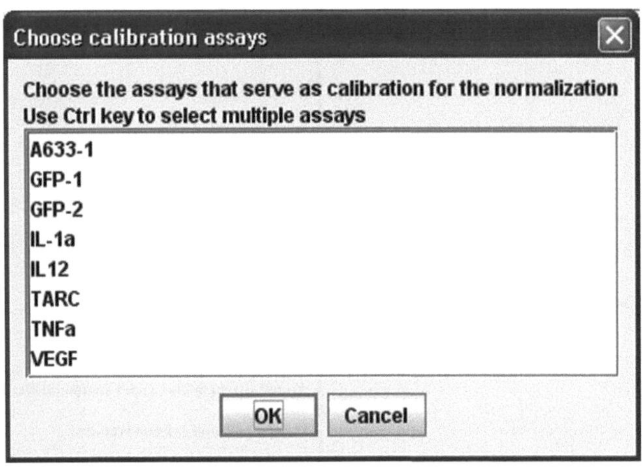

Fig. 4. ProMAT Calibrator select calibrants screen.

a browser window will appear with diagnostic images and tables. Those images and the HTML file are saved in the output directory regardless of which option is chosen so the results may be viewed later.

When ProMAT Calibrator is complete, the output directory will contain a few CSV files that contain data and JPG files that contain diagnostic images.

- *adjusted_values.csv*: this file contains a table of the calibrated values for all input data along with the original spot intensities and the metadata to identify each spot. In this table, the spot intensities are adjusted according to two models: a diagnostic model (*Adjusted.Intensity.Diagnostic.Model*) and a calibration model (*Adjusted.Intensity.Calibration.Model*). The diagnostic model is used to identify systematic patterns in the data, whereas the calibration model adjusts the data for chip-level effects using the selected calibrant spots.
- *variances.csv*: this file contains a table of the within-array and between-array variances (on the log scale, if the option to log transform the data was selected) for each antigen/imager setting pair.
- ProMAT Calibrator also creates new versions of the experiment information and array layout files in the output directory. This allows the user to immediately run ProMAT using the output from ProMAT Calibrator.
- *array_means_*.jpg*: this file provides graphs of the array means for the original values and adjusted values (versus spot print order) so that the effects of the data calibration can be seen for each assay.
- *calibrant_*.jpg*: plots of spot intensity versus print order of the measured values and adjusted values for each of the calibration spots.

3.4. Standard Curves and Concentration Estimates

3.4.1. Overview

After data calibration, standard curves may be estimated and used to predict sample protein concentrations. In order to generate a standard curve for every assay, we incubate a subset of chips with a serial dilution of a mixture of purified antigens of known concentrations. These chips are processed in the same manner as those treated with the biological samples; ideally they are randomized throughout the experiment with the samples, so that there is no systematic difference in the chips used for the samples and those used for the standards in the chip printing, incubation or image analysis. Once a standard curve is calculated from the spot intensities and protein concentrations of the standards, the protein concentrations of the samples can be estimated by referencing their spot intensities to their corresponding standard curve.

Several standard curve models are used to represent the relationship between concentration and spot intensity. In general, we expect the standard data to have a lower bound corresponding to background noise at zero concentration and then to rise monotonically with concentration until reaching an upper bound because of either instrument or assay saturation.

A four-parameter logistic is one of the most commonly used standard curve models because it follows the expected S-shape of the curve with lower and upper bounds. In cases where the upper or lower bound is not observed in the standard data, power or linear models may also be used. We also commonly model standard curves using monotonic splines because of their flexibility and accuracy (5). Spline models are a set of piecewise polynomial curves that are smoothly joined at specified *x*-values called knots.

When multiple curve types are selected, ProMAT fits all the selected models to the data and then uses a PRESS statistic to determine which curve best fits the data (see Note 5). The advantage of this approach is that it provides an objective criterion for choosing one standard curve estimate over another, but it takes longer to compute and there may be benefits to using the same model for all assays or for the same assay on different days of analysis.

ProMAT also calculates a statistical confidence interval for each standard curve estimate. The confidence interval is a useful diagnostic of the quality of the assay. This interval also plays a role in the estimation of corresponding concentration prediction intervals. ProMAT provides two ways to calculate confidence and prediction intervals: analytic bounds and Monte Carlo simulation. See Note 6 for a comparison of the two methods.

When predicting protein concentrations, ProMAT aggregates replicate spot values within a chip and then references the aggregate value to the standard curve. The aggregation reduces the effects of sampling variability, and thus reduces the concentration prediction error. The uncertainty in a concentration prediction has two sources: the uncertainty in the estimated standard curve and uncertainty in the sample intensity measurement. Replicate spot intensity measurements will decrease the effects of both sources.

3.4.2. Using ProMAT

When ProMAT is started, the main screen for data and parameter entry appears (Fig. 5). In this screen, the locations of all data files are specified as well as other parameters that control the analysis. To prepare for analysis in ProMAT, first create the slide layout and experiment information files as described in Subheading 3.2. Alternatively, if ProMAT Calibrator is used first, the output includes the required data, slide layout and experiment information files.

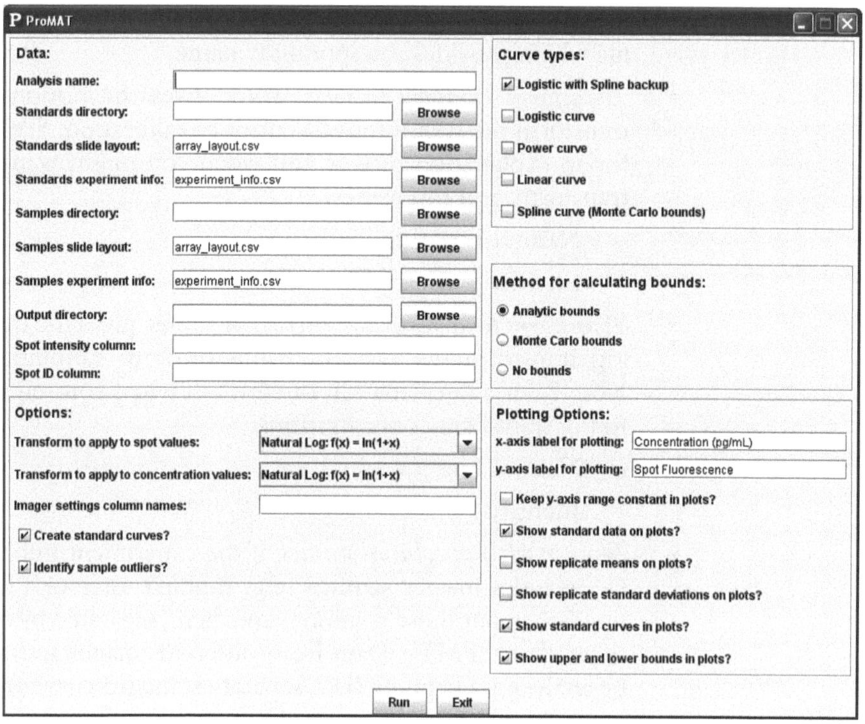

Fig. 5. ProMAT Screen.

3.4.2.1. ProMAT Parameters

- *Analysis name*: (optional) this will be used to name all output files.

Data

- *Standards directory*: directory containing the data for the standard chips.
- *Standards slide layout*: the slide layout file for the standards data described in Subheading 3.2.
- *Standards experiment info*: the experiment information file for the standards data described in Subheading 3.2.
- *Samples directory*: directory containing the data for the sample chips. This is optional and if left blank, ProMAT will generate standard curves and create figures, then exit.
- *Samples slide layout*: the slide layout file for the sample data described in Subheading 3.2. Only used if samples directory is filled in.
- *Samples experiment info*: the experiment information file for the sample data described in Subheading 3.2. Only used if samples directory is filled in.
- *Output directory*: the directory in which the output will be created.
- *Spot intensity column*: the name of the column to use for spot intensity (see Note 3).

- *Spot ID column*: the name of the column in the slide layout file which provides the spot assay name.

Options

- *Transform to apply to spot values*: gives the option to log transform the spot intensities prior to calibration. See Note 4 for an explanation of why and when you might want to log transform spot intensities.
  - Natural Log $f(x) = \ln(1 + x)$
  - Identity $f(x) = x$
- *Transform to apply to concentration values*: gives the option to log transform the concentrations prior to calibration. See Note 7 for an explanation of why and when you might want to log transform concentrations.
  - Natural Log $f(x) = \ln(1 + x)$
  - Identity $f(x) = x$
- *Imager settings column names*: if the experiment information file includes imager settings (e.g. multiple laser or PMT settings) then put those column names here, separated by commas (e.g., Laser, PMT). Data from different imager settings will be analyzed separately (i.e., separate standard curves for each imager setting) and then replicate protein concentrations at different imager settings will be combined in the last step using a weighted average.
- *Create standard curves?* If this box is unchecked, ProMAT will read and plot the data, as usual, but will not create standard curves or estimate protein concentrations. In this case, ProMAT will still extract and organize the raw intensity data from multiple microarray data files, but will not convert these values into antigen concentrations.
- *Identify sample outliers?* If this box is checked, ProMAT will identify outliers in the sample data and exclude them from analysis.

Curve types

- Choose one or more curve models to fit to the data. For each assay, the model that best fits the data will be chosen from those selected.
- *Logistic with Spline backup* (attempts to fit a logistic curve and, in cases where the algorithm does not converge and fitting this curve is not possible, provides a spline model instead)
- *Logistic curve*
- *Power curve*
- *Linear curve*
- *Spline curve*

Method for calculating bounds

Choose one option to determine how confidence and prediction intervals are calculated.

- *Analytic bounds*: these are the fastest to compute but cannot be used with spline curves and have greater uncertainty near curve asymptotes.
- *Monte Carlo bounds*: these take longer to calculate than analytic bounds but are more accurate. See Note 6 for further explanation and comparison of analytic and Monte Carlo bounds.
- *No bounds*: when bounds are not needed or a quick, preliminary analysis of the data is desired, selecting this option will reduce ProMAT run time.

Plotting options

These options affect how the standard curve data are graphed by ProMAT. These graphs are useful for a rapid evaluation of data quality, but these options do not alter the curve fitting or sample concentration estimates in any way.

- *x-axis label for plotting*: the label on the *x*-axis (concentration parameter, such as pg/ml or molarity)
- *y-axis label for plotting*: the label on the *y*-axis (spot intensity)
- *Keep y-axis range constant in plots?* Checking this box means that the *y*-axis will have the same minimum and maximum in all plots – useful for comparing different standard curves to each other.
- *Show standard data on plots?* When this box is checked, the individual data points used to fit the standard curve will be plotted as black points on the plots.
- *Show replicate means on plots?* When this box is checked, mean values for each set of replicate concentration points are shown on the plots as light blue points.
- *Show replicate standard deviations on plots?* When this box is checked, the standard deviation for each set of data points (i.e., for one concentration value) is plotted above and below the mean.
- *Show standard curves on plots?* When this box is checked, the standard curve is plotted as a black line.
- *Show upper and lower bounds in plots?* When this is checked the bounds are shown in blue. No bounds are graphed if the "No bounds" option (see previous section) is selected. In this case, selecting this option has no effect on the data analysis or the graphs.

After selecting the desired options, click *Run* to start the analysis. When ProMAT is finished, a message is displayed showing the directory in which the output is stored (that is, the output directory specified by the user). For a full analysis (that includes standard curves estimation and concentration prediction), two types of diagnostic plots will be created in the output directory. The first plot is a standard curve with confidence bounds on the fitted curves, which is discussed in more detail in the next paragraph.

The second plot type is shown in Fig. 6. The lower right graph shows the standard curve (black line) and the standard data from which the curve was estimated. The blue lines are the upper and lower confidence bounds. On the left side is a histogram of sample spot intensities for this assay. This allows the user to easily determine if the range of sample spot values falls within the usable range of the standard curve (estimated by red dashed lines). The upper graph shows the percent coefficient of variation for the curve (the *x*-axes of the standard curve and coefficient of variation are matched). A single HTML file (created in the output directory) allows the user to easily browse through the second type of

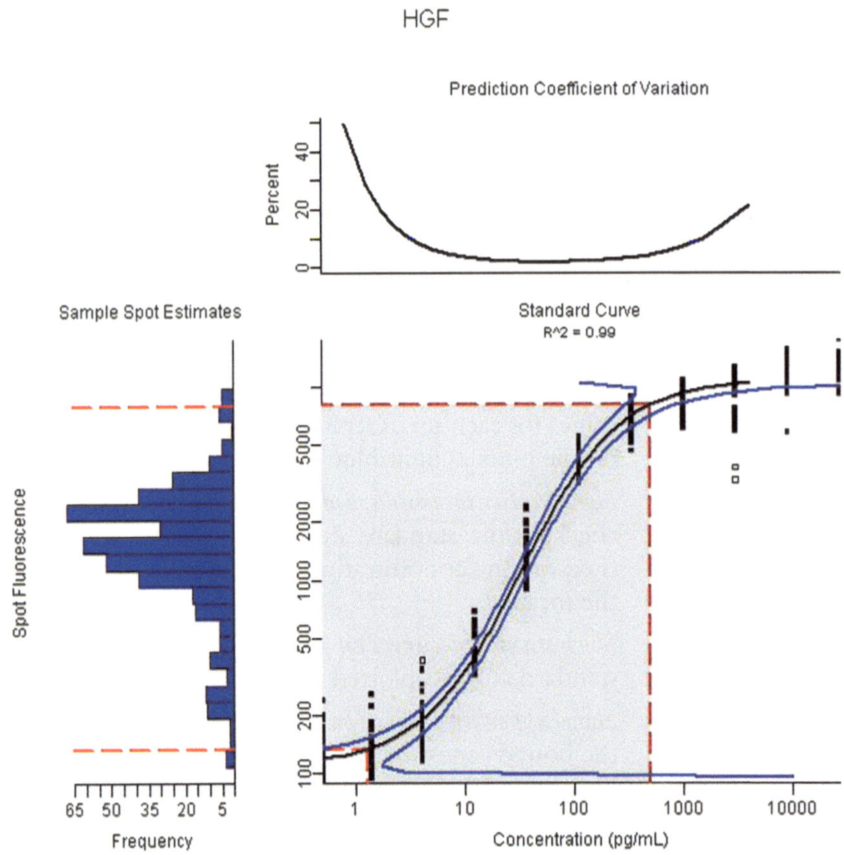

Fig. 6. Diagnostic figure produced by ProMAT. The *lower right panel* shows the estimated standard *curve* (*black line*), prediction bounds (*blue lines*) and standard data (*black points*). The *lower left panel* shows how the spot intensity values align with the standard *curve* and the *upper panel* shows the prediction coefficient of variation.

diagnostic plots for all assays, thus providing a convenient way to review all of the standard curves and corresponding sample intensity values for a single study.

In addition to the diagnostic plots, four data tables are created in the output directory: *predicted_concentrations.csv*, *standard_data.csv*, and *standard_curve_statistics.csv*.

- *predicted_concentrations.csv*: contains predicted concentrations and prediction bounds for each sample observation, where the replicates have been pooled prior to prediction. This file includes the spot and sample IDs and dilution plus the number of spots determined to be outliers which were excluded from the concentration prediction.
- *standard_data.csv*: contains the data used to fit the standard curves, and is provided for user reference. This file includes all the columns found in the array layout and experiment information files (e.g., *sample.id*, *dilution*, *max.concentration*) plus the data file names, spot intensities and the calculated actual concentration which is *max.concentration* dilution*.
- *standard_curve_statistics.csv*: contains information about each standard curve including equation and goodness-of-fit statistics such as R^2 and mean-squared error. (The ProMAT documentation includes a full list of the data in this file.) This file also provides data on the sensitivity (i.e., the lower limit of detection), calculated by two different methods which are described in the ProMAT documentation.
- *spot_intensity_variances.csv*: contains estimates of the variance of spot replicates for both the standards and samples data for each curve. If you choose to log transform your data prior to curve fitting, then these variances are on the log scale.

3.5. Tracking of Metadata and Data Processing Steps

3.5.1. Overview

We have found that ProMAT reduces our data analysis time by about a factor of 10, and provides a variety of quality metrics that go far beyond what would be obtained with standard analyses using spreadsheets. However, it does not provide the capability to organize data and metadata in a way that will facilitate the comparison of results over extended time periods or multiple experiments, which is crucial when conducting studies using thousands of samples that cannot be analyzed in a single day's experiment. This capability is important, because large-scale validation studies with thousands of samples will likely be required to truly define the utility of a biomarker panel for the detecting rare diseases such as cancer (6). Ideally, tracking of these data would be done in a manner similar to the "minimum information about a microarray experiment" (MIAME)-criteria developed for the DNA microarrays (7).

For this purpose, we integrated ProMAT with BioArray Software Environment (BASE) (4) to create the ELISA-BASE (3) system.

BASE is a MIAME-compliant data management tool developed for DNA gene expression microarrays. We have extended BASE for use with ELISA microarrays and incorporated ProMAT as a plugin (a plugin for ProMAT Calibrator is currently being developed).

When using ELISA-BASE, the researcher imports experimental data in a format very similar to that which is used to run ProMAT (i.e., image analysis files plus a file of metadata such as samples, antigen and detection mixtures and dilution information). The ELISA experiment importer creates all database objects needed to organize the data, including adding any sample IDs that do not already exist in the database. There is also an antigen and detection mixture importer, which helps the user to specify the concentrations used to make these mixtures (which, in the case of the antigen mixture, is needed to generate standard curves and to estimate sample protein concentrations).

The ProMAT plugin to ELISA-BASE is very similar to the standalone ProMAT version and produces identical output and diagnostic images. However, the database implementation allows the user to combine and/or filter datasets before analysis, and also tracks the data provenance (e.g. all previous data processing steps as well as the parameters to those steps), which can be useful for future analyses.

3.5.2. Using ELISA-BASE

3.5.2.1. Antigen/Detection Mix Concentration Importer

Information on individual antigens, detection antibodies, and mixtures of these reagents, are stored in the Samples table in ELISA-BASE. Mixtures are stored as pooled "samples" of their constituent parts (e.g., a standard antigen mix references the individual antigens from which it was pooled), thus providing a link to all of the metadata associated with each reagent used in a particular study. Antigen mixes and detection antibody mixes must also provide antigen or antibody concentration data for generating the standard curves. Thus, ELISA-BASE provides the Antigen/Detection Mix Concentration Importer to create these pooled samples with all the necessary concentration data.

Note that each antigen mix or detection mix only needs to be entered into BASE once. Thereafter, it can be used in as many experiments as necessary. This is a useful feature, because we commonly make up these reagents in bulk, aliquot and freeze them, and use the same mix for an extended period of time. Thus, this feature allows us to track changes in individual reagents and mixtures over time. If any changes in standard curves are observed between studies, we are able to determine if these problems are associated with changes in reagents. Importing these data on the reagents needs to be done before importing the experimental data, which is discussed below.

1. Create a CSV file (see Note 1) where the first column is the external ID (in ELISA-BASE) of the antigen or detection

antibody, and the second column is the concentration (do not include units) in the mixture. The first row may contain column names but it is not necessary.

2. In ELISA-BASE, go to *View → Samples* and find your antigen mix or detection mix sample (or click *New* to create a new sample). Click on the sample name to open that sample.
3. Click the *Import* button.
4. Choose the Antigen/Detection Mix Concentration Importer from the drop-down list then click *Next*.
5. Parameters:
   - *Type: Antigen or Detection Antibody Mix*: choose one depending on the type of mixture you are creating.
   - *File containing concentrations*: click Browse..., then click on the Upload file... button and upload the file created in step 1.
   - *Header*: choose true if the file has column names and false otherwise.
6. Click the *Next* button then the *Finish* button.

If the plugin was successful, the sample now shows that it was pooled from multiple samples, corresponding to the IDs provided. Also there will be data in either the *AntigenConcentration* or *DetectionConcentration* annotation.

You may now refer to the antigen or detection mix in an experiment information file when importing a new set of data. The *sample.id* column of the experiment information file should contain the external ID (not the name) of the antigen mix for the appropriate arrays. The *detection* column should contain the external ID of the appropriate detection antibody mix.

3.5.2.2. ELISA Experiment Importer

The next step is to import the experimental data into ELISA-BASE, which is done using the ELISA Experiment Importer plugin. See Note 8 on how to configure the data importer for the data format.

Data Setup

1. Zip the image analysis data files into a single file. This may be done using WinZip or WinRAR on Windows or with the *zip* command on the Windows or Linux command line.
2. Create an experiment information file as described in Subheading 3.2. If using GPR or ScanArray files, you do not need to specify Laser and PMT because these can be read automatically from the data files, by checking the option *Use Laser and PMT values from data file headers* when running the ELISA Experiment Importer.

Running the Plugin in ELISA-BASE

1. Choose the experiment into which you want to import data by clicking on its name in the Experiments list. Then click *Import ELISA Experiment Importer* and then *Next*.

2. The plugin parameters are displayed:
   - *Zip archive of data files*: Select (or upload) the zip archive created in the Data Setup step.
   - *Experiment info file*: Select (or upload) the experiment information file.
   - *Slide number column name*: The name of the column in the experiment information file that gives the slide identifier (e.g., Slide or slide.number).
   - *Use Laser and PMT values from data file headers?* If *true* is selected and the data files are GPR or ScanArray format, then the Laser and PMT values will be read from the data file headers rather than the experiment information file.
   - *Spot intensity value*: Choose a formula to define your spot intensity values.
   - *Label*: Choose the fluorescent label used (e.g. Cy3, Cy5).
   - *Scan name prefix*: Used to prefix the name of all scan items in the database created by the plugin.
   - *Hybridization name prefix*: Used to prefix the name of all hybridization items in the database created by the plugin.
   - *Create samples?* If *true* is selected then new Sample items will be created for samples listed in the experiment information file if those items are not found in the database.
3. Click *Next* and then *Finish* to run the program.

On completion, the experiment will have a Raw Bioassay for each array in the dataset and a new root Bioassay Set with the spot intensity value chosen by the user. The data are organized to be able to run the ProMAT plugin immediately although the user may choose to filter or transform the data prior to further analysis.

3.5.2.3. ProMAT Plugin

The ProMAT Plugin to ELISA-BASE creates standard curves and estimates protein concentrations. The analysis it performs is the same as that in the standalone ProMAT tool (described above), and the parameters are largely the same.

1. From the bioassay set that you wish to analyze with ProMAT, go to the *Run Analysis* tab and select the ProMAT Plugin. Note that the bioassay set chosen must have been created by the ELISA Experiment Importer (or be a child of such a bioassay set).
2. If there is more than one configuration in the drop-down list, choose the one corresponding to your spot intensity data files, and then click *Next*.
3. The next screen allows you to enter the ProMAT parameters, which are the same as those described for the ProMAT

tool in Subheading 3.4.2, with the exception of one added parameter:

- *Standard curve data*: choose one or more antigen mixtures found in this experiment to create the standard curves.

4. Click *Next* and then *Finish* to begin ProMAT analysis.

On completion, a dialog box will give a file location of a zip archive, which contains the ProMAT results. This zip file may be located by selecting *View* → *Files* after closing the dialog. The zip file of ProMAT results contains all of the output files described in Subheading 3.4.2, which have been zipped together for convenient downloading.

4. Notes

1. In Excel, to save a spreadsheet as a CSV file, choose *File* → *Save As* and choose CSV in the *Save As Type* box.
2. We deal with chip-level calibrants in ProMAT Calibrator, although in theory a calibrant may be used at different levels, from the experiment or slide level down to the spot level. A calibrant has two important qualities that must be considered when choosing or developing an assay to perform this function. The calibrant assay must not interfere with the other assays used, and the calibrant and target assays must be similarly affected by the processing factors to be tracked. The experimental process used to choose our chip-level calibrant is documented in (2).
3. Most microarray image analysis tools have multiple ways of calculating spot fluorescence and thus provide multiple measures for each spot. We prefer to use the median of the spot pixels as it is most resistant to bias because of sampling pixels outside the spot.
4. In microarray spot fluorescence data (as in many types of sensor readings) the variance between replicates increases as values increase. This is undesirable when fitting a statistical model to the data, as most models assume the variance is constant. We typically log-transform the spot intensities to make the variances approximately the same at all concentration levels.
5. The PRESS statistic assesses the predictive capabilities of a model by re-fitting the model while leaving out each value once, then predicting the left-out value (8). Thus if there are n data points $\{(x_1, y_1), \ldots, (xn, yn)\}$ then the algorithm has n iterations, where model $M\text{–}i$ is the model produced using all the data except the ith value, and ei is the difference between yi and the predicted value for xi:

$$e_i = y_i - M_{-i}(x_i).$$

Then the PRESS statistic is the sum of the squared error values:

$$\text{PRESS} = \sum_{i=1}^{n} \left(y_i - M_{-i}(x_i)\right)^2.$$

The model with the smallest PRESS statistic has the best fit according to this criterion.

6. ProMAT provides two methods for estimating concentration uncertainties: propagation of error (PE) and Monte Carlo (MC) simulation. PE allows us to derive a closed form equation for the variance of a concentration estimate from the standard curve equation and an estimate of spot replicate variability. PE uncertainties are fast and efficient to calculate. MC uncertainties bounds are calculated by simulating concentration as a function of spot intensity, and include the estimated replicate spot variability in the simulation. We have found MC uncertainty bounds to be much more accurate than PE bounds (5), although they require more computation time. One reason is demonstrated in Fig. 7. Both parts show the same four-parameter logistic standard curve. On the left are PE bounds and on the right MC bounds. As the bounds approach the asymptotes of the logistic curve, the PE bounds diverge because the standard deviation approaches infinity more quickly than the curve does.
7. Using serial dilutions of the standard mixture results in concentration values that are very close together on the lower end of the scale and very far apart on the higher end when

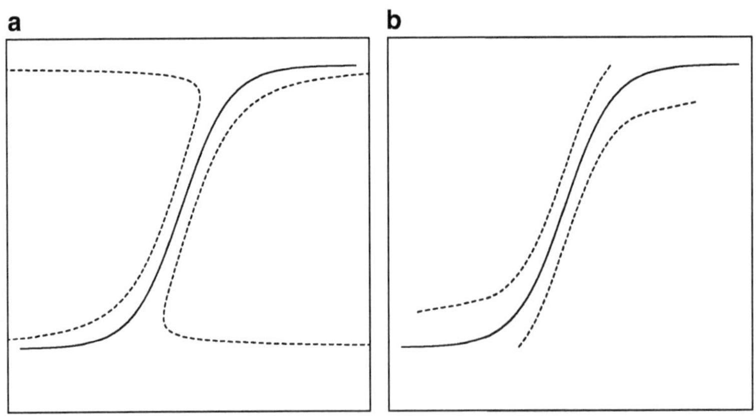

Fig. 7. Four-parameter logistic *curve* with uncertainty bounds (*dotted lines*). The *left graph* shows bounds calculating using propagation of error, the *right graph* shows bounds calculated using Monte Carlo.

using a linear scale. If we were to fit the standard curve to the data as-is, the data at the higher concentrations would have much more weight when calculating the fitted curve than those at the lower concentration levels, which is undesirable. We prefer the data to be evenly spaced so that all dilutions have equal impact on the standard curve that is calculated, and this can be accomplished by applying a log transform to the concentration values.

8. The ELISA Experiment Importer uses BASE's built in Raw Data Importer plugin, which relies on the user to create a configuration for their data file format. Before using the ELISA Experiment Importer for the first time, go to Administrate → Plugins → Plugin definitions and choose *Raw data importer*, then click *New configuration*.... The configuration tool will walk you through specifying the file format and allow you to test the configuration against a data file to verify that it works.

Acknowledgements

This work was funded by the National Institute of Biomedical Imaging & Bioengineering (R01 EB006177) and by the National Cancer Institute (U01 CA117378).

References

1. White A.M., Daly, D.S., Varnum, S.M., Anderson, K.K., Bollinger, N., and Zangar, R.C. (2006) ProMAT: protein microarray analysis tool. *Bioinformatics* **22**, 1278–1279.
2. Zangar R.C., Daly, D.S., White, A.M., Servoss, S.L., Tan, R., and Collett, J.R. (2009) ProMAT Calibrator: a tool for reducing experimental bias in antibody microarrays. *Journal of Proteome Research* **8**, 3937–3943.
3. White A.M., Collett, J.L., Seurynck-Servoss, S.L., Daly, D.S., and Zangar, R.C. (2009) ELISA-BASE: an integrated bioinformatics tool for analyzing and tracking ELISA microarray data. *Bioinformatics* **25**, 1566–1567.
4. Saal, L.H., Troein, C., Vallon-Christersson, J., Gruvberger, S., Borg, Å., and Peterson, C. (2002) BioArray Software Environment (BASE): a platform for comprehensive management and analysis of microarray data. *Genome Biology* **3**, software0003.1–software0003.6.
5. Daly D.S., Anderson, K.K., White, A.M., Gonzalez, R.M., Varnum, S.M., and Zangar, R.C. (2008) Predicting protein concentrations with ELISA microarray assays, monotonic splines and Monte Carlo simulation. *Statistical Applications in Genetics and Molecular Biology* 7, Article 21.
6. Zangar, R., Daly, D., and White, A. (2006) ELISA microarray technology as a high-throughput system for cancer biomarker validation. *Expert Review of Proteomics* **3**, 37–44.
7. Brazma, A., Hingamp, P., Quackenbush, J., Sherlock, G., Spellman, P., Stoeckert, C., Aach, J., Ansorge, W., Ball, C.A., Causton, H.C., Gaasterland, T., Glenisson, P., Holstege, F.C., Kim, I. F., Markowitz, V., Matese, J.C., Parkinson, H., Robinson, A., Sarkans, U., Schulze-Kremer, S., Stewart, J., Taylor, R., Vilo, J., and Vingron, M. (2001) Minimum information about a microarray experiment (MIAME)-toward standards for microarray data. *Nature Genetics* **29**, 365–371.
8. Myers R.H. (1990) *Classical and modern regression with applications.* Boston, MA: PWS-KENT Publishing Company.

Chapter 14

Proteomics Databases and Repositories

Lennart Martens

Abstract

With the advent of more powerful and sensitive analytical techniques and instruments, the field of mass spectrometry based proteomics has seen a considerable increase in the amount of generated data. Correspondingly, the need to make these data publicly available in centralized online databases has also become more pressing. As a result, several such databases have been created, and steps are currently being taken to integrate these different systems under a single worldwide data-sharing umbrella. This chapter will discuss the importance of such databases and the necessary infrastructure that these databases require for efficient operation. Furthermore, the various kinds of information that proteomics databases can store will be described, along with the different types of databases that are available today. Finally, a selection of prominent repositories will be described in more detail, together with the international ProteomExchange consortium that is aimed at uniting all the different databases in a global data sharing collaboration.

Key words: Proteomics, Mass spectrometry, Identifications, Database, Repository, ProteomExchange

1. Introduction

The advent of high-throughput analytical methodologies (1), the development of improved instrumentation (2), and the availability of extensive protein sequence databases (3, 4) have all contributed to the maturation of mass spectrometry based proteomics into a robust platform that can generate vast amounts of data (5). As the capacity to produce data increased by several orders of magnitude, the importance of sharing published data with the larger community became apparent (6–8). In response to this overall requirement, several such databases have been created (9) since. Interestingly, the primary focus of these databases varies somewhat, leading to unique characteristics in each case. This variety, and the current lack of an efficient means of data exchange

Cathy H. Wu and Chuming Chen (eds.), *Bioinformatics for Comparative Proteomics*, Methods in Molecular Biology, vol. 694, DOI 10.1007/978-1-60761-977-2_14, © Springer Science+Business Media, LLC 2011

between the existing databases, can lead to substantial confusion for both data submitters and data users alike.

To provide a comprehensive guide to the interested user (as either submitter or consumer), this review chapter will begin by outlining the underlying need for these databases, as well as the key infrastructure they require. The various kinds of information that proteomics databases can store will be described next, followed by the different types of databases that are available today. Finally, a description of the most prominent databases will be given, concluding with the current plans for the worldwide exchange of data across the different resources in existence today.

2. Why Do We Need Proteomics Databases?

The life sciences have a long-standing tradition in terms of the public availability of data, which is remarkable in light of the closed-access culture that continues to pervade many of the neighbouring fields (e.g. chemistry). One of the first data types to enjoy widespread public release in the life sciences concerned the three-dimensional protein structures deposited in the Protein Data Bank (PDB) (10). Part of the motivation for public access to the structures was provided by the effort and cost involved in obtaining a single structure, thus making public dissemination of any structure very cost effective (11). Over the years, many other data types in the life sciences have followed this example, with the human genome sequence as one of the most highly publicized examples (12). Thanks to this pervasive spirit of data sharing, researches today can freely access genomes (3), RNA microarray data (13), protein sequences and their annotations (4), protein structures (14), protein interactions with proteins, nucleotides and small molecules (15, 16), and even chemical compounds of biological interest (17). For the interested reader, a thorough review of all the publicly accessible resources in the life sciences can be found in a review by Vizcaíno et al. (18).

Interestingly, as outlined in the introduction, the field of mass spectrometry based proteomics owes its existence as a high throughput analytical platform to this free availability of genome sequencing data. Indeed, the identification of peptides and proteins is most commonly based on spectrum matching to known protein or peptide sequences (6). Additionally, although mass spectrometry data can in principle be generated quickly and at a reasonable cost, publicly sharing proteomics data remains interesting from a cost-benefit point of view because of the heterogeneity in approaches (1), instruments (2) and identification algorithms (19) applied across the field, and the largely complementary analysis perspectives obtained by each of the possible combinations (20–22). Finally, current analysis techniques are typically applied on a single tissue or cell-type, resulting in tissue-specific findings that often also yield complementary findings (23). Taken together, the exploration of the proteome of a complex organism can be seen as a compound task, in which a variety of different analyses can resolve a complete picture, whereas an individual analysis is likely to fail to present a comprehensive result. For these reasons, the aggregation of data produced across different groups worldwide provides a cost efficient means to ultimately derive complete (tissue) proteomes.

Yet even for simpler organisms or highly specialized datasets, public availability of data can be very useful. One possible use case is the overall reuse, reanalysis, and validation of data (6), along with data interpretation and algorithm development (8).

Overall, the necessity for public data sharing is perhaps most eloquently justified in the words of Thomas Jefferson, who remarked that "Information, no matter how expensive to create, can be replicated and shared at little or no cost".

3. Key Infrastructure for Proteomics Databases

The first and foremost requirement for any centralized database is the availability of a means for efficient data dissemination. It is no coincidence that the invention of the printing press coincided with the advent of The Enlightenment, when the basis for modern science was laid. Obviously, the internet has provided an extremely powerful and convenient infrastructure for data dissemination, and most of the resources in the life sciences can therefore be found on the web (18).

Once the means of dissemination is thus established, a database must ensure that it can accept incoming data from submitters, and that it can deliver its contents to prospective users in a readable and comprehensive format. In the ideal situation, both submission and dissemination of data occur in the same format, which is standardized across all submitters and consumers. In the case of proteomics databases however, each resource has developed a unique set of formats for this task, resulting in a confusing situation for submitters and consumers. Submitters often find that the data format they use internally in the lab does not correspond to the desired input format for the database they wish to submit to, necessitating data format conversions that may be difficult to perform in the absence of trained informatics staff, or that may result in a loss of information (7). Data consumers on the other hand, encounter data format conversion issues as soon as they try to integrate information from multiple databases, especially when these individual databases do not yet exchange their data holdings automatically. For proteomics data, standards development has been in active progress over the last years, and relevant standard formats have recently emerged for the transfer of data to and from central databases (24). Additionally, the necessity to contact different databases to obtain a comprehensive coverage of the available public data will soon be a thing of the past. The ProteomExchange consortium (see Subheading 7 below) is currently actively working toward automated data exchange between key databases in the field.

Finally, shared data always needs to be placed in context – a set of mass spectra and derived peptide or protein identifications by themselves are largely meaningless. Minimal information on the origins and processing of the data needs to be available along with the actual experimental findings. To this end, it is important that data submitters are aware of, and adhere to, minimal reporting requirements that have been created by the wider community. Such guidelines have already been created in the field of proteomics, initially based on the efforts by individual journals (25), but increasingly complemented by requirements defined by broad community consultation (26).

4. Information Stored in Proteomics Databases

Mass spectrometry based proteomics databases can store a variety of data types. This section describes these various types of information, and briefly discusses the different forms these data may take.

In general, two main types of information are combined in a proteomics experiment: the experimental data recorded by the instrument, and the results inferred from this data. A third data type discussed here concerns the experimental metadata, that captures any and all relevant information about sample, experimental protocol, data processing and interpretation, and author contact details.

4.1. Data

The primary data obtained in mass spectrometry based proteomics consists of mass spectra. In its simplest incarnation, a mass spectrum is a combination of mass-over-charge (m/z) values, and their respective intensities. Additional information can include information about a precursor peak (for fragmentation spectra) and spectrum elution time (for liquid-chromatography based online separations). The mass spectra recorded by the instrument are typically stored in a vendor- or even instrument-specific formatted file, which typically employs some form of binary data encoding and/or compression (7). As a result, these files are not readily accessible to users, as detailed knowledge about the encoding scheme is required to decode the data. Access to these files can be obtained programmatically through so-called vendor libraries, or graphically through the proprietary instrument software. In both cases, the required libraries or software is typically shipped with the instrument for free, but will have to be purchased by prospective data consumers that do not own an instrument; in some cases, a separate purchasing option may not even be available at all. Despite the accessibility problems, these binary formats do present certain advantages: they contain the most complete source of information available after analysis, and they tend to support fast reading.

Although "power users" tend to favour these binary files, most end users will struggle to interact with these files, resulting in the common practise to derive text-based peak lists from the binary files prior to processing and identification (7). Because of spectral processing, these peak lists typically contain far less information than their counterpart spectra in the original binary file. Three steps are frequently applied in this conversion step: denoising, charge deconvolution, and deisotoping (6). Despite the reduction in detail of the data, these peak lists remain the most commonly used data format for submission of mass spectrometer data to search engines. From a data sharing point of view, the peak lists are quite convenient as well, because they tend to be relatively small compared to the full binary data files they originate from (7), and because their text format makes them easily human- and computer-readable. However, the lack of meta-information in these files does greatly limit their actual usefulness as a data transport format.

To find the middle way between the verbose, encoded binary files, and the minimalist and highly accessible peak lists, the recently released mzML standard format (27) combines the ability to store detailed data and metadata, at any level of processing, in an easily accessible and readable format. The base format for mzML is text-based XML, and although the spectra are still encoded inside mzML, internet-standard base-64 encoding is used for this purpose. Base-64 encoding and decoding routines are ubiquitously supported in contemporary programming languages and this encoding therefore does not affect the accessibility of the format.

4.2. Results

Mass spectra are typically used to infer peptide or protein sequences by aligning the experimental spectra to theoretical spectra computed from protein and peptide sequences from sequence databases (6, 28). These peptides or proteins are assigned scores, which may or may not be matched against cut-offs (29), and which can be further manually or automatically validated (6, 30, 31). If peptides are identified from spectra (which is the typical case in modern high-throughput methodologies) an extra step is required, in which the peptides need to be matched to proteins. This so-called protein inference step is in fact one of the most difficult operations in the interpretation of mass spectrometry data (6, 32). Taken together, the identified peptides and the inferred proteins constitute the results output from a typical proteomics experiment. Depending on the search engine used to obtain these data, these results can take a variety of formats, many of which are text based.

In essence, the minimal information obtained is a set of protein accession numbers, each of which is backed by a list of peptide sequences that were identified for that protein. Yet although the peptide sequences are universal, employing the single-letter notation amino acid alphabet, the protein accession numbers are linked to a particular database of origin. Concretely, this means that the same protein (say, human beta actin) has a different accession number in each different database (e.g. P60709 in UniProtKB/Swiss-Prot, ENSP00000349960 in Ensembl, IPI00021439 in IPI, and NP_001092 in RefSeq). This diversity of formal identifiers for a single protein creates a considerable problem for comparative proteomics studies, and databases somehow have to cope with this strategy. As outlined in Subheading 5 below, different types of databases come up with different solutions for this problem, but briefly, the approaches fall apart in two broad categories: (a) use a single protein database for protein identification or inference, thus unifying all accession numbers in the same namespace; and (b) translating all protein accession numbers across all namespaces, to ensure that all known identifiers for the protein are mapped to this protein. The latter task is far from trivial, and as a result several independent systems have been devised and set up for this specific purpose (4, 33, 34). The mzIdentML format has been created as an XML-based data standard to communicate peptide and protein identification data, along with metadata describing the processing of the results.

In addition to the identification of peptides and proteins, proteomics experiments are increasingly geared toward quantifying the observed proteins (35). A large variety of experimental methods and software applications are correspondingly becoming available (36). From the proteomics databases point of view however, quantitative procedures are not yet well established enough to allow consensus data capture. Rather, specific solutions for specific data types have been proposed by independent research groups (37). The incorporation of quantitative data into proteomics databases is currently being actively pursued, and it is highly likely that this type of data will be standardized across the field over the coming years, much like mass spectral and identification data formats have been.

4.3. Metadata

Experimental data and its inferred results by themselves do not provide a lot of useful information to an end user. Indeed, for a researcher interrogating the data, it is important to know the origin of the sample and the experimental context used to obtain the results. If this information is captured as free text, formatted and written according to the specific style and habits of the individual author (as is the case for a Materials and Methods section in a journal article, for instance), this information is both available and interpretable for humans. However, computers struggle quite with such so-called free text, to the point that it is an active field of research to create algorithms that provide incremental improvements in computer "reading" of such text. If it is therefore sufficient to make sample, protocol, instrument, and processing information available, free text annotation will suffice. However, if it is also deemed necessary to have computers to read and process this information, a better system for annotating metadata is required. One important justification for making annotations computer-readable is the ability to quickly and efficiently search the annotations for data of interest, say all data derived from human hepatocytes using electrospray ionization (ESI) instruments. To support such queries, the vocabulary used for annotating the various aspects of the experiment must be fixed. This can be achieved by agreeing on a common vocabulary, but in practise, this turns out to be quite difficult. As an example, consider the various ways in which people might report a "time-of-flight" analyzer: "time of flight", "TOF", "tof", "T.O.F", "t.o.f.", etc. To computers these are all very different things, whereas humans tend to think of them as one and the same thing. The final solution therefore adapts to the way humans think about concepts – a concept (such as "time-of-flight") is assigned a unique name (or accession number) (say "MS: 100041") and this number is equally assigned to all the synonyms. Now, we can have submitters annotate their data as derived from an "MS: 100041" analyzer, which is an unambiguous description that reflects the abstract, concept-based thinking of humans. Whether that accession number is accompanied by the name "TOF" or "t.o.f." is no longer relevant to the computer – it knows what "MS: 100041" is. A commonly used example of this fixed vocabulary is the NCBI taxonomy, where the number "9606" stands for "human" (or "Homo sapiens", or "homo sapiens", or any of the other common synonyms).

A further innovation can be added to such a list of defined and numbered terms: they can be linked up in complex hierarchies. Again, taxonomy provides a ready example; "9606" (human) is linked to the parent genus "Homo" (9605), which is in turn further linked (over several steps) to "Eukaryota" (2759). This linking of concepts results in a controlled vocabulary (CV) or an ontology. The main difference between these two lies in their depth of coverage in a field – an ontology is considered a comprehensive representation of an entire domain, whereas a CV is often much more constricted and specialized.

By using CV or ontology terms to annotate data, we gain several key advantages over free text: (a) the annotation is readily computer readable, and can be queried automatically; (b) the links between concepts allows for very powerful queries, for instance by searching for an exact match for a term, as well as for matches to any of the terms children (e.g. searching for data from "brain" retrieves data from "cerebellum" as well); and (c) because of the links between concepts, the task for a submitter is greatly simplified as well – annotating data as "cerebellum" will implicitly provide the far-reaching context that this is a part of the brain, for instance. This system thus provides clear benefits for data consumers and data submitters alike, a and this is an important reason for the importance of such CVs and ontologies in the abovementioned standard formats for mass spectral data (mzML) and identification results (mzIdentML). Apart from annotating the data and the inferred results, this same approach is used to annotate the overall experimental context by some repositories as well.

5. Different Types of Proteomics Databases

As described in Subheading 6 below, there are various different proteomics databases available today. Interestingly, these databases typically have dissimilar objectives, which are reflected in the accepted and disseminated data in each case. This section provides an overall framework of database types that allows current and future databases to be classified easily. It is worth noting however that certain databases can be classified as more than one of the types outlined below.

5.1. Research-Oriented Databases

These databases are primarily built to serve as a storehouse for data dedicated to a particular research goal. A typical example is research into peptide fragmentation characteristics (38), or into peptide detection biases (39). The main purpose of the database is thus to provide a starting point for further work, which is often reflected in the data stored. Research databases aimed at understanding peptide fragmentation might for instance forgo detailed sample annotation or protein identifications, as the focus lies specifically on the mass spectra and their inferred peptides. Similarly, a database aimed at peptide detection biases might not store mass spectra (which typically contribute the bulk of the data volume) but rather only peptide identifications and inferred proteins.

5.2. Pipeline-Oriented Databases

As outlined above, the identification of proteins from mass spectra is a relatively complex process, involving several steps of data processing and inference. Software pipelines have therefore been set up to automate such analyses, and developers can relatively easily append a database at the end of their pipeline. The main benefit of this approach is the uniform origin of the data, but at the same time, this can also be perceived as the greatest drawback of such databases. This because processing and identification algorithms often introduce biases in the output (19–21), which can either confound downstream analysis, or can lead to over fitting.

5.3. Annotation-Oriented Databases

These databases capture proteomics results to refine or extend the annotation of existing sequence databases. Typical examples include the annotation of biologically relevant post-translational modifications (PTMs), the verification and correction of predicted open reading frames (ORFs), and the delineation of splice sites. As mentioned, these databases tend to focus on the results rather than on the mass spectral data, unless they implicitly employ quality control criteria that involve careful evaluation of the match between a peptide sequence and its mass spectrum of origin.

5.4. Transmission-Oriented Databases

This category of highly specialized databases is aimed specifically at solving the data dissemination issue that is particularly acute for the raw, binary output files of the mass spectrometer. A single liquid chromatography (LC) run on a modern electrospray mass spectrometer for instance, can generate a file that is several hundred gigabytes in size. Communicating the results of a single study (often comprising tens of such runs) to the scientific community thus quickly becomes challenging. Transmission-oriented databases therefore employ sophisticated internet-oriented data sharing architectures that involve distributed protocols such as the well-known peer-to-peer architecture. These databases often resemble file systems more than they do databases, and the information contained in them tends to be highly heterogeneous. This makes the data difficult to query, especially when searching for details (e.g. all data pertaining to a particular protein). Data is therefore typically retrieved from these repositories on the level of an entire study.

5.5. Data Repositories

Data repositories are the entities that most people (somewhat erroneously) associate with the term "database". They exist as central databases that verbatim capture data that is submitted to them. Aimed at simply collating all publicly available information, they often lack clear research or annotation goals or associated pipelines, but because of their generic structure interested data consumers can easily use them for these purposes. Data in repositories tends to be stored in a highly structured and well-annotated way, allowing detailed queries to be run with ease. Most of the data submitted to repositories will be related to published papers, as direct requests from, or requirements by, scientific journals in the field are the primary motivators for authors to submit their data (40–42).

6. Prominent Proteomics Databases

A number of proteomics databases and repositories have sprung up over the past few years, but the most prominent efforts are listed specifically here, along with a brief description. The interested reader can find additional information in dedicated review articles, specifically aimed at enumerating and positioning a comprehensive list of databases (9, 43).

6.1. Global Proteome Machine Database

The first mass spectrometry based proteomics database to be published, the Global Proteome Machine Database (GPMD) (44) (http://gpmdb.thegpm.org) is a typical example of a pipeline-oriented database. Originally intended as an add-on data store for the online version of the open source X!Tandem search engine (45), the GPMDB has accumulated a wealth of data since its inception. As a result, its classification can now be made as a research database, as well as an annotation-oriented database. Indeed, the data accumulated in GPMDB is used to create spectral libraries, and inspire the development of a spectral library-searching tool, whereas the identified peptides are matched to the Ensembl genome database. GPMDB contains data from a variety of organisms, but is somewhat biased toward ion trap data on the instrument front.

6.2. PeptideAtlas

The PeptideAtlas project (http://www.peptideatlas.org), originally published in 2005 (46) was originally set up as an annotation-oriented database, although it simultaneously served as the endpoint for the Trans-Proteomic Pipeline (TPP) processing pipeline. Similar to GPMDB, the remit of PeptideAtlas has grown over time, and it has served as a research database for the development of spectral libraries and corresponding search tools as well (47). Furthermore, PeptideAtlas also now provides information about the detectability of peptides (48). One noteworthy aspect of PeptideAtlas is the occurrence of several "builds" of the system; a new such PeptideAtlas build is typically created on the basis of an individual organism (49, 50), although tissue-specific (51) Atlas builds have been created as well.

6.3. Proteomics Identifications Database

The Proteomics Identifications Database (PRIDE; http://www.ebi.ac.uk/pride), developed at the European Bioinformatics Institute (EBI), was the first fully structured proteomics data repository (52). The main purpose of PRIDE is to capture and disseminate proteomics data from across the field, without biases or editorial control in data origins, data processing approaches, and annotation. Having undergone dramatic growth, PRIDE is now one of the largest and certainly the most varied collection of proteomics data in the field. Based entirely on direct data submissions, PRIDE relies heavily on freely available and platform-independent, user friendly data submission tool called PRIDE Converter (53). PRIDE contains details ranging from mass spectra to identified proteins, as well as extensive metadata at all levels of detail. A BioMart for data retrieval is available as well (54).

6.4. Human Proteinpedia

The Human Proteinpedia project (http://www.humanproteinpedia.org) falls neatly in the annotation-oriented database category (55). Intended to complement the curated Human Protein Reference Database (HPRD) (56) with community-provided annotation, this database collects submitted proteomics data and uses the findings to directly annotate the protein entries in HPRD with observed post-translational modifications and (subcellular) localization. In contrast to the previously described resources, Proteinpedia is specifically dedicated to data derived from human samples.

6.5. NCBI Peptidome

A recent addition to the field, the NCBI's Peptidome database (57) (http://www.ncbi.nlm.nih.gov/peptidome) can be considered as a sibling of the abovementioned PRIDE database. A true repository, Peptidome accepts submitted data without assuming editorial control and stores and disseminates this data in a well-defined and structured way. Similar to the PRIDE database, this generic repository can be used for a large variety of downstream activities, including research and annotation. The current volume of data at Peptidome remains somewhat restricted, but this is certain to change substantially over time. It is worth noting that the PRIDE and Peptidome teams have already committed themselves to a full exchange of publicly available data, which will ensure that data submitted to one repository can be found in the other as well. This collaboration forms the keystone of the ProteomExchange collaboration outlined in Subheading 7.

6.6. Tranche

The Tranche databases (58) (https://proteomecommons.org/tranche) occupies a special place in this list, as it is the only example of a transmission-oriented database. Built around a sophisticated peer-to-peer protocol and a robust distributed server architecture, the Tranche project currently provides the best available means to share the large data volumes associated with raw, binary data files. It is used in this capacity by databases such as PRIDE and PeptideAtlas that deposit submitted raw data into Tranche and link out to the files. Recent developments in Tranche have allowed the annotation of a data set with metadata, although the actual data files can take many forms and in general do not support detailed searching.

7. Proteom Exchange and the Future for Proteomics Databases

The existence of a variety of proteomics databases has led to a fragmentation of the available data, and causes unnecessary confusion for eager submitters willing to share their data. The data fragmentation further complicates matters for data consumers, as they have to browse several databases, each fitted with a custom interface and output format, to collate a full picture of the publicly available data for their topic of interest. To alleviate this situation, the ProteomExchange consortium was set up (59) to guide the data flow through the databases in a clear and comprehensive way. Currently in the final phases of development, ProteomExchange will rely on the PRIDE and Peptidome repositories as entry points for submitted data, with automatic replication for public between these databases. All submitted data will also be automatically sent to Tranche, and will be picked up form there by core members such as PeptideAtlas. A common ProteomExchange accession number will ensure that the data can be found easily, regardless of the database queried. ProteomExchange also comes with an automated notification system that will send structured messages via mail and RSS to announce the availability of new datasets to any interested party in the general public. As a result, the submission and consumption of proteomics data will become a much simpler and straightforward procedure, while allowing the consumer to choose the particular view on the data offered by the database of preference.

Once operational, ProteomExchange will provide a data dissemination infrastructure for proteomics similar to the International Nucleotide Sequence Database Collaboration (INSDC) that has turned genome-sequencing efforts into a stable and rich foundation for the growth of proteomics. It is to be hoped that the efforts now undertaken at the proteomics level will in turn enable and nurture future research fields to contribute to our understanding of life.

Acknowledgements

The author would like to thank Henning Hermjakob and Rolf Apweiler for their support.

References

1. Gevaert K., Van Damme P., Ghesquière B., Impens F., Martens L., Helsens K. et al. (2007) A la carte proteomics with an emphasis on gel-free techniques. *Proteomics* **7**, 2698–2718.
2. Domon B. and Aebersold R. (2006) Mass spectrometry and protein analysis. *Science* **312**, 212–217.
3. Hubbard T., Aken B., Ayling S., Ballester B., Beal K., Bragin E. et al. (2009) Ensembl 2009. *Nucleic Acids Res* **37**, D690–D607.
4. The UniProt Consortium (2009) The Universal Protein Resource (UniProt) 2009. *Nucleic Acids Res* **37**, D169–D174.
5. Aebersold R. and Mann M. (2003) Mass spectrometry-based proteomics. *Nature* **422**, 198–207.
6. Martens L. and Hermjakob H. (2007) Proteomics data validation: why all must provide data. *Mol Biosyst* **3**, 518–522.
7. Martens L., Nesvizhskii A.I., Hermjakob H., Adamski M., Omenn G.S., Vandekerckhove J. et al. (2005) Do we want our data raw? Including binary mass spectrometry data in public proteomics data repositories. *Proteomics* **5**, 3501–3505.
8. Prince J.T., Carlson M.W., Wang R., Lu P. and Marcotte E.M. (2004) The need for a public proteomics repository. *Nat Biotechnol* **22**, 471–472.
9. Mead J., Bianco L. and Bessant C. (2009) Recent developments in public proteomic MS repositories and pipelines. *Proteomics* **9**, 861–881.
10. Bernstein F.C., Koetzle T.F., Williams G.J., Meyer E.F.J., Brice M.D., Rodgers J.R. et al. (1977) The Protein Data Bank: a computer-based archival file for macromolecular structures. *J Mol Biol* **112**, 535–542.
11. Berman H. (2008) The Protein Data Bank: a historical perspective. *Acta Crystallogr* **64**, 88–95.
12. Lander E.S., Linton L.M., Birren B., Nusbaum C., Zody M.C., Baldwin J. et al. (2001) Initial sequencing and analysis of the human genome. *Nature* **409**, 860–921.
13. Parkinson H., Kapushesky M., Shojatalab M., Abeygunawardena N., Coulson R., Farne A. et al. (2007) ArrayExpress – a public database of microarray experiments and gene expression profiles. *Nucleic Acids Res* **35**, D747–D750.
14. Berman H., Henrick K., Nakamura H. and Markley J. (2007) The worldwide Protein Data Bank (wwPDB): ensuring a single, uniform archive of PDB data. *Nucleic Acids Res* **35**, D301–D303.
15. Chatr-aryamontri A., Ceol A., Palazzi L., Nardelli G., Schneider M., Castagnoli L. et al. (2007) MINT: the Molecular INTeraction database. *Nucleic Acids Res* **35**, D572–D574.
16. Kerrien S., Alam-Faruque Y., Aranda B., Bancarz I., Bridge A., Derow C. et al. (2007) IntAct – open source resource for molecular interaction data. *Nucleic Acids Res* **35**, D561–D565.
17. Degtyarenko K., de Matos P., Ennis M., Hastings J., Zbinden M., McNaught A. et al. (2008) ChEBI: a database and ontology for chemical entities of biological interest. *Nucleic Acids Res* **36**, D344–D350.
18. Vizcaíno J., Mueller M., Hermjakob H. and Martens L. (2009) Charting online OMICS resources: a navigational chart for clinical researchers. *Proteomics Clin Appl* **3**, 18–29.
19. Kapp E.A., Schütz F., Connolly L.M., Chakel J.A., Meza J.E., Miller C.A. et al. (2005) An evaluation, comparison, and accurate benchmarking of several publicly available MS/MS search algorithms: sensitivity and specificity analysis. *Proteomics* **5**, 3475–3490.
20. Reidegeld K.A., Muller M., Stephan C., Bluggel M., Hamacher M., Martens L. et al. (2006) The power of cooperative investigation: summary and comparison of the HUPO Brain Proteome Project pilot study results. *Proteomics* **6**, 4997–5014.
21. Klie S., Martens L., Vizcaíno J.A., Côté R., Jones P., Apweiler R. et al. (2008) Analyzing large-scale proteomics projects with latent semantic indexing. *J Proteome Res* **7**, 182–191.
22. Mueller M., Vizcaíno J.A., Jones P., Côté R., Thorneycroft D., Apweiler R. et al. (2008) Analysis of the experimental detection of central nervous system related genes in human brain and cerebrospinal fluid datasets. *Proteomics* **8**, 1138–1148.
23. Martens L., Muller M., Stephan C., Hamacher M., Reidegeld K.A., Meyer H.E. et al. (2006) A comparison of the HUPO Brain Proteome Project pilot with other proteomics studies. *Proteomics* **6**, 5076–5086.
24. Martens L., Orchard S., Apweiler R. and Hermjakob H. (2007) Human Proteome Organization Proteomics Standards Initiative: data standardization, a view on developments and policy. *Mol Cell Proteomics* **6**, 1666–1667.
25. Carr S., Aebersold R., Baldwin M., Burlingame A., Clauser K. and Nesvizhskii A. (2004) The

need for guidelines in publication of peptide and protein identification data: Working Group on Publication Guidelines for Peptide and Protein Identification Data. *Mol Cell Proteomics* **3**, 531–533.

26. Taylor C.F., Binz P., Aebersold R., Affolter M., Barkovich R., Deutsch E.W. et al. (2008) Guidelines for reporting the use of mass spectrometry in proteomics. *Nat Biotechnol* **26**, 860–861.
27. Deutsch E. (2008) mzML: a single, unifying data format for mass spectrometer output. *Proteomics* **8**, 2776–2777.
28. Sadygov R.G., Cociorva D. and Yates J.R. (2004) Large-scale database searching using tandem mass spectra: looking up the answer in the back of the book. *Nat Methods* **1**, 195–202.
29. Nesvizhskii A.I., Vitek O. and Aebersold R. (2007) Analysis and validation of proteomic data generated by tandem mass spectrometry. *Nat Methods* **4**, 787–797.
30. Keller A., Nesvizhskii A.I., Kolker E. and Aebersold R. (2002) Empirical statistical model to estimate the accuracy of peptide identifications made by MS/MS and database search. *Anal Chem* **74**, 5383–5392.
31. Helsens K., Timmerman E., Vandekerckhove J., Gevaert K. and Martens L. (2008) Peptizer: A tool for assessing false positive peptide identifications and manually validating selected results. *Mol Cell Proteomics* **7**, 2364–2372.
32. Nesvizhskii A.I. and Aebersold R. (2005) Interpretation of shotgun proteomic data: the protein inference problem. *Mol Cell Proteomics* **4**, 1419–1440.
33. Babnigg G. and Giometti C.S. (2006) A database of unique protein sequence identifiers for proteome studies. *Proteomics* **6**, 4514–4522.
34. Côté R.G., Jones P., Martens L., Kerrien S., Reisinger F., Lin Q. et al. (2007) The Protein Identifier Cross-Reference (PICR) service: reconciling protein identifiers across multiple source databases. *BMC Bioinformatics* **8**, 401.
35. Panchaud A., Affolter M., Moreillon P. and Kussmann M. (2008) Experimental and computational approaches to quantitative proteomics: status quo and outlook. *J Proteomics* **71**, 19–33.
36. Mueller L.N., Brusniak M., Mani D.R. and Aebersold R. (2008) An assessment of software solutions for the analysis of mass spectrometry based quantitative proteomics data. *J Proteome Res* **7**, 51–61.
37. Siepen J.A., Swainston N., Jones A.R., Hart S.R., Hermjakob H., Jones P. et al. (2007) An informatic pipeline for the data capture and submission of quantitative proteomic data using iTRAQTM. *Proteome Sci* **5**, 4.
38. Klammer A.A., Reynolds S.M., Bilmes J.A., MacCoss M.J. and Noble W.S. (2008) Modeling peptide fragmentation with dynamic Bayesian networks for peptide identification. *Bioinformatics* **24**, i348–i356.
39. Mallick P., Schirle M., Chen S.S., Flory M.R., Lee H., Martin D. et al. (2007) Computational prediction of proteotypic peptides for quantitative proteomics. *Nat Biotechnol* **25**, 125–131.
40. Anonymous. (2008) Thou shalt share your data. *Nat Methods* **5**, 209–209.
41. Anonymous. (2007) Democratizing proteomics data. *Nat Biotechnol* **25**, 262.
42. Anonymous. (2007) Compete, collaborate, compel. *Nat Genet* **39**, 931.
43. Mead J.A., Shadforth I.P. and Bessant C. (2007) Public proteomic MS repositories and pipelines: available tools and biological applications. *Proteomics* **7**, 2769–2786.
44. Craig R., Cortens J. and Beavis R. (2004) Open source system for analyzing, validating, and storing protein identification data. *J Proteome Res* **3**, 1234–1242.
45. Craig R. and Beavis R.C. (2004) TANDEM: matching proteins with tandem mass spectra. *Bioinformatics* **20**, 1466–1467.
46. Desiere F., Deutsch E.W., Nesvizhskii A.I., Mallick P., King N.L., Eng J.K. et al. (2005) Integration with the human genome of peptide sequences obtained by high-throughput mass spectrometry. *Genome Biol* **6**, R9.
47. Lam H., Deutsch E.W., Eddes J.S., Eng J.K., King N., Stein S.E. et al. (2007) Development and validation of a spectral library searching method for peptide identification from MS/MS. *Proteomics* **7**, 655–667.
48. Deutsch E., Lam H. and Aebersold R. (2008) PeptideAtlas: a resource for target selection for emerging targeted proteomics workflows. *EMBO Rep* **9**, 429–434.
49. Van P.T., Schmid A.K., King N.L., Kaur A., Pan M., Whitehead K. et al. (2008) *Halobacterium salinarum* NRC-1 PeptideAtlas: toward strategies for targeted proteomics and improved proteome coverage. *J Proteome Res* **7**, 3755–3764.
50. Loevenich S.N., Brunner E., King N.L., Deutsch E.W., Stein S.E. et al. (2009) The Drosophila melanogaster PeptideAtlas facilitates the use of peptide data for improved fly proteomics and genome annotation. *BMC Bioinformatics* **10**, 59.
51. Deutsch E.W., Eng J.K., Zhang H., King N.L., Nesvizhskii A.I., Lin B. et al. (2005)

Human Plasma PeptideAtlas. *Proteomics* **5**, 3497–3500.

52. Martens L., Hermjakob H., Jones P., Adamski M., Taylor C., States D. et al. (2005) PRIDE: the proteomics identifications database. *Proteomics* **5**, 3537–3545.
53. Barsnes H., Vizcaíno J.A., Eidhammer I. and Martens L. (2009) PRIDE Converter: making proteomics data-sharing easy. *Nat Biotechnol* **27**, 598–599.
54. Jones P., Cote R., Cho S., Klie S., Martens L., Quinn A. et al. (2008) PRIDE: new developments and new datasets. *Nucleic Acids Res* **36**, D878–D883.
55. Mathivanan S., Ahmed M., Ahn N.G., Alexandre H., Amanchy R., Andrews P.C. et al. (2008) Human Proteinpedia enables sharing of human protein data. *Nat Biotechnol* **26**, 164–167.
56. Mishra G.R., Suresh M., Kumaran K., Kannabiran N., Suresh S., Bala P. et al. (2006) Human protein reference database – 2006 update. *Nucleic Acids Res* **34**, D411–D414.
57. Slotta D.J., Barrett T. and Edgar R. (2009) NCBI Peptidome: a new public repository for mass spectrometry peptide identifications. *Nat Biotechnol* **27**, 600–601.
58. Falkner J.A., Hill J.A. and Andrews P.C. (2008) Proteomics FASTA archive and reference resource. *Proteomics* **8**, 1756–1757.
59. Hermjakob H. and Apweiler R. (2006) The Proteomics Identifications Database (PRIDE) and the ProteomExchange Consortium: making proteomics data accessible. *Expert Rev Proteomics* **3**, 1–3.

Chapter 15

Preparing Molecular Interaction Data for Publication

Sandra Orchard and Henning Hermjakob

Abstract

It is now becoming more usual for journals to request the submission of the data accompanying an article to an appropriate public repository. Such users may access the data in a format appropriate for display and reanalysis. It is commonly accepted that molecular interaction databases will hold all the large-scale interaction datasets and enrich this with lower throughput data. Previously this small-scale interaction data has been archively curated from the literature but, increasingly, deposition of such information is also being seen as an integral part of the publication process. This chapter acts as a brief guide to preparing both large- and small-scale data for publication and gives a range of different submission options.

Key words: Human proteome organisation, Proteomics standards initiative, International molecular exchange consortium, Data standardization

1. Introduction

The systematic mapping of molecular interaction data, in particular protein-protein interactions, has proven a rich source of information for many groups, providing valuable insights into the understanding played by a protein in processes, pathways, and in particular cell types. Protein-protein interaction databases play a necessary role in capturing, collating, and redistributing the wealth of interaction data that is published in the literature each year. The scope of these databases has increased over time, with the degree of annotation captured becoming richer and some repositories, such as the IntAct molecular interaction database (1), are beginning to capture all possible interactions within a cell or organism, including those made by protein-small molecules and protein-nucleic acids. To increase both data coverage and data quality in these resources, it is essential that direct deposition

Cathy H. Wu and Chuming Chen (eds.), *Bioinformatics for Comparative Proteomics*, Methods in Molecular Biology, vol. 694, DOI 10.1007/978-1-60761-977-2_15, © Springer Science+Business Media, LLC 2011

by data producers as part of the publication process is encouraged and made as simple a process as possible.

1.1. HUPO and the Proteomics Standards Initiative

Before 2004 the user was served by an ever-increasing number of data resources, all of which collected interaction information from the literature, and in some cases, from direct submissions from research workers but an individual wishing to download the information from a number of sources was faced with parsing multiple, separately constructed databases, each with its own individual structure and data format. Merging the data into a single repository then required further effort, as did identifying those papers that had been redundantly curated by more than one database. It was at this point that several of these resources were brought together by the Human Proteome Organisation to tackle these problems and provide an improved service for the user.

The Human Proteome Organisation (HUPO) was formed in 2001 to consolidate national and regional proteome organizations into a single worldwide body (2). The Proteome Standards Initiative (PSI) was established by HUPO with the remit of standardizing data representation within the field of proteomics to the end that public domain databases can be established where all such data can be deposited, exchanged between such databases, or downloaded and used by laboratory workers (3). Each workgroup within the HUPO-PSI has produced a series of documents and resources to aid in the process of data standardization and exchange.

In 2004, the Molecular Interaction workgroup published Level 1.0 of the PSI-MI XML interchange schema, with accompanying controlled vocabularies, jointly developed by both major producers of protein interaction data and by a number of database resources (4). Version 1.0 of the format focused exclusively on protein interactions, and was widely implemented and supported by both software tool development and data providers. As a direct result of requests from users, database groups and data providers the original PSI-MI format was considerably extended to increase the interactor types that could be described within the format to encompass all biomolecules. The description of both experimental conditions and experimental features on participating molecules, such as the description of purification tags or deletion or point mutations, was considerably enhanced and made more flexible. The abilities to describe kinetic as well as modeled interaction parameters were also added. PSI-MI XML2.5 was published in 2007 and has remained stable since then (5).

The PSI-MI XML2.5 format allows a detailed representation of fully annotated interaction records both for interdatabase and database-end user data communication. However, to support many use cases, such as fast Perl parsing or loading into Microsoft Excel, that only require a simple, tabular format of interaction records, the MITAB2.5 format was defined as part of PSI-MI 2.5. The MITAB2.5 format only describes binary interactions, one pair of interactors per row in a simple tab-delimitated format.

Controlled vocabularies (CVs) are used throughout the PSI-MI schema to standardize the meaning of data objects. Their use ensures that the same term used throughout a description by a data producer, instead of a synonym or alternative spelling, and also that the interpretation of the meaning of that term remains consistent between multiple data producers and users. To achieve this, all terms have definitions and, where appropriate, are supported by one or more literature references. The controlled vocabularies have a hierarchical structure, in the form of a direct acyclic graph (DAG), higher-level terms being more general than lower-level descriptors, allowing annotation to be performed to an appropriate level of granularity while also enabling search tools to return all mapped objects to both parent and child terms, if required. The Molecular Interaction CV (MI) (4, 5) may be accessed either on the Open Biomedical Ontologies website (http://www.obofoundry.org) or via the Ontology-Lookup Service at the European Bioinformatics Institute (http://www.ebi.ac.uk/ols). It is produced and maintained by the HUPO-PSI and is used for the annotation of molecular interaction data.

Almost all major interaction data producers now make data available in PSI-XML2.5 format and many also in MITAB2.5. Any data submitted to an IMEx database, will be available in both formats and can then be accessed by the increasing number of visualization and analysis tools using these formats.

1.2. The Minimum Information about a Molecular Interaction Experiment

The minimum information about a molecular interaction experiment (MIMIx) guidelines provide a checklist for anyone preparing interaction data, be it as little as a single interaction within a paper describing the characterisation of a protein, for either publication in a peer-reviewed article, deposition in an interaction database or displaying a large dataset on a website (6). MIMIx represents a compromise between the depth of information necessary to describe all relevant aspects of an interaction experiment and the reporting burden placed on scientists who generate the data. It is of increasing importance that databases, and the datasets they contain, are maintained to, at least, MIMIx compatibility, as increasingly tools and services are being written on the assumption that this minimum level of information be supplied. For example, the R statistics package (7) is compromised if databases have not included information on interaction directionality (e.g., bait–prey relationships), which is a MIMIx requirement. Any submission to an interaction database should be prepared to MIMIx specifications.

1.3. The International Molecular Exchange Consortium

The International Molecular Exchange Consortium (IMEx) currently consists of four full members (IntAct (1), MINT (8), DIP (9), MatrixDB (10), and MPIDB (11)) with a number of other databases either maintaining observer status or applying to join. All data submitted to any one of these databases will be shared with the consortium members, such that users need only access one site to find all the requisite information fulfilling their query.

1.4. Data Publication

Increasingly journals are looking to encourage the deposition of datasets in the public domain, where it is possible for network biologists to access and download in an appropriate format. IMEx member databases handle such submissions, providing accession numbers to be cited in the accompanying article, which will allow access to the data in all participating resources.

2. Methods

2.1. Choice of Database

All IMEx member databases will accept protein-protein interaction data, and some, such as IntAct will accept all forms of interaction data and MatrixDB also protein-small molecule information. While data resources such as IntAct and DIP will accept all data, other data resources have areas of specialist, for example MPIDB is a microbial protein interaction database and MatrixDB gathers extracellular matrix interaction data. Potential submitters may access a full list of databases and their submission details on the IMEx website (www.imexconsortium.org) or go directly to the individual resource.

2.2. Deposition of Small-Scale Data

Small-scale datasets are of enormous value to interaction databases, as the interactions they contain tend to be confirmed by multiple methodologies and often provide details missing in larger datasets such as binding sites or kinetic data. The simplest method of depositing interaction data is to provide a copy of the manuscript, either before or during the journal review process, to the database where the interaction can be loaded by an experienced curator. The information will be held in confidence until the publication of article when it will be released into the public domain. Before submission, however, authors are encouraged to read the MIMIx guidelines and ensure that the data in the paper reaches these minimum requirements. In particular, authors should ensure that all molecules participating in interactions are fully and unambiguously described, for example, should cDNAs be transcribed into host cell system it should be made very clear the organism from which it originated. Use of an accession number from a public domain database will give both source organism and sequence length (see Note 1).

2.3. Deposition of Medium-Scale Data

Medium-scale datasets, generally consisting to tens to several hundreds of interactions (be these two-hybrid bait-prey pairs, or several one-bait multiple-prey combinations) are routinely described in a single publication and would constitute a single deposition. Most biologists prefer to store such data in an Excel™ spreadsheet and this is an appropriate method for submission to an interaction database. Again, submitters should refer to the MIMIx guidelines and ensure that unambiguous interactor identifiers should be used, such as public database accession numbers (see Note 1). If the author has used their own internal identifier system this may be included on the spreadsheet, and will be retained in the entry, but the sequence should also be mapped to an external database resource. Details of expression constructs should also be given, such as tags or mutations, which may well be described most simply in free text or diagram in an accompanying document such as the paper itself (see Note 2). In all cases, it is recommended that the submission be accompanied by a copy of the manuscript so that any small-scale experiments can also be included in a deposition.

Alternatively, the submitter may wish to make use of a preformatted Excel sheet provided by the IMEx consortium, which is available from http://imex.sourceforge.net/MIMIx/index.html. The workbook should be opened using Microsoft Excel 2000 or later on a computer running a Windows operating system. Unfortunately, the workbook will not work correctly on a non-Windows operating system or on an Office clone such as OpenOffice.org. This is because of the extensive use of Visual Basic for Applications (VBA) code in the spreadsheet that relies heavily on code libraries that are only available under Windows.

The first time the workbook is opened, a Security Warning dialogue box will be shown. The workbook includes macros written in VBA. These have been signed using a Digital Signature signed by Thawte (http://www.thawte.com/digital-certificates) on behalf of the EBI. The purpose of this certificate is to guarantee that the code embedded in the workbook has not been modified outside the EBI and is virus free. The exact appearance and wording of these dialogues may differ depending on the specific Microsoft operating system and the version of Microsoft Excel that is being used. Once the user clicks on the Details link on the Security Warning dialogue box a new dialogue will open, indicating that the digital signature originates from the EBI. Once the certificate has been viewed, and the digital signature has been issued by Thawte Code Signing CA and has been issued to the European Bioinformatics Institute the user can then proceed. If the certificated is accepted permanently by checking the "Always trust macros from this publisher" box and clicking on the Enable macros button, this dialogue box does not reappear.

Each page in the workbook should then be filled in. The filling of the Database, Experimental role, Biological role, Interaction detection, and Participant detection fields are facilitated by drop-down menus, which contain suggestions for the most appropriate terms (see Note 3). These fields will update, if necessary, whenever the workbook accesses the internet. The workbook should then be submitted to the database of choice.

2.4. Submission of Large-Scale Amounts of Data

High-throughput data producers, those experimenters working in a continual pipeline with bioinformatic support, may wish to use the XML format (5) which may be continually sent to an IMEx-member database throughout the period of data generation and made public when appropriate.

Potential users should download and install the Java Development Kit (JDK) version 5.0 or 6.0 from http://java.sun.com/. There are several download packages available for the JDK. Choose the one that is labeled as JDK 5.0/6.0 Update X rather than the (much larger) packages including the NetBeans IDE and/or Java EE as these components are not necessary for the protocol. Detailed installation instructions for a particular system can be found on this Website as well.

Downloading and installing Subversion from http://subversion.tigris.org/ will allow the user to check out the examples source code (if using Windows or Mac OS you may want to download the binaries). For the example, the command line client is used, so there is no need to download a third-party client. Maven (http://maven.apache.org/download.html) may be used to run the examples.

2.5. Obtaining the Examples

A command line terminal should be opened, allowing the user to navigate to the folder of choice and execute the following code:

svn checkout – http://psidev.svn.sourceforge.net/svnroot/psidev/psi/mi/xml/psimixml-examples

This will create a folder called psimixml-examples, which contains the example project. The layout of the sources follows the Maven standards (http://maven.apache.org/guides/introduction/introduction-to-the-standard-directory-layout.html). The example we are going to work with is located at:

psimixml-examples/src/main/java/org/hupo/psi/mi/xml/example/SimpleExample.java

The class already contains some code to generate an XML file. The first part of the example creates some example data. This is the section that can be modified to create your own data. The second part contains the actual writing of the created data to an XML file. The example contains comments about the steps in more detail.

All the code of the class is inside a *main* method, which makes this class executable. This is not recommended for production code, but for example purposes we want to execute from the command line.

Maven is used to execute the class from the command line. To do so, run the following command:

```
mvn -P exec-simple
```

Once run, an XML file should appear in the psimixml-examples/target folder, containing the data created in the example. A semantic validator for PSI-MI files (http://www.ebi.ac.uk/intact/validator/start.xhtml) (Kerrien et al., in preparation) is also available. The validator checks the correct use of PSI-MI ontologies in a data file, plus applies additional semantic consistency rules written to conform to IMEx curation standards. The code should be modified to generate individual PSI-MI XML2.5 files. The schema will require the use of PSI-MI controlled vocabulary terms. Once the file validates, it may be submitted to a database of choice.

3. Notes

1. The most common cause of data loss when published interaction data is archively curated is the use by authors of ambiguous molecule identifiers or the lack of originating organism when clones are described. Interactors should always be referenced to an external public domain resource, even if an author-derived identifier is the main descriptor within the paper. For proteins this should be a protein sequence database such as UniProtKB, and when appropriate, identification can be as detailed as specifying a particular isoform by a database accession number. For genes, Ensembl (http://www.ensembl.org), Ensembl Genomes (http://www.ensemblgenomes.org/), or Entrez Gene (http://www.ncbi.nlm.nih.gov/sites/entrez) identifiers are appropriate resources; nucleic acids may be identified by a DDBJ/EMBL/GenBank identifier and small molecules by a ChEBI accession number (http://www.ebi.ac.uk/chebi) (12). If a molecule is missing from the source database, you should include the sequence/structure in your submission, and if possible the database will arrange for its inclusion in the reference resource.
2. If features, such as mutations, are to be mapped to a sequence, the numbering should reflect the current version of the sequence given in any resource unless the author is publishing an alternative sequence in the manuscript.
3. New techniques, or derivations of existing techniques, may not yet be adequately described by the current version of the HUPO-PSI CV. New terms may be requested on the SourceForge tracker at https://sourceforge.net/tracker/?group_id=65472&atid=612426.

References

1. Aranda, B., Achuthan, P., Alam-Faruque, Y., Armean, I., Bridge, A., Derow, C., Feuermann, M., Kerrien, S., Khadake, J., Kerssemakers, J., Leroy, C., Menden, M., Michaut, M., Montecchi-Palazzi, L., Neuhauser, S.N., Orchard, S., Perreau, V., Roechert, B., Tashakkori, A., van Eijk, K., Hermjakob, H. (2010) The IntAct molecular interaction database in 2010. *Nucleic Acid Res* **38**, D525–D531.
2. Anon. (2005) Proteomics' new order. *Nature* **437**, 169–170.
3. Orchard, S., Hermjakob, H. (2008) Standardising proteomics data: the work of the HUPO proteomics standards initiative. *J Proteomics Bioinform* **1**, 3–5.
4. Hermjakob, H., Montecchi-Palazzi, L., Bader, G., Wojcik, J., Salwinski, L., Ceol, A., Moore, S., Orchard, S., Sarkans, U., von Mering, C., Roechert, B., Poux, S., Jung, E., Mersch, H., Kersey, P., Lappe, M., Li, Y., Zeng, R., Rana, D., Nikolski, M., Husi, H., Brun, C., Shanker, K., Grant, S.G., Sander, C., Bork, P., Zhu, W., Pandey, A., Brazma, A., Jacq, B., Vidal, M., Sherman, D., Legrain, P., Cesareni, G., Xenarios, I., Eisenberg, D., Steipe, B., Hogue, C., Apweiler, R. (2004) The HUPO PSI's molecular interaction format – a community standard for the representation of protein interaction data. *Nat Biotechnol* **22**, 177–183.
5. Kerrien, S., Orchard, S., Montecchi-Palazzi, L., Aranda, B., Quinn, A.F., Vinod, N., Bader, G.D., Xenarios, I., Wojcik, J., Sherman, D., Tyers, M., Salama, J.J., Moore, S., Ceol, A., Chatr-aryamontri, A., Oesterheld, M., Stümpflen, V., Salwinski, L., Nerothin, J., Cerami, E., Cusick, M.E., Vidal, M., Gilson, M., Armstrong, J., Woollard, P., Hogue, C., Eisenberg, D., Cesareni, G., Apweiler, R., Hermjakob, H. (2007) Broadening the horizon – level 2.5 of the HUPO-PSI format for molecular interactions. *BMC Biol* **5**, 44
6. Orchard, S., Salwinski, L., Kerrien, S., Montecchi-Palazzi, L., Oesterheld, M., Stümpflen, V., Ceol, A., Chatr-aryamontri, A., Armstrong, J., Woollard, P., Salama, J.J., Moore, S., Wojcik, J., Bader, G.D., Vidal, M., Cusick, M.E., Gerstein, M., Gavin, A.-C., Superti-Furga, G., Greenblatt, J., Bader, J., Uetz, P., Tyers, M., Legrain, P., Fields, S., Mulder, N., Gilson, M., Niepmann, M., Burgoon, L., De Las Rivas, J., Priesto, C., Perreau, V.M., Hogue, C., Mewes, H.-W., Apweiler, R., Xenarios, I., Eisenberg, D., Cesareni, C., Hermjakob, H. (2007) The minimum information required for reporting a molecular interaction experiment (MIMIx). *Nat Biotechnol* **25**, 894–898.
7. Chiang, T., Li, N., Orchard, S., Kerrien, S., Hermjakob, H., Gentleman, R., Huber, W. (2008) Rintact: enabling computational analysis of molecular interaction data from the IntAct repository. *Bioinformatics* **24**, 1100–1101.
8. Chatr-aryamontri, A., Ceol, A., Palazzi, L.M., Nardelli, G., Schneider, M.V., Castagnoli, L., Cesareni, G. (2007) MINT: the Molecular INTeraction database. *Nucleic Acids Res* **35**, D572–D574.
9. Xenarios, I., Salwinski, L., Duan, X.J., Higney, P., Kim, S., Eisenberg, D. (2002) DIP: the database of interacting proteins. A research tool for studying cellular networks of protein interactions. *Nucleic Acids Res* **30**, 303–305.
10. Chautard, E., Ballut, L., Thierry-Mieg, N., Ricard-Blum, S. (2009) MatrixDB, a database focused on extracellular protein–protein and protein–carbohydrate interactions. *Bioinformatics* **25**, 690–691.
11. Goll, J., Rajagopala, S.V., Shiau, S.C., Wu, H., Lamb, B.T., Uetz, P. (2008) MPIDB: the microbial protein interaction database. *Bioinformatics* **24**, 1743–1744.
12. Degtyarenko, K., de Matos, P., Ennis, M., Hastings, J., Zbinden, M., McNaught, A., Alcantara, R., Darsow, M., Guedj, M., Ashburner, M. (2008) ChEBI: a database and ontology for chemical entities of biological interest. *Nucleic Acids Res* **36**, D344–D350.

Chapter 16

Submitting Proteomics Data to PRIDE Using PRIDE Converter

Harald Barsnes, Juan Antonio Vizcaíno, Florian Reisinger, Ingvar Eidhammer, and Lennart Martens

Abstract

With the continuously growing amount of proteomics data being produced, it has become increasingly important to make these data publicly available so that they can be audited, reanalyzed, and reused. More and more journals are also starting to request the deposition of MS data in publicly available repositories for submitted proteomics manuscripts. In this chapter we focus on one of the most commonly used proteomics data repositories, PRIDE (the PRoteomics IDEntifications database, http://www.ebi.ac.uk/pride), and demonstrate how a new graphical user interface tool called PRIDE Converter (http://pride-converter.googlecode.com) greatly simplifies the submission of data to PRIDE.

Key words: Mass spectrometry, Proteomics, Data repository, Data conversion

1. Introduction

1.1. Background

Public availability of source data and results is the standard for most areas of proteomics, e.g., protein sequences in UniProt (1) (http://www.uniprot.org), protein structures in the Protein Databank (2) and protein modifications in UniMod (3) (http://www.unimod.org), and RESID (4) (www.ebi.ac.uk/RESID). Correspondingly, journals are increasingly requesting that mass spectrometry based proteomics data, accompanying submitted manuscripts, is also made publicly available (5, 6). However, before such data deposition can be made strictly mandatory, the submission process first has to become straightforward. Peptide and protein identifications using mass spectrometry (MS) face some additional challenges compared to the other data types in this respect: The data sets can be both very complex and very

Cathy H. Wu and Chuming Chen (eds.), *Bioinformatics for Comparative Proteomics*, Methods in Molecular Biology, vol. 694, DOI 10.1007/978-1-60761-977-2_16, © Springer Science+Business Media, LLC 2011

large. Additionally, the lack of standard formats for data exchange and storage further confounds matters, although recent standardization efforts by the Human Proteome Organization's Proteomics Standards Initiative (HUPO PSI) for MS driven proteomics, mainly centered on mzIdentML (previously known as analysisXML) and mzML (http://www.psidev.info) are likely to improve this situation in the near future.

Several repositories for proteomics MS data have been established, with PRIDE, GPMDB, PeptideAtlas, NCBI Peptidome and Proteinpedia (7) being the most prominent. Among these the PRIDE data repository (8–10) at the European Bioinformatics Institute (http://www.ebi.ac.uk/pride) stands out by combining several important properties. First of all, it represents an actual data repository, as it assumes no editorial control over and data is kept exactly as submitted. Secondly, it includes a convenient but powerful system that supports anonymous peer review of submitted data, while maintaining the submission as private. PRIDE stores three different types of information: peptide and protein identifications derived from MS or MS/MS experiments, MS and MS/MS mass spectra as peak lists, and any and all associated metadata. Both in terms of the amount of metadata that can be stored, and in the way that this information is structured, PRIDE exceeds the capabilities of other repositories. Because of these various strengths, discussed in more detail elsewhere (8–10), PRIDE has become one of the recommended locations for making MS proteomics data publicly available, e.g., (5, 6) and http://www3.interscience.wiley.com/homepages/76510741/2120_instruc.pdf. The remainder of this chapter focuses on how submitting data to PRIDE has been greatly simplified by the development of a graphical user interface data conversion tool called PRIDE Converter (11) (http://pride-converter.googlecode.com).

1.2. Tools for Submitting Data to PRIDE

In the past submitting data to PRIDE could be challenging, especially for wet-lab scientists without a background in bioinformatics or local informatics support. Apart from the inherent complexity of MS proteomics data, the main reason for these difficulties can be attributed to the long list of different data formats being used in the proteomics community, which somehow has to be dealt with when submitting the data to a repository.

When PRIDE was created an XML-based data format, referred to as PRIDE XML, was chosen as the default format for submissions. PRIDE XML is built around the HUPO PSI's mzData standard for mass spectrometry (http://www.ebi.ac.uk/pride/schemaXmlspyDocumentation.do) (9). However, converting proteomics data to PRIDE XML can be quite complex, and as a result, several tools for converting data into PRIDE XML have been developed over the past few years. The first of these were the ProteomeHarvest PRIDE Submission Spreadsheet (http://www.ebi.ac.uk/pride/proteomeharvest), which is a Microsoft Excel-based tool only suitable for small-scale submissions, and PRIDE Wizard (12) (http://www.mcisb.org/resources/PrideWizard), supporting Mascot search result files. A tool called ProCon (http://www.medizinisches-proteom-center.de/software) was developed more recently for converting data from the ProteinScape LIMS system to PRIDE XML.

Unfortunately, all have limited support for different proteomics data formats and the inability of some of the tools to handle larger data sets further reduces their usability. A new tool called PRIDE Converter (11) (http://pride-converter.googlecode.com) was therefore developed. It improves on the existing tools in three essential ways (1) it supports a large variety of input formats (see Table 1), (2) it is suitable for both small and large data submissions, and (3) having a wizard-like graphical user interface it is very intuitive and easy to use.

Table 1
The currently supported data formats in PRIDE Converter

Data format name	More information
Mascot DAT Files	http://www.matrixscience.com
Mascot Generic Files	http://www.matrixscience.com
X!Tandem	http://www.thegpm.org/TANDEM
Spectrum Mill	http://www.chem.agilent.com
Micromass PKL Files	http://www.matrixscience.com/help/data_file_help.html#QTOF
SEQUEST Result Files	http://fields.scripps.edu
SEQUEST DTA Files	http://fields.scripps.edu
OMSSA	http://pubchem.ncbi.nlm.nih.gov/omssa
Peptide- and ProteinProphet	http://peptideprophet.sourceforge.net http://proteinprophet.sourceforge.net
ms_lims 7	http://genesis.ugent.be/ms_lims
VEMS PKX Files	http://personal.cicbiogune.es/rmatthiesen
MS2	http://doi.wiley.com/10.1002/rcm.1603
mzData	http://www.psidev.info/index.php?q=node/80#mzdata
mzXML	http://tools.proteomecenter.org/wiki/index.php?title=Formats:mzXML
DTASelect	http://fields.scripps.edu

2. Materials

PRIDE Converter is built using Java 1.5, is platform independent and currently supports the conversion of 15 different input formats into PRIDE XML (see Table 1). It has a wizard-like graphical user interface divided into eight steps guiding the user from the specification of the input files to the final conversion result. In each step, the user is requested to provide appropriate metadata using controlled vocabulary terms that are retrieved using the Ontology Lookup Service (OLS) (13) (see Fig. 6). PRIDE Converter is open source, freely available, and support is available via the PRIDE support team at pride-support@ebi.ac.uk. For more details see the PRIDE Converter home page at http://pride-converter.googlecode.com.

3. Methods

The remainder of this chapter contains a step-by-step tutorial on how to use PRIDE Converter to convert proteomics data into valid PRIDE XML. Additional help can be found by clicking the help icons in the lower left corner of each frame of the running user interface or by visiting the PRIDE Converter home page, where example data and the tool itself can be downloaded. No further installation is required; simply download and unzip the file.

3.1. Starting PRIDE Converter

After unzipping the downloaded archive, double click on the PRIDEConverter-X.Y.Z.jar file (where X.Y.Z represents the version number) to start PRIDE Converter (see Note 1). If problems should occur, the PRIDE Converter home page contains tips and hints on how to handle the most common issues and this information will therefore not be repeated here. At each start up PRIDE Converter checks if a newer version of the converter has become available. It is always recommended to use the latest version, and one can easily import user settings, etc. from previous versions when upgrading. For more information about upgrading see the PRIDE Converter home page.

3.2. Data Source Selection

After PRIDE Converter has started the data source selection screen is shown, see Fig. 1. Here all the currently supported formats are shown, see Table 1 for the complete list, and additional information about each format is displayed when selecting a given data format. In this tutorial we use "SEQUEST Result Files" as example data, but the main procedure is identical for all formats.

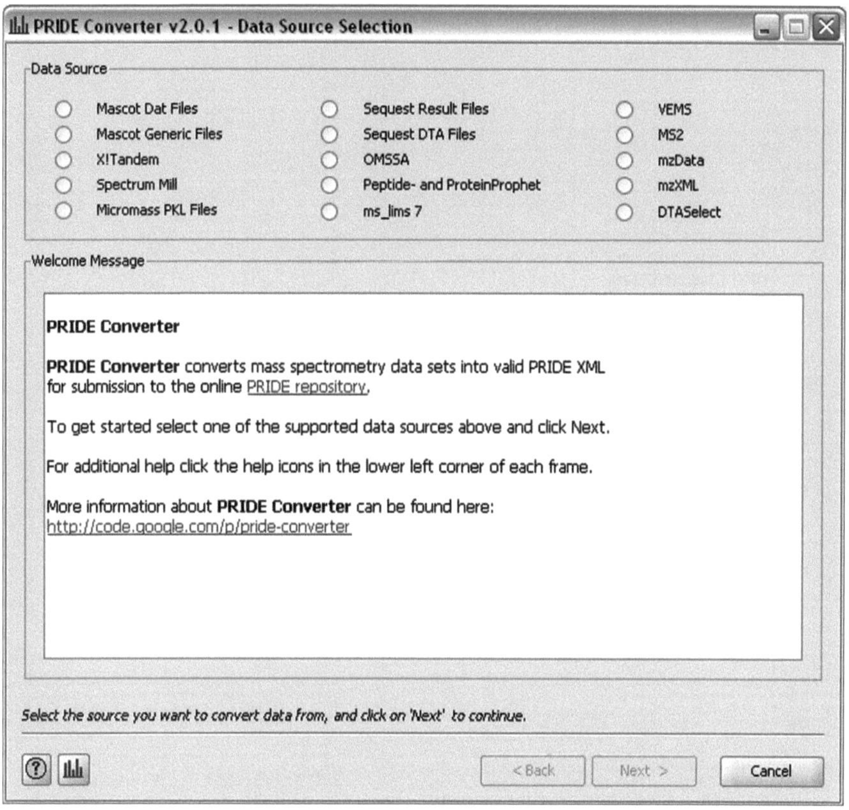

Fig. 1. Opening screen of PRIDE Converter, Data Source Selection, showing the supported data formats. Additional information about each format is displayed when selecting a given format.

(Example data files can be found on the PRIDE Converter home page.) Select "SEQUEST Result Files" in the list at the top and click on "Next >" to continue.

3.3. Step 1 of 8: Data File Selection

The wizard is divided into eight simple steps where the user provides different types of information at each step. The first step is the "File Selection" step, see Fig. 2, where the proteomics data files to be converted are selected. This step varies somewhat between the different formats, because of the various ways of storing the data. For SEQUEST result files the data is divided into two types: the spectrum files (dta files) and the identification files (out files). Other formats store all the information in one file, e.g., Mascot dat files, or in a database, e.g., ms_lims, but the selection process is fairly similar (see Note 2).

3.4. Step 2 of 8: Spectra Selection

The spectra to be included are selected in this step. In most cases it is best to include all spectra, but sometimes it makes more sense to only include a subset. This option is provided at the "Spectra

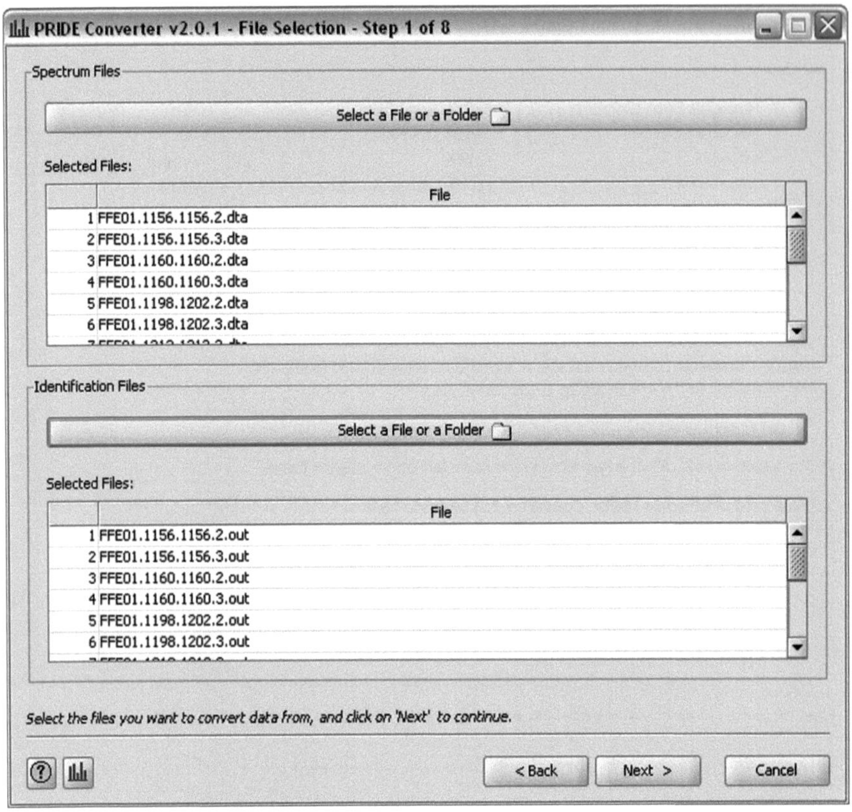

Fig. 2. First wizard step: Data File Selection. Here the files to be converted are selected. Note that this step differs somewhat for the different data formats, depending on how the data is stored, i.e., all the data in one file, the data separated into two files, the data stored in a database etc. In this case the two data file types option is shown, using SEQUEST data as the example.

Selection" step, see Fig. 3, where several alternatives for spectra selection are presented. The simplest alternatives are "Select All Spectra" or "Select Identified Spectra", i.e., include all spectra or just the identified ones. For more advanced selection, use, "Advanced Spectra Selection" or "Manual Spectra Selection". "Advanced Spectra Selection" enables the selection of subsets of spectra based on either filenames or identification IDs, whereas "Manual Spectra Selection" enables manual selection of spectra based on individual information about each spectrum. For manual selection first load the spectra by clicking the "Load Spectra" button. When using the more advanced spectra selection options it is always recommended to verify the selection before continuing.

For some data formats it is also possible to set boundaries for the identifications, either by a minimum peptide score level, a minimum Mascot confidence level or by adding a protein identification filter. The protein identification filter is mainly provided

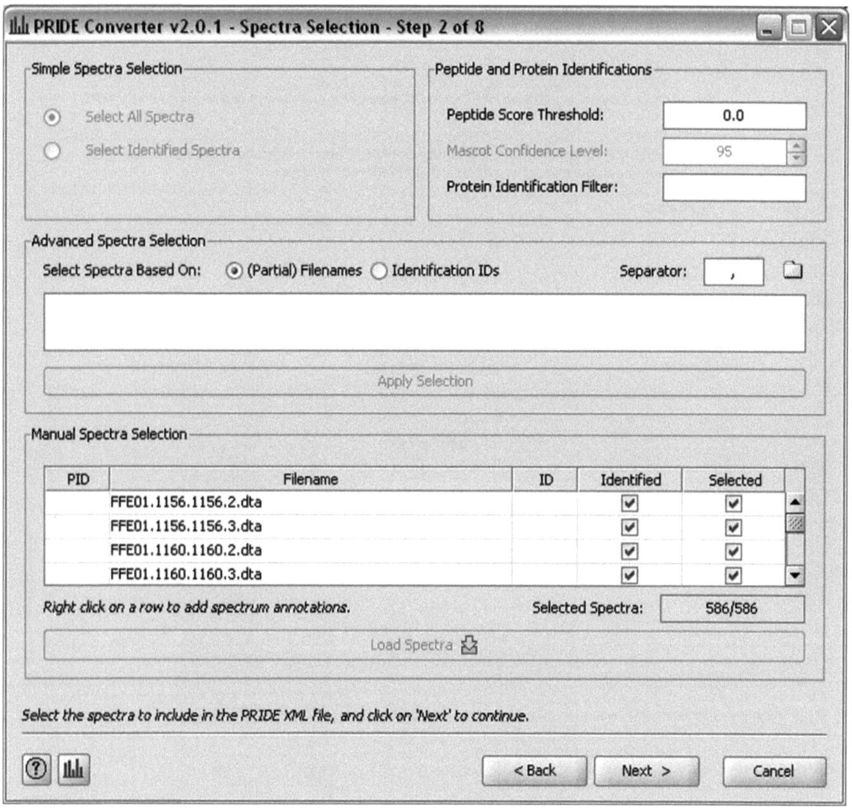

Fig. 3. Second wizard step: Spectra Selection. Here the spectra to be used are selected. Note that this step differs somewhat depending on the data format used. Also note that spectra selection is not available for all data formats. Here SEQUEST result files are used as the example data. See text for more details.

to be able to remove protein hits from decoy databases, but other uses might also be imagined (see Note 3).

One last point to mention is the possibility to include individual spectrum annotations. Particular information about a given spectrum can be included either as plain text user parameters or, as recommended, by using controlled vocabulary terms (for more details about controlled vocabulary terms, see Step 4). Both types of terms are added by right-clicking on the given row in the spectrum table and selecting the "View/Change Spectrum Annotations."

Note that the options available at this step are dependent on the data format selected, and that spectra selection might not be available for all formats.

3.5. Step 3 of 8: Experiment Properties

At this step the main properties of the experiment are described, including title, description, experiment label, and an optional project name, see Fig. 4. The project name is used as a way of organizing related experiments in a hierarchical structure.

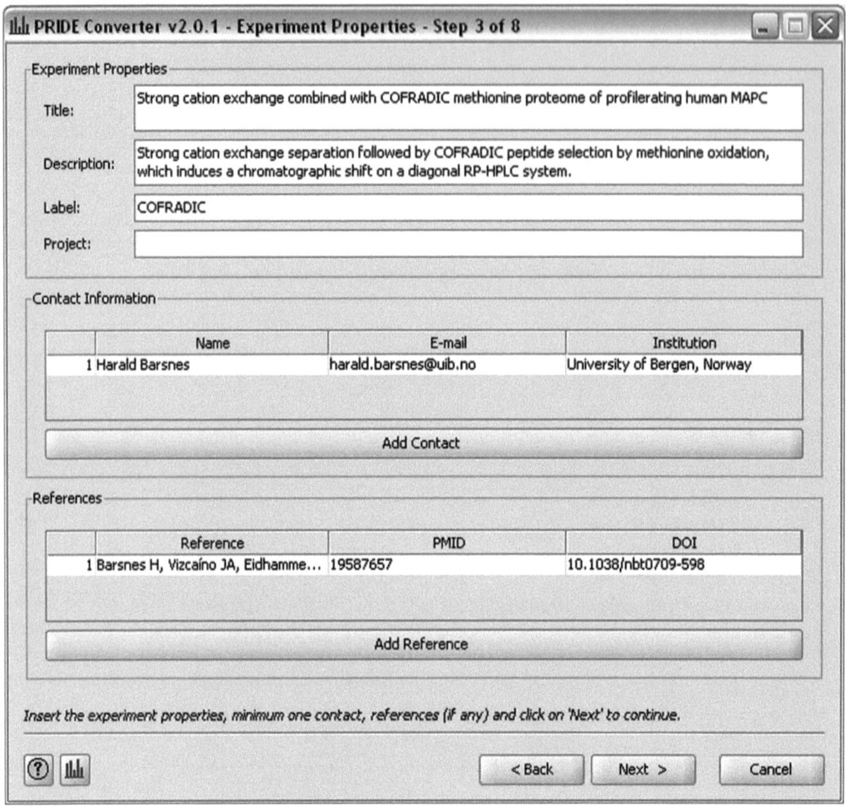

Fig. 4. Third wizard step: Experiment Properties. In this step information about the experiment is inserted, including contact information and references if any. See text for more details.

It is not mandatory to provide a project name when submitting related experiments but it is highly recommended as it will allow efficient retrieval of data across the related experiments. The information provided at this step is used throughout PRIDE to describe the experiment, and in some of the searching features. Contact information is important, and makes it easy for people to contact the owners of the data set. A minimum of one contact has to be provided. If the submitted data set is published as part of a scientific paper, references to this paper should also be included. The added references will be highly visible for people viewing the data via the PRIDE web page and most likely result in more people referencing the papers when using the data.

3.6. Step 4 of 8: Sample Properties

Information about the sample being used for the analysis is inserted in the "Sample Properties" step, see Fig. 5. There are two main options: single sample and multiple samples. Adding multiple samples makes it possible to include quantification data, where iTRAQ (14) is the currently supported format (see Note 4). Because most users will have a single sample the following will

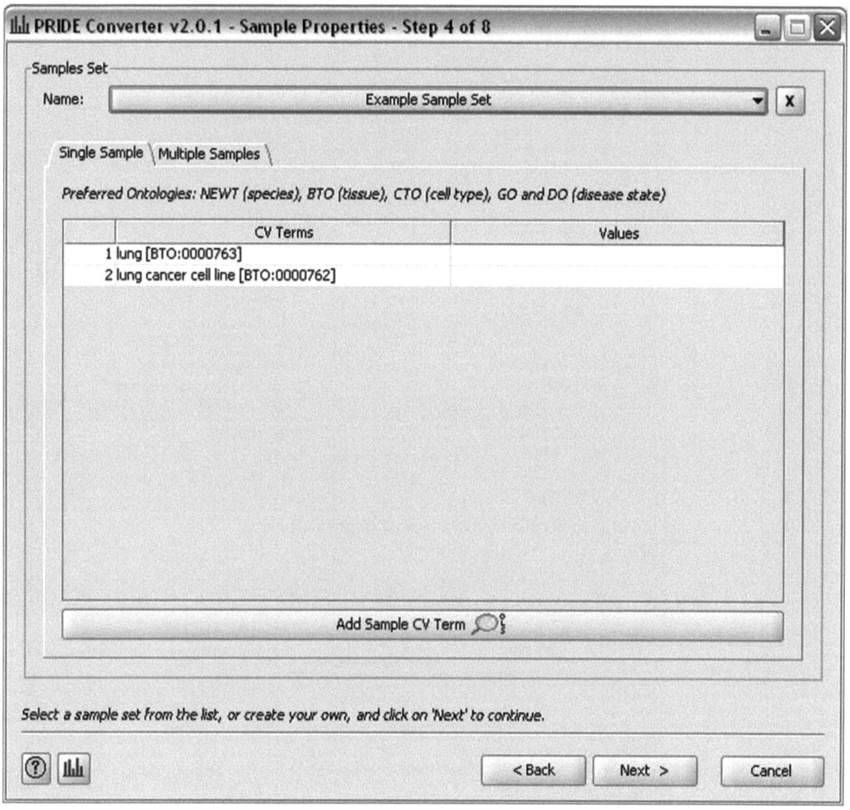

Fig. 5. Fourth wizard step: Sample Properties. Here the properties of the sample used in the experiment are described. See text for details.

focus on how to annotate single samples. However, the process is very similar for multiple samples (see Note 4). At the top of the frame one can choose from a set of already created sample sets. Simply select the appropriate sample set in the drop down menu. If none of the available sample sets are satisfactory, new ones are easily created either from scratch, by selecting "Create a new sample set…" at the bottom of the drop down menu, or by extending one of the existing sample sets.

A sample is described using so-called controlled vocabulary (CV) terms, enabling all users to describe the data using the same terms. Using CV terms has many advantages, which are explained in more detail elsewhere (15), the most important advantage in this context is that it simplifies data mining across different data sets. The use of CV terms provides annotation of metadata in a structured way, because CV terms are organized in a hierarchical structure and all terms have definitions. For finding CV terms PRIDE Converter uses an online connection to the OLS (13) (http://ols-dialog.googlecode.com and http://www.ebi.ac.uk/ontology-lookup), see Fig. 6. To find a CV term click on the

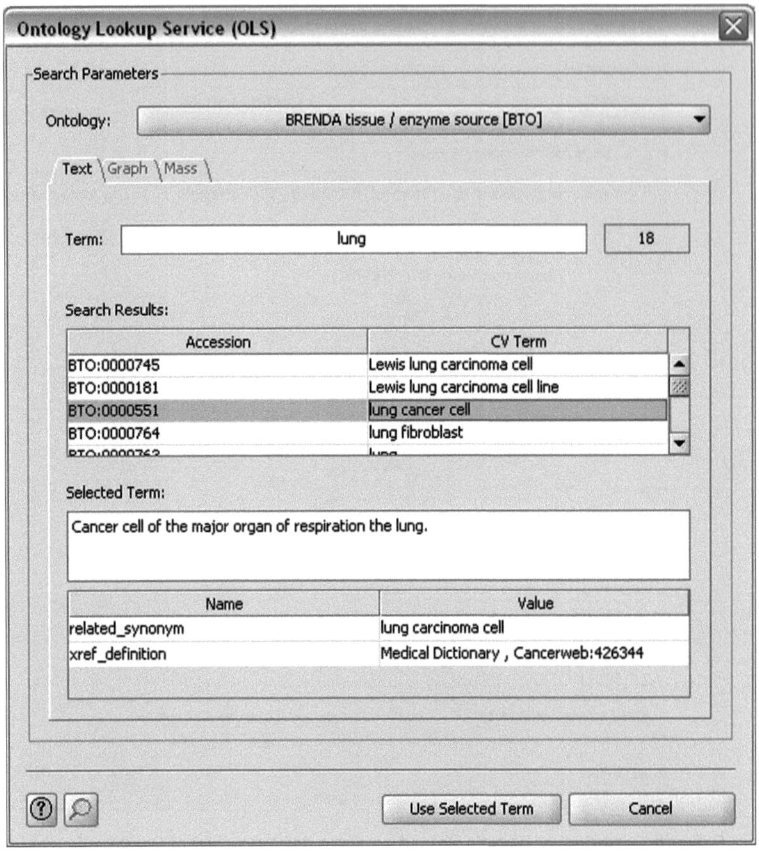

Fig. 6. Ontology Lookup Service (OLS) dialog. Finding and using the correct CV (controlled vocabulary) terms is made easier using the Ontology Lookup Service. See text for details.

"Add Sample CV Term" button, and the OLS dialog will appear. After selecting the relevant ontology, the correct CV term can easily be found by typing in the first few letters of the term into the autocompleting search field. A list of matching terms will appear, and additional information is displayed when selecting a term in the list. When the correct term has been identified, add it to the sample annotation by clicking the "Use Selected Term" button (see Note 5). Added terms can easily be altered, either by right-clicking on the term and selecting "Edit" from the popup menu, or by double clicking on the term. A minimum of one sample CV term must be provided.

3.7. Step 5 of 8: Protocol Properties

"Protocol Properties," see Fig. 7, are annotated in much the same way as "Sample Properties." Again one can select an already existing protocol, or create a new protocol, and as before the properties are annotated using CV terms, but in this case each protocol step can contain more than one CV term. To add a protocol step,

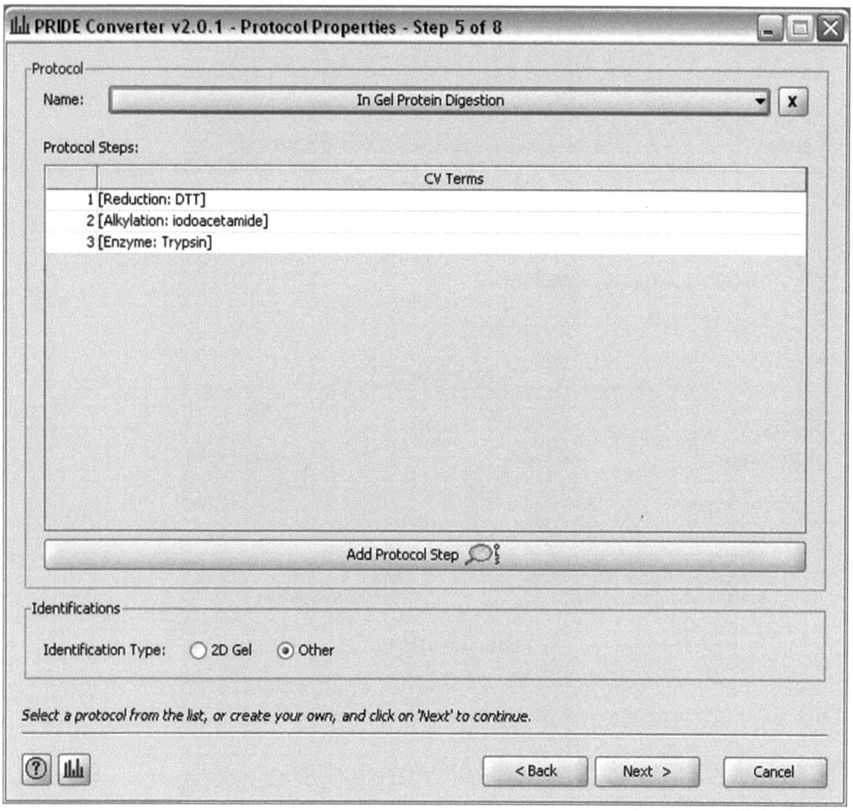

Fig. 7. Fifth wizard step: Protocol Properties. Describing the protocol steps used for the experiment. See text for details.

click on the "Add Protocol Step" button. A new dialog then appears where multiple CV terms describing the given step can be added. Click the "Ontology Lookup Service" button to find a CV term in the same way as previously described. Note that it is also possible to attach values to these CV terms, e.g., the name of the enzyme used. After selecting all the terms for a given protocol step, click the "OK" button to add the step to the protocol description. Multiple steps are added by repeating the procedure, and similar to the CV terms used for sample description, added terms can be altered either by right clicking on the term and selecting "Edit" from the popup menu, or by double clicking on the term itself. A minimum of one protocol step has to be provided.

At the bottom of the Protocol Properties frame there is an option of selecting the identification type. The current options are "2D Gel" and "Other," where the latter will be the preferred option for most users. "2D Gel" can be used if at a later stage 2D gel images should be added to the PRIDE XML file, see Note 6.

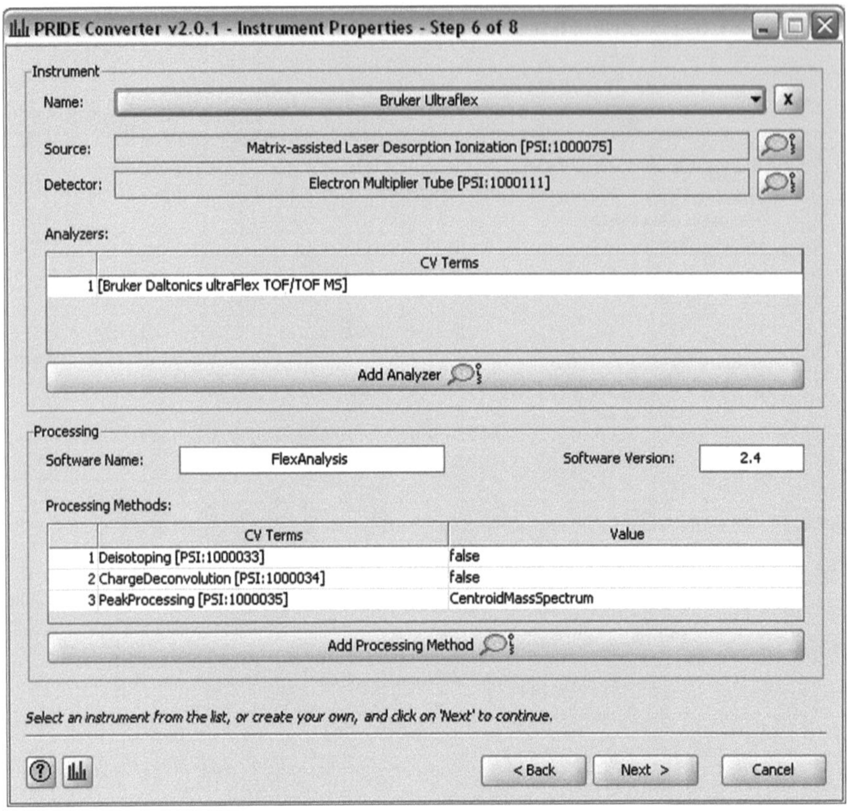

Fig. 8. Sixth wizard step: Instrument and Processing Properties. Details about the instrument and the processing methods used are inserted at this step. See text for details.

3.8. Step 6 of 8: Instrument Properties

This frame has two main parts, the instrument details at the top and the processing details at the bottom, see Fig. 8. Both are described using CV terms, and the process is very similar to the overall annotation procedure already explained. The choice of selecting an already created instrument or creating a new one is available using the drop down menu at the top. All fields are mandatory, including the addition of at least one analyzer and at least one processing method.

For some data formats, instrument information is included in the data files, and if included, the information is extracted and inserted into the respective fields and tables. When this is the case it is important to verify that the acquired information is correct and complete. Please provide any missing information if necessary.

3.9. Step 7 of 8: User Parameters

Additional information about the data set can be annotated by user parameters, see Fig. 9. User parameters store plain text information provided by the submitter in a nonstructured way. By clicking the "Add User Parameter" button, name and value pairs (using plain text) can be added to further annotate the data set. There is no limit on the number of parameters, and including user parameters is not mandatory.

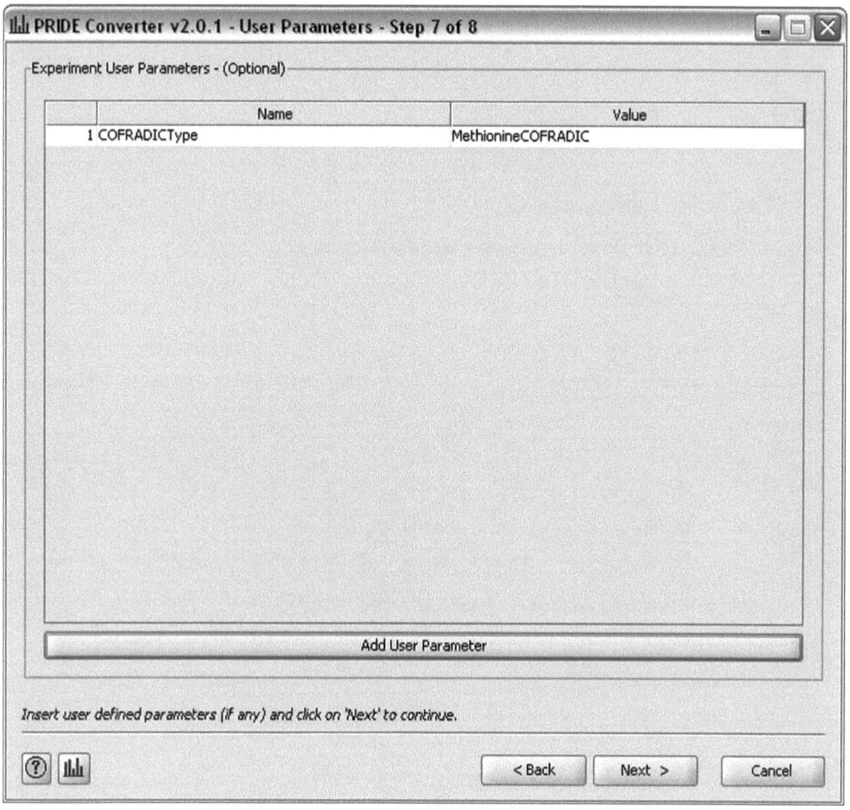

Fig. 9. Seventh wizard step: User Parameters. Additional description of the experiment can be added as plain text user parameters. See text for details.

3.10. Step 8 of 8: Output Properties

The final wizard step concerns the setting of the "Output Properties," see Fig. 10. The frame consists of four parts. At the top the output folder is provided. This is the location where the resulting PRIDE XML file will be saved. To change the location, click the folder icon to the right of the text field. Next is the resubmission section. When resubmitting a data set, check the resubmission box and provide the accession number of the original submission. Otherwise leave the box unchecked.

A section containing a set of format specific parameters follows. The first one can be used to mimic the Mascot web result by rounding the score and the threshold before comparison. The second option, if selected, causes PRIDE Converter to expect a comma instead of a period as the decimal symbol. If not checked, a period is assumed to be the decimal symbol used in the data files, and this is the standard for all data formats. Finally, the location of the OMSSA installation folder can be set. Because it contains required information about the peptide modifications, setting this field is mandatory when converting OMSSA omx files.

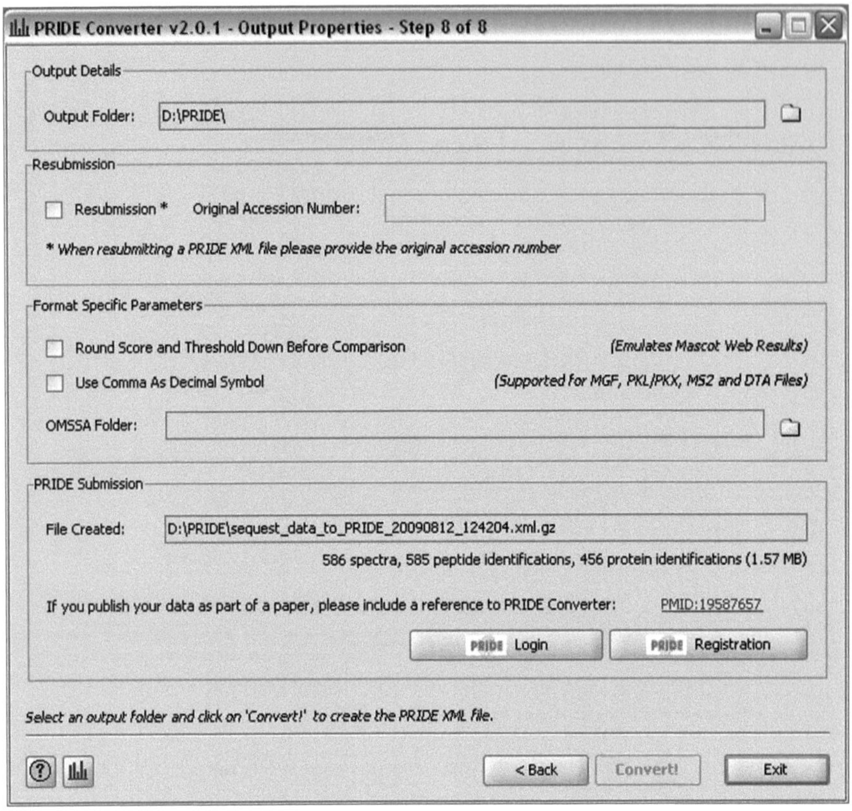

Fig. 10. Eighth (and final) wizard step: Output Properties. The final details before converting are set at this step. Note that the screenshot shows the step after the creation of a PRIDE XML file. See text for more details.

The last section contains information about the converted file, including the complete file name and the file properties, i.e., the file size and the number of peptide and protein identifications. To start the conversion process, simply click the "Convert!" button.

3.11. Conversion Process

When the "Convert!" button has been clicked PRIDE Converter starts converting all the selected files into one valid PRIDE XML file. However, depending on the data format used the converter may need additional input. One typical example of this concerns the details about detected peptide modifications. When a modification is detected PRIDE Converter tries to map it to a set of default PSI-MOD (16) CV terms. These default mappings cover all the most common modifications, but if a given modification is not automatically mapped, the user is prompted to do so manually using the OLS. Selected mappings will be added to PRIDE Converter's default mappings and proposed the next time the given modification is detected. Given that a modification mapping might not be unique, e.g., C* can correspond to different modifications for different data formats, the user will have to verify the modification each time the file is converted. However, only the first occurrence of the modification has to be confirmed.

For some data formats, mainly Mascot dat files, one also gets the choice of how to map detected protein isoforms. Three different options are given (1) always select the first isoform found, (2) do a manual selection for each peptide-protein mapping, or (3) provide a list of peptide to protein mappings. Selecting the first option requires the least amount of additional effort, but in some cases the more advanced options are needed.

After verifying the modification mappings (and handling the protein isoforms if required), the conversion process continues and a progress bar with detailed information about the progress is shown. The conversion process can be cancelled at any time, but please note that the actual cancellation might take some time depending on the underlying processes being run. If the conversion is not cancelled, and no errors occur (see Notes 7 and 8), an automatically validated PRIDE XML file will be created and a dialog containing information about the created file will be displayed, including the name and size of the file, and an overview of the contents of the file, i.e., the number of spectra and the number of identified peptides and proteins. It is advisable to verify that this information makes sense before continuing.

The created PRIDE XML file is now ready for submission. There are basically two ways of submitting your file to PRIDE: either by using the direct submission system on the PRIDE web page (http://www.ebi.ac.uk/pride) or by uploading the file to the PRIDE FTP server. The first option is only available for very small files (up to 15 MB unzipped), and accessing the submission site can be done by clicking the "PRIDE Login" button at the bottom of the frame. New users will have to register as users of PRIDE first by clicking the "PRIDE Registration" button and filling in the required information. If the created PRIDE XML file exceeds the limit for using the direct submission system, a message will be displayed urging the user to contact the PRIDE support team to get access details to the PRIDE FTP server.

For additional information or help please see http://pride-converter.googlecode.com or feel free to contact the PRIDE support team at pride-support@ebi.ac.uk.

4. Notes

1. PRIDE Converter can also be started from the command line: "java-jar PRIDEConverter-X.Y.Z.jar". Most failures to start PRIDE Converter will result in a "pride_converter.log" file located in the user home directory. For more information see the PRIDE Converter home page.

2. For most formats, several files can be combined into one PRIDE XML file. In this way the submitter can then choose how to organize their experiments. One PRIDE XML file corresponds to one experiment in the system after submission.

3. To use the protein identification filter to remove decoy database hits, simply insert the tag used to distinguish the decoy hits, e.g., "decoy_", into the protein identification filter text field. During conversion all proteins having an accession number containing this tag will then be excluded from the resulting PRIDE XML file. Note that only one filter can be used per conversion.

4. Multiple samples are supported by selecting the "Multiple Samples" tab, and the samples are annotated using CV terms in much the same way as for single samples. The main difference is the ability to have more than one sample. Click on "Add Sample," and click the "Ontology Lookup Service" button to add CV terms to this particular sample. Each sample must also be given a unique name. To add iTRAQ labels to the samples, select the iTRAQ reagent type in the "Quantification" column. When at least one sample has been annotated with an iTRAQ reagent type the "Quantification Parameters" section becomes enabled, where the iTRAQ parameters can be set. At present, only 4-plex iTRAQ is supported.

5. If unsure about which ontology to use, go to the OLS web page (http://www.ebi.ac.uk/ontology-lookup) and search for the desired term using the "Search in all ontologies" option.

6. Adding 2D gel images can currently not be done using PRIDE Converter directly. Contact the PRIDE support team after submitting the file in order to submit your gel images.

7. If an error occurs when using PRIDE Converter, an error message describing the problem will be written to the "ErrorLog.txt" file located in the "PRIDE Converter"/Properties folder. For most errors a dialog containing details about the problem will also be displayed. For more information about the most common issues see the PRIDE Converter home page.

8. In some cases the default memory size is not big enough to perform the conversion and an "Out of Memory" error will be shown when trying to convert the file or files. If this occurs the memory size can be increased by editing the "JavaOptions.txt" file in the "PRIDE Converter"/Properties folder. For more details see the PRIDE Converter home page.

References

1. UniProt Consortium. (2008) The universal protein resource (UniProt). *Nucleic Acids Res.* **36**, D190–D195.
2. Berman, H., Henrick, K., Nakamura, H., and Markley, J. L. (2007) The worldwide Protein Data Bank (wwPDB): ensuring a single, uniform archive of PDB data. *Nucleic Acids Res.* **35**, D301–D303.
3. Creasy, D. M. and Cottrell, J. S. (2004) UniMod: protein modifications for mass spectrometry. *Proteomics* **4**, 1534–1536.
4. Garavelli, J. (2004) The RESID database of protein modifications as a resource and annotation tool. *Proteomics* **4**, 1527–1533.
5. Editors. (2007) Democratizing proteomics data. *Nat Biotechnol.* **25**, 262.
6. Editors. (2008) Thou shalt share your data. *Nat Methods* **5**, 209.
7. Mead, J. A., Bianco, L., and Bessant, C. (2009) Recent developments in public proteomic MS repositories and pipelines. *Proteomics* **9**, 861–881.
8. Jones, P., Côté, R. G., Cho, S. Y., Klie, S., Martens, L., Quinn, A. F., Thorneycroft, D., and Hermjakob, H. (2008) PRIDE: new developments and new datasets. *Nucleic Acids Res.* **36**, D878–D883.
9. Jones, P., Côté, R. G., Martens, L., Quinn, A. F., Taylor, C. F., Derache, W., Hermjakob, H., and Apweiler, R. (2006) PRIDE: a public repository of protein and peptide identifications for the proteomics community. *Nucleic Acids Res.* **34**, D659–D663.
10. Martens, L., Hermjakob, H., Jones, P., Adamski, M., Taylor, C., States, D., Gevaert, K., Vandekerckhove, J., and Apweiler, R. (2005) PRIDE: the proteomics identifications database. *Proteomics* **5**, 3537–3545.
11. Barsnes, H., Vizcaíno, J. A., Eidhammer, I., and Martens, L. (2009) PRIDE Converter: making proteomics data-sharing easy. *Nat Biotechnol.* **27**, 598–599.
12. Siepen, J. A., Swainston, N., Jones, A. R., Hart, S. R., Hermjakob, H., Jones, P., and Hubbard, S. J. (2007) An informatic pipeline for the data capture and submission of quantitative proteomic data using iTRAQ. *Proteome Sci.* **5**, 4.
13. Côté, R. G., Jones, P., Martens, L., Apweiler, R., and Hermjakob, H. (2008) The Ontology Lookup Service: more data and better tools for controlled vocabulary queries. *Nucleic Acids Res.* **36**, W372–W376.
14. Zieske, L. (2006) A perspective on the use of iTRAQ reagent technology for protein complex and profiling studies. *J Exp Bot.* **57**, 1501–1508.
15. Martens, L., Palazzi, L. M., and Hermjakob, H. (2008) Data standards and controlled vocabularies for proteomics. *Methods Mol Biol.* **484**, 279–286.
16. Montecchi-Palazzi, L., Beavis, R., Binz, P. A., Chalkley, R. J., Cottrell, J., Creasy, D., Shofstahl, J., Seymour, S. L., and Garavelli, J. S. (2008) The PSI-MOD community standard for representation of protein modification data. *Nat Biotechnol.* **26**, 864–866.

Chapter 17

Automated Data Integration and Determination of Posttranslational Modifications with the Protein Inference Engine

Stuart R. Jefferys and Morgan C. Giddings

Abstract

This chapter describes using the Protein Inference Engine (PIE) to integrate various types of data – especially top down and bottom up mass spectrometer (MS) data – to describe a protein's posttranslational modifications (PTMs). PTMs include cleavage events such as the n-terminal loss of methionine and residue modifications like phosphorylation. Modifications are key elements in many biological processes, but are difficult to study as no single, general method adequately characterizes a protein's PTMs; manually integrating data from several MS experiments is usually required. The PIE is designed to automate this process using a guess and refine process similar to how an expert manually integrates data. The PIE repeatedly "imagines" a possible modification set, evaluates it using available data, and then tries to improve on it. After many rounds of refinement, the resulting modification set is proposed as a candidate answer. Multiple candidate answers are generated to obtain both best and near-best answers. Near-best answers are crucial in allowing for proteins with more than one supported modification pattern (isoforms) and obtaining robust results given incomplete and inconsistent data.

The goal of this chapter is to walk the reader through installing and using the downloadable version of PIE, both in general and by means of a specific, detailed example. The example integrates several types of experimental and background (prior) data. It is not a "perfect-world" scenario, but has been designed to illustrate several real-world difficulties that may be encountered when trying to analyze imperfect data.

Key words: PTM, MCMC, Simulated annealing, Proteomics, Top-down, Bottom-up, Data integration, PIE

1. Introduction

1.1. PTMs in Biology

Proteins underlie many of the processes that sustain life. They function as catalysts in cellular reactions, as signaling network components coordinating cellular processes, or simply as scaffolds that provide necessary cellular structure. The closer we study

Cathy H. Wu and Chuming Chen (eds.), *Bioinformatics for Comparative Proteomics*, Methods in Molecular Biology, vol. 694, DOI 10.1007/978-1-60761-977-2_17, © Springer Science+Business Media, LLC 2011

proteins, the more complex their function, structure, and regulation seem. One key aspect of that complexity is the modulation of protein behavior by chemical changes made co- or posttranslationally (1, 2). Some of these chemical changes alter the sequence of a protein after its translation at the ribosome, removing a number of amino acids from one or both ends. Others involve chemical groups that are added to or subtracted from proteins – often by specialized enzymes that are themselves modified proteins. Although modification occurs both during and after translation, we refer to all modifications hereafter as posttranslational modifications, or PTMs.

One reason PTMs are important in cellular systems is they allow for rapid responses to changing environmental conditions. For example, in bacteria like *E. coli*, a chemotactic circuit that senses nutrients in the environment. This circuit uses methylation/demethylation of receptor proteins to change their sensitivity to ligands in the environment, and uses phosphorylation of soluble proteins like CheY and CheA to signal changing nutrient conditions. The downstream results of these modifications are changes in the behavior of flagellar motors, affecting swimming (3, 4). Without PTMs, an organism would be dramatically limited in its ability to respond quickly to changing environmental conditions.

PTMs also play critical roles in human health and disease. On histone proteins, around which DNA is wound, control which genes are expressed and when. Misregulation of histone modification can be extremely deleterious (5). Cancer is modulated by p53 and other proteins, involving regulation through phosphorylation (6). Immune responses are also frequently modulated by PTMs through toll-like receptor pathways (7). The details of if, when and how proteins are modified is central to uncovering how disease processes work and determining potential therapeutic actions.

Given the important role of PTMs, a key goal of proteomics research has been to develop approaches and methods that can maintain fragile PTMs during handling, and then tease apart the subtle signals that indicate the location and type of PTMs on proteins (8). Although considerable progress has been made, there remain substantial hurdles to realization of a fully automated approach to identifying PTMs.

1.2. Challenge of Identifying PTMs

Proteins exhibit diverse chemical properties and PTMs further increase that diversity. This is useful biologically, but it makes the development of techniques to analyze modified proteins challenging. Mass spectrometry (MS) is the only widely successful method that can analyze most proteins and identify their chemical composition in their in vivo form (9).

MS measures molecular masses with such accuracy that it is possible to distinguish the difference in mass between a protein containing the amino acid Asparagine (N), weighing in at 114.1038 atomic mass units (Daltons or Da), versus one containing aspartic acid (D), weighing 115.0886 Da. With many mass spectrometers capable of accuracy of ppm, that means a 50 kDa protein can be measured within ±0.25 Da, to resolve the 0.9 Da difference between N and D residues. Similarly, when a protein is modified by the addition of chemical groups such as methyl (adding 14.0269 Da) or phosphoryl (adding 79.9799 Da), the engendered change in protein mass can clearly be detected (Fig. 1).

An accurate intact protein mass contains a great deal of information about a protein, but it cannot be easily interpreted and does not tell us everything we want to know:

- It doesn't tell us which residues have which chemical adducts.
- Intact masses are not unique. Different modifications or modifications sets with the same mass are said to be isobaric. For example, three methyl adducts are isobaric with a single acetyl adduct (within 0.05 Da).
- Intact proteins are hard to manipulate prior to and during mass spectrometry. They often dislike staying in solution and may not ionize or "fly" well in the mass spectrometer (10).

To address the difficulties with top down analysis of proteins, a number of approaches have been devised. The most common method is termed "bottom-up" proteomics. In this approach, the

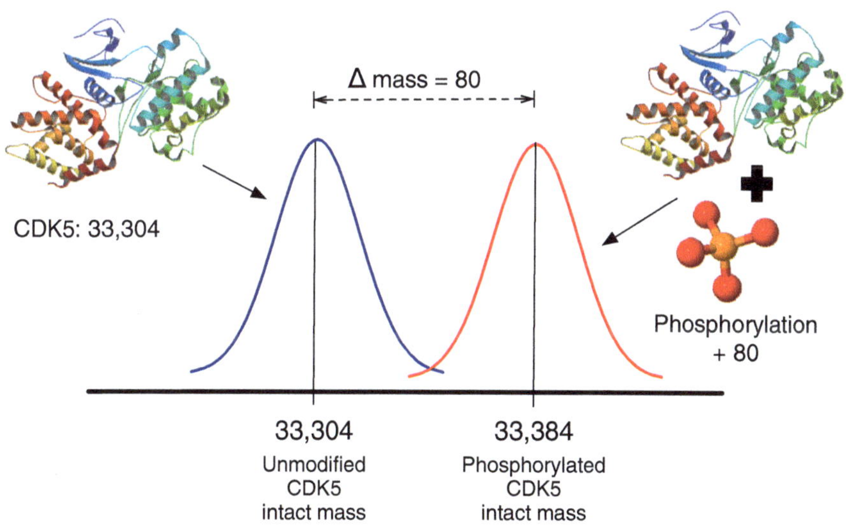

Fig. 1. Intact mass change caused by phosphorylation. Modification of a protein will influence its measured mass value. Here the mass of an unmodified (CDK5) protein sequence is 33,304 Da. The mass of the modified protein is 33,384 Da. The difference, +80 Da, is easily discernible by many mass spec approaches as the difference expected because of a phosphorylation modification. However, it could be because of some other combination of modifications that add up to 80 Da.

intact proteins are digested into much smaller pieces, "peptides," each of which can be individually analyzed by mass spectrometry. Not only can we measure the masses of these peptides, but also fragment them within the mass spectrometer, and measure the masses of the fragmentation products. This process is termed tandem mass spectrometry (MS/MS) (Fig. 2). With commonly available MS/MS search software, we can then use the peptide mass and/or its MS/MS spectrum to determine what kind of PTM was present and even which specific residue it was present on (e.g., (11–13)).

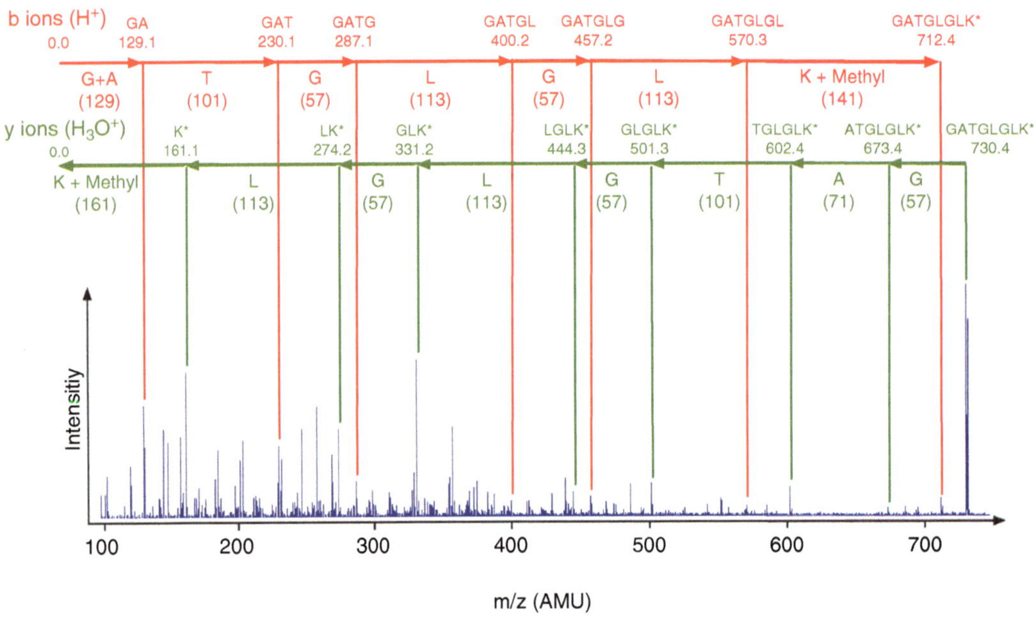

Fig. 2. MS/MS can reveal specific modifications and their position. A representative MS/MS spectra for a peptide is shown. Ideally, a single randomly placed peptide bond is broken in each peptide molecule, resulting in two collections of fragments, one of pieces from the N terminus side of the break, the other from the C terminus side. Fragments of all possible substring lengths are generated. The N terminus fragments for this peptide would be V-, VK-, VKD*-, VKD*L-, … and the C terminus fragments would be R-, RV-, RVG-, RVGP-, etc. Each fragment has a (different) mass, and results in a sequence of peaks on the MS/MS spectra. The resulting ladders of mass peaks (one from each end of the peptide) can be assigned to increasingly long fragments: 175.12 = R, 274.19 = R + V, 331.21 = R + V + G, etc. As with intact masses, shifts caused by a modification will be detected. Here a beta-methylthiolation modification has occurred on the aspartic acid residue third from the N-terminal end. Instead of the expected mass of V + K + D at 343, a peak at mass 389 was found. This represents VKD + 46, a mass shift that is attributed to the presence of modification. Unfortunately, the interpretation of MS/MS spectra can get complicated. Peptides may be broken in more than one place, and/or at bonds besides those between amino acids, producing distinct ion types. This, along with contaminants or heterogeneous spectra, often results in extra peaks – those not labeled in the diagram. Additionally, some expected fragmentation sites might not be seen at all, resulting in a single step of the sequence ladder including two or more amino acids, such as occurs for VK in this example. If there are many modifications and/or the specific sequence of the peptide is not known before hand, it can be difficult to interpret this spectra, especially because modifications can add to the variability in fragmentation patterns seen.

Another approach is to start with an intact mass measurement, and then take it a few steps further by fragmenting the intact protein within the mass spectrometer by one of a variety of methods, then measuring the pieces. In some cases, pieces may be isolated and further fragmented, leading to a process of MS^n, where n reflects the number of successive fragmentation steps. This is termed the "top-down" approach (14–16).

The challenge with these approaches is that they each have inherent limitations for detecting PTMs. In the bottom-up approach, coverage of a protein by a complete set of overlapping or abutting peptides is almost never achieved, and so there are gaps. Further, if there are distinct PTM variants of a protein, sorting out the combinatorics becomes very challenging once the protein is digested, because observations on distinct parts of the protein are unlinked from one another. In addition, it is often difficult to definitively identify the site and type of PTM within an MS/MS fragmentation spectrum.

The top-down approach faces its own challenges, such as the difficulty of isolating and analyzing intact proteins, and the difficulty of interpreting the complex top-down spectra (17). The top-down method is typically performed using electrospray ionization as the ion source, which produces ion species in multiple charge states. The process of "deconvoluting" the multiple species and their multiple charge states produced by top-down fragmentation is not fully solved.

These impediments have led many groups to adopt hybrid approaches, using a combination of strategies that complement each other, such as "top-down/bottom-up" (TDBU) proteomics (18–20). In TDBU, bottom-up data usually provide definitive protein identities, along with a partial map of specific modification sites and types, whereas the top-down data usually provide insight on the overall state of the protein (e.g., is there one methylation or two at any given time on the protein?). This combination has proved powerful to more completely elucidate the modification state of proteins.

However, hybrids like TDBU have a major challenge: integrating the data from disparate mass spectrometry experiments and approaches into a cohesive picture of the protein's original *in vivo* state. It is not a trivial problem. Each measurement holds a piece of information about the protein's state, but it is usually incomplete. Worse, sometimes the data are conflicting. For example, if two isoforms of a protein are present, one with a methylated residue and one without, we may get some bottom up peptides covering that site from each – some with and some without the presence of a methylation. Usually a combination of ad hoc search methods and manual analyses must be applied to distill these data sets into a final picture of the protein's modification state. This is a significant limitation to the more widespread adoption of such hybrid approaches. Tools to aid in the automation of this process are just now starting to emerge (21), and it is this task we have addressed through the development of the Protein Inference Engine (PIE) (Manuscript in preparation, Jefferys and Giddings).

The PIE is designed to rapidly and automatically integrate disparate types of proteomic measurements into a conclusive picture of the modification state of the protein. It is highly modular, with each module allowing it to incorporate a distinct type of information. Presently there are modules that use intact mass measurements, peptide MS/MS measurements, residue-specific probabilities of various modification types, expert knowledge, and specialized PTM site predictions, such as those from programs like (22–26). The program can readily accommodate conflicting information, and if there are multiple PTM solutions (e.g., multiple isoforms), the program will output multiple high-scoring solutions.

PIE uses a guess refine methodology based on Markov Chain Monte Carlo (MCMC) (27) simulated annealing (28) to explore possible answers to the question "Given the data I have, what PTMs do I have and where are they located on my protein?" Guesses are candidate answers, each a specific modification set such as ("1 AA truncated from the n-terminal end of the protein," "a methylation on A #2 – the new N terminus," and "a methylation on AA #36.")

Given that an average protein has around 300 residues (29), and allowing for a minimal set of just ten different adduct modifications, there is an impossibly large search space of 10^{300} candidate guesses. This cannot be searched exhaustively in our lifetimes – or for that matter, within the expected lifetime of the universe. However, MCMC allows for exploring such huge combinatoric spaces. MCMC based approaches have long history in the physical sciences (27, 30) and although not as widespread in biological contexts, they have been successful used to explore very difficult search spaces. One well-known software package is `Mr Bayes` (31).

Each data module D_i in the PIE models a data type i and can evaluate guesses for a particular PTM state. Using the data it understands, each module returns a score S_D for its own data type. The scores from all data modules are combined into a total score $S = S_{D1} \times S_{D2} \times \ldots$ This total score represents how well a given guess is supported by all available data, but the individual module scores are retained to allow comparison of the relative value of different data sets.

PIE makes a guess, looks at nearby guesses, and chooses one as a potential new best guess. By analogy to a physical surface, each such new guess is referred to as a step. The scores of the old guess, S_{old}, and the new guess S_{new} are compared, and the ratio S_{old}/S_{new} is used to determine whether to refine the guess and keep the new one, or to keep looking for a better new guess.

After many guesses (100,000 steps or so), the PIE reports the highest scoring candidate it found during the search. Determining the needed run length (number of steps) is necessary (see Note 1) but once determined each run of this length ends with at least a good candidate for the highest *possible* scoring modification set. This candidate is a prediction for the modification set or sets most supported by the integrated data as represented by the scoring modules, which are the answers we seek.

Running multiple searches allows interpreting the collection of best candidates and comparing the optimal and near optimal answers. Finding near-optimal answers is required to identify the single highest supported modification set, but these answers also inform us about potential multiple protein isoforms and help interpret incomplete or inconsistent data.

2. Materials

This section describes the components needed to run the PIE for your own data or with the example described in Subheading 3. This includes a brief description of installing the PIE, a tour of the data modules used in the example and explicit listings of the sample data and analysis parameters needed. All data analysis was carried out with version 0.3 of the PIE using the data modules and example data distributed with this version. PIE is available from http://bioinfo.med.unc.edu/Downloads/. Some of the details described in this chapter may change as interfaces are improved and data models are extended or added, so the documentation distributed with the PIE should be considered the authoritative reference.

2.1. Installing the PIE

The PIE 0.3 should run on any system that supports Java 5, although we have only tested extensively on Mac OS X (10.5). This version operates from the command line and has the following prerequisites:

- Java 1.5 or greater (http://www.java.com/en/download/manual.jsp).
- A user familiar with the basic concepts of mass spectrometry-based proteomics.
- A text editor to format input information.
- Some way to view and manipulate the results (e.g., R or Excel).
- That you agree to the noncommercial license pie is offered under.

Note that a graphical user interface is under development and should be available by the time this is published. Check our website at http://bioinfo.med.unc.edu for more information.

2.2. The PIE Distribution

PIE is distributed from http://bioinfo.med.unc.edu/Downloads/ as a compressed file. Download the latest version (highest numbered) file and unpack it. The resulting directory will be named `PIE-`*version*, but will be referred to here simply as `PIE`.

`PIE/bin/` – The pie application files, including `pie.jar`.
`PIE/data` – Template data files.
`PIE/demo` – Example runs including input data and results.
`PIE/doc` – Documentation.
`PIE/R` – Sample scripts for plotting graphs using the R statistics package (see Note 2).

2.3. Installing and Running PIE

PIE can be run directly from the distribution directory, or installed to run like an application. We briefly describe here how to run PIE directly. Instructions for installing and running PIE as an application are included with the distribution in the *`INSTALL.html`* document.

The PIE is written in Java and packaged in *`PIE/bin/pie.jar`*. As an executable jar-packaged program, this can be run on any system supporting java by using the *`java -jar`* command. Two arguments are required: the *`pie.jar`* application file and a run.properties based parameter file:

```
> java -jar "/path/to/PIE/pie.jar" "/path/to/myRun.properties"
```

A template *`run.properties`* file will be copied and modified for use as the second parameter, as discussed in Subheading 3. This will provide all the information needed to run PIE, including the names of the data files to be read and integrated.

2.4. Input Data

The PIE has a modular design, allowing it to integrate multiple data types by specifying a scoring module for each data type. None of the data types are required for PIE to run, because it is expected that data will be incomplete and contain errors and inconsistencies. However, there must be sufficient information content in the input data to produce useful output.

High-quality data likely to produce a reliable, single high scoring result has the following features:

- The target is a single protein or protein fragment with at most a few dominant patterns (isoforms).
- The target protein sequence is known, excepting N-, C-terminal truncations, and/or pre-specified individual amino acid substitutions.

- A list of all modifications to be considered is provided to the PIE.
- A high-resolution intact mass is available for the target isoform. This can be partially replaced by a constraint on the total number of modifications.
- Available peptide or MS/MS data with significant coverage of the target protein.

See Note 3 for some approaches that can be used when these conditions are not met (see also Note 4).

The data needed for the operation of the PIE fall into four categories:

- *Molecule data:* basic mass information amino acids, modifications, etc.
- *Experimental data:* evidence-based, specific to an analyzed protein or isoform.
- *Prior data:* average expectations, beliefs and background theoretical distributions.
- *Runtime data:* Parameters describing how to run the PIE.

Most molecular, experimental, and prior data are provided to the PIE through separate text files formatted as tables with predefined columns. In general, the first row gives column names and each following row represents one independent data element, such as a detected fragment. The row order is unimportant. Each column is a property of a row, such as the protein a fragment (row) is associated with or it's mass. Additionally, rows may be blank or start with a "#" symbol to indicate they are comments, not data.

Additional data is provided through the run.properties file (Listings 7 and 8). This file provides both general application settings – such as default input and output directories – as well as parameterizing each data-scoring module. All settings in the file are specified by key=value parameters. It is divided into three main sections: Section 1 is Data and Data Models (Listing 7). Within this section, each scoring module is identified by a "ModuleName" and has its own subsection, including an `isModuleNameScoring=`*`parameter`* turning it on or off and a moduleNameDataFile=parameter giving the file name it reads, if needed. Section 2 provides parameters to the underlying MCMC statistical engine. These do not generally need modifying and will not be discussed the third and final section, Run and Reporting Parameters, (Listing 8) is concerned with results: what to do during a run and how to report answers.

To make it convenient to specify the locations of files as simple file names without the entire path, the input parameters allow specifying up to three default input directories. Files are loaded first from the `defaultDataDir=`*parameter*, then from `experimentSetDataDir=`*parameter*, then from the `experimentDataDir=`*parameter*, and finally from the local directory. A file found in more than one directory will be loaded only from the *last* directory it is found in. The default configuration is to read all data from the `defaultDataDir`, leaving the other directories unspecified. However, hierarchically organized data is convenient when analyzing multiple proteins from large common data sets.

2.5. Molecule Data

The PIE needs to know the masses for amino acids, adduct modifications, and water. Each is provided in a separate table-based file, respectively ***aminoacid.txt***, ***modifications.txt*** and ***molecules.txt***. Each row in these files represents a different molecule, and each column describes some basic property of that molecule such as a (globally) unique name, aliases, and various measures for atomic masses. Only the average mass column is currently used. The monoisotopic and most abundant isotopic mass (MAIM) columns are present to allow for a planned extension, see Note 5.

The adduct modification file *modifications.txt* (Listing 1) is special in that besides mass data for the adduct modifications; it defines the modifications the PIE will search for. Data in this file was taken from `Proclame` (32), but is also available from sites like http://www.unimod.org/ (33). Adduct modifications are defined in terms of functional groups that may bond with a protein creating PTMs, so their masses must account for molecular gains or losses during binding. For example, a methyl group is 16.04 Da, but both the protein and the methyl must loose a hydrogen (2 × 1.01 Da) to form a covalent bond, making the net mass change of a methylation equal to 14.02 Da. Adducts with multiple modes of binding (different net mass changes) need to be listed multiple times. Only modifications described in this file will be searched for! Changes to the modifications list require changing the modification-based prior data as discussed later in Subheading 2.7, and

Listing 1. Default modification list.

Name	Abbreviation	Code	DeltaAvg	DeltaMono	DeltaMAI
Phosphorylation	Phos	P	79.9799	79.9663	80.0
Methylation	Meth	M	14.0269	14.0156	14.0
Acetylation	Acet	A	42.0373	42.0106	42.0
Oxidation	Oxid	O	15.9994	15.994915	16.0
Amidation	Amid	I	-0.9847	-0.98402	-1.0
Deamidation	Deam	i	0.9847	0.98402	1.0
Farnesylation	Farn	N	204.356	204.188	204.0
Formylation	Form	F	28.0104	27.9949	28.0
Myristoylation	Myrs	m	210.36	210.198	210.0
Palmitoylation	Palm	L	238.414	238.23	238.0
Selenocysteine	SelC	E	62.9606	63.9216	63.0

the number of modifications selected has consequences for the accuracy and running time for the PIE (see Note 6).

2.6. Experimental Data

Experimental data is the main source of information used by the PIE to select and localize modifications. It can presently use three types of experimental data: an accurate intact mass, a set of fragments matched to the target protein via a program like `GFS` (34) or `MASCOT` (12), and a set of MS/MS (sequenced) fragments with exact modification positioning information (35). Each data type requires a separate data module with accompanying text file (for the provided L16-A example they are in the distribution under `pie/demo/L16/experimental/`). Each experimental data can describe multiple protein targets, but only one protein at a time can be integrated with the current version of the PIE. A brief description of these modules and data files are described here; more complete documentation is available in the user manual.

The PIE reads protein sequence information from the *`targets.fasta`* file. This file is in standard FASTA file format (http://www.ncbi.nlm.nih.gov/blast/fasta). The protein name is read from the definition line text up to the first space, and which protein to use is specified by the `targetProteinName`=*parameter*.

Intact mass data from high-resolution mass spectrometry, e.g., from an FTICR or orbitrap instrument, is read from the *`intact.txt`* file (Listing 2). This data is evaluated using the `IntactMassScoring` data module. For each different protein `Name`, this file provides the total experimental mass, `MassAvg`, and the approximate mass `Error` (in ppm or absolute mass units). More accuracy means more resolving power when deciding between nearly equivalent answers. A mass accuracy of better than 10 ppm is ideal, though the PIE may give reasonable answers at 20 or even 50 ppm, depending on the situation. As previously mentioned, this and all other mass measurements must be of the same type, in this example they represent average isotope mass measurements (see Note 5).

Fragment data, such as that produced by matching peptide masses (e.g., peptide mass fingerprints) to proteins via a program like `GFS` (34) or `MASCOT` (12), is read from the *`fragments.txt`* (Listing 3). This data is evaluated using the `FragmentScoring` data module. To represent fragments, each line in the *`fragments.txt`* file is a separately matched peptide with an experimental mass (`FragMass`), the `Protein` it is matched to, the `Start` and `End` of that match, and the implied sequence (`AminoSequence`). If the peptide-matching program

Listing 2. Intact modification.

Name	MassAvg	MassMono	MassMai	Error	ErrorModel
L16-A	15222.1977	15222.1977	15222.1977	10.0	PPM

Listing 3. Unlocalized (matching precursor mass based) peptide information.

Protein	Start	End	FragMass	AminoSequence	ModList	Score
L16-A	6	13	1075.340	TKFRKMHK		51
L16-A	8	15	1059.311	FRKMHKGR		75
L16-A	16	33	1854.049	NRGLAQGTDVSFGSFGLK		37
L16-A	18	33	1583.764	GLAQGTDVSFGSFGLK		81
L16-A	18	33	1583.764	GLAQGTDVSFGSFGLK		81
L16-A	18	37	1967.199	GLAQGTDVSFGSFGLKAVGR		56
L16-A	40	54	1758.096	LTARQIEAARRAMTR	Methylation	42
L16-A	44	54	1316.553	QIEAARRAMTR	Methylation	86
L16-A	44	57	1614.940	QIEAARRAMTRAVK	Methylation	85
L16-A	62	80	2278.774	IWIRVFPDKPITEKPLAVR		87
L16-A	66	80	1710.061	VFPDKPITEKPLAVR		58
L16-A	81	113	3694.293	MGKGKGNVEYWVALIQPGKVLYEMDGVPEELAR	Oxidation	88
L16-A	114	126	1399.736	EAFKLAAAKLPIK		66
L16-A	118	126	924.191	LAAAKLPIK		70
L16-A	118	135	1933.422	LAAAKLPIKTTFVTKTVM		64
L16-A	127	135	1027.255	TTFVTKTVM		70

Listing 4. Localized (bottom-up MS/MS sequence based) fragment information.

fragmentNum	Protein	AminoSequence	ModList	InitAlignPos	Score
1	L16-A	GLAQGTDVSFGSFGLK		18	1.0
2	L16-A	VFPDKPITEKPLAVR		66	1.0
3	L16-A	GNVEYWVALIQPGK		86	1.0
4	L16-A	VLYEMDGVPEELAR	(8|Oxidation)	100	1.0

predicts any modifications, these are included as `ModList`. A (logarithmic) match `Score` indicating the quality of the match (e.g., a `Mascot` score) makes it easier to results from multiple peptides that overlap, repeat, or conflict. Where matched peptides are available, the PIE will use them to guide choice and placement of modifications implied in top-down data, including using peptides without modifications to guide where modifications are probably not found (see Note 7).

Data from sequenced MS/MS spectra, such as from peptides or middle-down fragments, is read from the *`localizedFragments.txt`* data file (Listing 4). The `LocalizedFragmentScoring` data module reads this file, including any modified amino acids identified by the sequencing, and evaluates it using an exponential model based on discrete differences. Each line in the file represents a separate sequenced spectra. Columns specify the `Protein` name that a sequence is matched to, the `InitAlignPos` (Start) positions for that sequence relative to the matched protein, the determined `AminoSequence` of the peptide/fragment, and the list (`ModList`) of modifications along with their location. A (logarithmic) `Score` of the peptide match allows the integration of conflicting information, such as when multiple PTM isoforms are present.

2.7. Prior Data

Experimentation is the only real way to determine and localize modifications, but real-world data are often incomplete and sometimes contradictory. When evaluating and interpreting less than perfect data, an expert relies on prior knowledge and experience. For example, an expert might know that phosphorylation adducts are common whereas β-methylthiolation adducts are very rare, and use that to guide the assignments. The PIE uses a set of prior data models to accomplish a similar feat. Background expectation data can be obtained from resources like Uniprot (http://www.uniprot.org/). Five separate prior data models are used: three models supply background information on the expected distribution and locations of adduct modifications, one model describes cleavages, and one allows specific rule-based biases to be applied.

The cleavage model is implemented by the `cleavageScoring` module, based on a simple open and continue model similar to affine gap scoring in sequence alignment (36). Evaluation based on four parameters from the *run.properties* file: the `nLoss`= parameter is the likelihood of the first AA being truncated from the n-terminal end, the `nLossMore`= parameter is the likelihood of each additional n-terminal AA cleaved beyond the first, `cLoss` and `cLossMore` are the same, but reference cleavage from the other end of the protein. Values near 1 lead to many cleavages, smaller values (0.5 and less) lead to few cleavages. This is a relatively primitive model, which easily handles the common "loss of N-terminal methionine" modification. This simple approach is a possible extension point, with models based on database scanning and on signal peptide prediction envisioned.

The three modification distribution models consist of a `modCountScoring` module that applies a distribution of the expected number of modifications based only on *run.properties* parameters, as well as the `modTypeScoring` and the `modLocationScoring` modules that use text files based on database scanning to predict and localize modifications. The `modTypeScoring` module uses the *modCount.txt* file that contains a row for each possible modification and a `Count` column, giving the (unnormalized) weight of how often a modification is expected (see Listing 5). The `modLocationScoring` module uses the *modLocation.txt* file, which is a table of weights giving, for each possible modification (row) and amino acid (column) combination, an (unnormalized) weight of how often that specific modified amino acid is expected (Listing 6). The default values are taken from dbPTM (37), but may often need to be modified. By default the program adds a +1 (a pseudocount) to all weights, to allow for novel modifications. To prevent any chance that a given AA/modification pairing will be suggested, a weight of "−1" be specified.

The `ModCountScoring` data module uses a center and spread model using `modRate`= as the expected number of adduct modifications and `modDelta`= as the average absolute value this

Listing 5. ModType.txt: relative weights for modifications.

PTM_Type	Count	PTM_Type	Count
Acetylation	2071	Methylation	746
Amidation	2150	Myristoylation	113
Deamidation	38	Palmitoylation	222
Farnesylation	62	Phosphorylation	22500
Formylation	32	Selenocysteine	2
Oxidation	1074		

Listing 6. ModLocation.txt: relative weights for modifications by AA.

PTM_Type	A	R	N	D	C	G	E	Q	H	I
Acetylation	424	7	-	6	5	60	10	-	-	-
Amidation	431	52	106	3	73	127	11	21	14	74
Deamidation	-	-	30	-	-	-	-	8	-	-
Farnesylation	-	-	-	-	62	-	-	-	-	-
Formylation	-	-	-	-	-	1	-	-	-	-
Oxidation	-	2	11	10	-	-	-	-	-	-
Methylation	10	180	22	-	40	-	29	22	14	0
Myristoylation	-	-	-	-	-	108	-	-	-	-
Palmitoylation	-	-	-	-	210	1	-	-	-	-
Phosphorylation	-	0	-	19	3	-	-	-	41	-
Selenocysteine	-1	-1	-1	-1	2	-1	-1	-1	-1	-1

PTM_Type	L	K	M	F	P	S	T	W	Y	V
Acetylation	-	792	240	-	14	432	64	-	2	15
Amidation	358	51	83	398	49	37	38	33	72	119
Deamidation	-	-	-	-	-	-	-	-	-	-
Farnesylation	-	-	-	-	-	-	-	-	-	-
Formylation	-	3	28	-	-	-	-	-	-	-
Oxidation	-	106	-	-	779	17	18	5	124	2
Methylation	7	407	4	6	5		-	-	0	-
Myristoylation	-	5	-	-	-	-	-	-	-	-
Palmitoylation	-	9	-	-	-	0	2	-	-	-
Phosphorylation	-	-	-	-	-	16590	3472	-	2375	-
Selenocysteine	-1	-1	-1	-1	-1	-1	-1	-1	-1	-1

amount this may be wrong by (Listing 7). If `modDelta` small, then only about `modRate` modifications will be predicted. If the error is large, then `modRate` is used as a guide, but easily ignored (see also Note 8).

The rule based module, `ruleScoring` will be expanded in future to allow for more convenient configuration, but currently allows setting value parameters for two specific conditions: How likely we think n-terminal acetylation is and how likely we think it is for both amidation and deamidation modification to occur in the same candidate. Values less than 1 are less likely than average, values greater than 1 are more likely than average. The purpose of this data module is to include odd bits of prior belief that might apply in a given situation.

Listing 7. Data and data models parameter file section.

```
# The data directories
#=====================
defaultDataDir = "/Users/srj/pieExperiments/ribosome/L16/A-15222"
experimentSetDataDir = ""
experimentDataDir = ""

# Molecular Data Files
#=====================
molDataFile = "molecules.txt"
aaDataFile = "aminoacids.txt"
modDataFile = "modifications.txt"

# Experimental Data and Scoring Parameters
#=========================================
proteinFastaFile = "targets.fasta"
targetProteinName = "L7/L12-A"

isIntactScoring = true
intactDataFile = "intact.txt"

isLocalizedFragmentScoring = true
localizedFragmentDataFile = "localizedFragments.txt"

isFragmentScoring = true
fragmentDataFile = "fragments.txt"
fragmentScoringAlgorithm = "deltaMass"

# Prior Data and Scoring Parameters
#==================================
isModLocationScoring = true
modLocationDataFile = "modLocation.txt"

isCleavageScoring = true
cLoss = 0.1
cLossMore = 0.5
nLoss = 0.5
nLossMore = 0.5

isModCountScoring = true
modRate = 0.0
modDelta = 100.0

isModTypeScoring = true
modTypeDataFile = "modType.txt"

isRuleScoring = true
```

2.8. Results

Parameters in the `run.properties` file tell the PIE how to run and how to output results, (Listing 8). A single run of the PIE usually consists of multiple searches, controlled by setting `runCount` = *value*. During each search, the number of steps taken to search for the best answer is controlled by `maxSteps` = *value*. As PIE runs, it will periodically output results to the console, after every `consoleUpdate` = *steps*. The `startSeed` parameter can be used to make the PIE behave deterministically by producing identical output for any given input data. If not specified, the PIE uses a random starting seed.

The PIE generates three files, a summary, a run detail, and a log file. By default these are generated into a subdirectory created at runtime in the directory specified though the `outputDir` = *parameter*. This directory is created if `isAutoOutputDir` = *true* (otherwise only the specified output directory is used). The subdirectory will be named using the date and time the PIE was run, such as `2009_10_12__21_55_00_032`. This prevents accidentally writing over previous data. Both the summary and detailed result files are tab-delimited text files, with a header line identifying the contents of each column. The log file is simply a narrative of what the PIE does as it happens. The names of these files are controlled by parameters, but can be left

Listing 8. Run and reporting parameter file section.

```
# MCMC Reporting parameters
#============================
outputDir = "/Users/srj/pieExperiments/target/"
isAutoOutputDir = true
detailedResultsFile = "pieDetails.txt"
summaryResultsFile = "pieSummary.txt"
logFile = "pie.log"
logFilterLevel = "DEBUG_LOW"

# Rerun information
#===================
startSeed = 0

# Main Runtime parameters
#========================
everyN = 5000
consoleUpdate = 25000
maxSteps = 100000
runCount = 2
```

at their "`pieSummary.txt`," "`pieDetails.txt`," and "`pie.log`" defaults.

The level of detail in the `pie.log` file can be controlled by changing the `logFilterLevel`. The default setting is generally adequate, recording an outline of what the PIE does including copies of all nondata messages. Changing the level from "`INFO`" to "`DEBUG`" or "`DEBUG_LOW`" will provide additional detail in the log file.

As pie runs, it will also output results to a detail file, after every `everyN` = steps. The detail file, *`pieSummary.txt`* is mainly useful when tuning MCMC parameters and is not discussed further. More detail can be found in the user documentation.

The *summary file* (Listing 9) is the main result file. It contains one entry for each search, `runCount` = data rows. Each line is one probable answer, a high scoring modification set candidate consistent with the data. Information reported includes `BestStep`, the step on which the highest scoring answer was found, `BestScore`, its total score, and the modifications predicted (`ModPos`, AA, and `ModName`). It also contains a separate `Best...Score` column for the score generated by each of the

Listing 9. Sample results summary file.

Steps	BestStep	BestScore	BestIntactMassScore	BestLocalizedFragmentScore
100000	32594	4.647362642103665E35	0.06784048759671008	1.6615349947311448E35
100000	49608	5.2017253454104814E36	0.6678687988002835	3.3230699894622897E35
100000	96999	5.180270237980864E36	0.6678687988056137	3.3230699894622897E35
100000	89018	2.6013295153132742E36	0.6678687988002835	1.6615349947311448E35

BestFragmentScore	BestModLocationScore	BestCleavageScore	BestModCountScore
169.34143123618742	1.0	0.5	1.0
119.11829455784581	1.0	0.5	0.8333333333333334
118.62697761168084	1.0	0.5	0.8333333333333334
119.13967573105404	1.0	0.5	0.8333333333333334

BestModTypeScore	BestRuleScore	CleavedLength	RawLength	NTerm	CTerm
0.4869393580246913	1.0	135	136	1	135
0.47222562396707823	1.0	135	136	1	135
0.4722256239670782	1.0	135	136	1	135
0.4722256239670782	1.0	135	136	1	135

TotalMass	ModCount	ModPos	AA	ModName
15220.106099999999	2	10	M	Formylation
15222.122	3	6	K	Acetylation
15222.122000000001	3	60	K	Acetylation
15222.122	3	2	P	Oxidation

ModPos	AA	ModName	ModPos	AA	ModName
125	K	Acetylation			
60	K	Methylation	107	P	Oxidation
107	P	Oxidation	131	K	Methylation
12	K	Methylation	60	K	Acetylation

individual component scores, allowing for detailed interpretation of results. Interpretation is covered in Subheading 3.4.

3. Methods

To illustrate the use of the PIE, we will follow a step-by-step analysis, using a synthetic example derived from real data pertaining to an isoform of the L16 ribosomal protein, shown in Fig. 17.3.

The experimental data consist of an accurate intact mass, about 50% coverage by matching peptide precursor masses (none showing modifications), and several matching ms/ms peptides, including one which has an oxidation localized to a specific residue.

The analysis takes place in four stages: *Setup, convergence, profiling, and interpretation.* During *setup*, the data and control parameters used by the PIE are collected and configured. During *convergence*, several small runs of the PIE are performed to determine an approximate convergence length. This length controls how much the PIE focuses on finding only the one best answer vs. finding more of the near-best answers. During *profiling*, the convergence length is used to generate an answer profile of candidate answers. Our example was chosen to include several common complications in obtaining this profile. The final stage, *interpretation* of the distribution, has the goal of examining the best and nearly best answers in the profile to provide both the sets of predicted modifications, and also information such as whether one or multiple isoforms may be present, how good the data is overall, and how valuable each of the integrated data sets are individually.

Each stage requires a number of steps to be performed sequentially, requiring various command line interactions with PIE and with the computer system on which PIE is running. For the example that follows, we provide both a general description of each step and explicit commands to perform it. However, even though PIE can be run on any system supporting Java, the details of the required interactions with that system will vary. In the interest of space, the explicit example commands are targeted only at a Un*x-based system such as Mac OS X or Linux. If you are using another system, it should be simple to extrapolate the needed command. For additional information, see the user manual.

3.1. Setup

Setting up a PIE run involves collecting and editing data files needed as input, and then setting `key=value` parameters in the configuration file used to control how PIE runs. For the most part, our example will use prepared files provided with the PIE distribution that are already formatted correctly. We will discuss only aspects of the files that are relevant to our examples. General information, such as the purpose and meaning of parameters not discussed here, is contained in Subheading 2 above and in the user's manual that accompanies the distribution. To do the example, you can follow these steps:

1. Download the PIE distribution as described in Subheading 2.
2. Create an experiment directory to work in and make it the current directory.

   ```
   > mkdir /Users/jefferys/ribosomeProject/L16/A-15222
   > cd /Users/jefferys/ribosomeProject/L16/A-15222
   ```

   Note: We use *L16/A-15222* as our working directory, based on the target protein and the intact mass of an imaginary primary isoform. The path */Users/jefferys/ribosomeProject/* should be replaced with the one used on your system. From here on this working directory is assumed.

3. Prepare experimental data for use by the PIE.

   Experimental data collected by the user must be correctly formatted to be read by PIE. For simplicity, we use the preformatted data files from `PIE/demo/L16/experimental/`:

   targets.fasta – Our target protein sequence, L16-A, is an entry in the provided file.

   intact.txt – Our (average) intact mass is 15222.19 Da, with error ~10 ppm.

   fragments.txt – A set of trypsin digest peptide masses that each matched one of the masses in a putative digestion of the sequence (see Fig. 3).

   localizedFragments.txt – Matched MS/MS peptides that can provide localization information on any modifications, such as the oxidation adduct in the example.

4. Copy the correctly formatted experimental data files to the working directory.

   ```
   >cp/path/to/PIE/demo/targets.fasta.
   >cp/path/to/PIE/demo/L16/experimental/intact.txt.
   >cp/path/to/PIE/demo/L16/experimental/fragments.txt.
   >cp/path/to/PIE/demo/L16/experimental/localizedFragments.txt.
   ```

Fig. 3. The target L16 isoform and experimental data an artificial L16-A protein isoform is shown in *green*, with three adduct modifications represented by *triangles*: acetylation at 2-L (*blue*), methylation at 49-R (*purple*), and oxidation at 107-P (*red*). There is also one AA truncation at the N terminus. This mock target is the answer that the PIE is seeking from the data. The two sets of *bars* above the target represent synthetic fragment (*grey*) and MS/MS (*black*) data. The MS/MS data identifies AA 107-P as having an oxidation modification (*red line*), and many others that are unmodified. The peptide data shows no modifications, but does identify additional unmodified regions of the protein. Multiple overlapping peptides with different match scores are represented; the darker the grey, the better the match. Confidence increases (*darker grey*) when peptides overlap. Although providing moderate coverage, the bottom-up data is significantly incomplete. It lacks any indication of the acetylation or methylation modifications, and allows for cleavage of up to six N-terminals AA before contradicting any peptide data. Using such incomplete data how PIE infers adduct modifications and terminal cleavages from an intact mass, and how incomplete data sets can support multiple answers.

For this example, we are using preformatted example data files. Generating the necessary files from experimental data is generally a straightforward task. Note the "." is not a period, but represents the working directory, and is required.

5. Copy the molecular mass data files to the working directory; Modify as necessary.

   ```
   >cp/path/to/PIE/data/molecule/aa.txt.
   >cp/path/to/PIE/data/molecule/molecule.txt.
   >cp/path/to/PIE/data/molecule/modification.
   txt.
   ```

   No modifications to these files are necessary for our example as it uses the default set of modifications, and is designed for average isotopic mass measurements.

6. Copy the prior data template files; Modify as necessary.

   ```
   >cp/path/to/PIE/data/molecule/modType.txt.
   >cp/path/to/PIE/data/molecule/modLocation.
   txt.
   ```

   We are using the default set of modifications to look for, so we do not need to add or subtract any rows from these files. However, because L16 is a ribosomal protein, we suspect that phosphorylations are less common than average, so we reduced the weighting for this adduct ten-fold (see Note 9). Edit the `modType.txt` file to the values shown in **boldface**

   `Phosphorylation` **2250**

7. Copy the run.properties template file.

   ```
   >cp/path/to/PIE/data/run.properties.
   ```

This is the file that will control details of how the PIE runs.

8. Edit the following parameters in the "Data and Data Models" section of the *run.properties* file. Edit the *run.properties* file to the values shown in **boldface**:

   ```
   defaultDataDir=
   "/Users/jefferys/ribosomeProject/L16/
   A-15222"
   targetProteinName="L7/L12-A"
   modRate=2.0
   modDelta=7.0
   ```

   The first edit tells the PIE which default directory to read from. The second identifies the protein name, and hence PIE will know which lines to read from the experimental data files. The last two edits adjust the data parameters for the `modCount` data-scoring module to match expectations for our project – see Note 8. We are otherwise using default file names and settings, including the cleavage and rule scoring modules.

9. Edit the following parameters in the "Run and Reporting" section of the *run.properties* file.

   ```
   outputDir="/Users/jefferys/ribosomeProject/
   L16/A-15222"
   ```

   This just defines the output directory to be the same as the input directory. As we are leaving `autoOutputDir` true, each PIE run will generate its own subdirectory. We are also not changing the default run parameters: `everyN=1,000`, `consoleUpdate=2,500`, `maxSteps=10,000` and `runCount=2`. These are set for the test run we execute next.

10. Execute a test run of the PIE.

    ```
    >java -jar "/path/to/PIE/bin/pie.jar" "./run.
    properties"
    ```

11. Verify that everything went well.
    - Should have run for a few seconds, writing to the screen as it went.
    - Should have exited without reporting an error. (Low memory errors are discussed in Note 10.)
    - *2009_12_17--18_23_35_951* should have been created, containing three files: *pieSummary.txt*, *pieDetails.txt*, and *pie.log*.
    - The *pieSummary.txt* file should contain two lines, with separate "best..." scoring columns for the total score and for each data model used. In our case, this is 1 total + 5 prior + 3 experimental = 9 columns.

- The *pieDetails.txt* file should contain two sets of 10 lines each.
- The *pie.log* file is useful if errors occur and the expected results are not generated.

12. Give the run.properties file and the timestamp directory better names.

    This is just to assure that when we refer to them later that we will know what the run was.

```
> mv 2009_12_17--18_23_35_951 test-2atE4
> mv run.properties test-2atE4.properties
```

3.2. Convergence

After setting up all the data, we need to experimentally determine one parameter, the convergence length. The goal is to adjust the number of steps taken by the MCMC walk to not just in find one best answer, but to obtain a useful profile of answers that contains the near-best answers as well. Efficiently generating a useful answer profile involves a trade-off between the number of candidates reported and the time spent searching for each candidate. Short runs will probe deeper into the lower scoring answers. Longer runs will focus coverage around just the top answers. Although exactly how run count and run length interact when generating answer profiles differs for every data set, the PIE's run time always increases proportionally with both. Because running time is significant (15 min or so with computers circa 2009) we want the smallest parameter values we can get away with. Empirically determining an initial value for the run length parameter is done by fixing the run count at 10, and then generating and comparing several small profiles with different run lengths (see Fig. 4 and Note 1). This results in overall time savings when trying to generating a useful answer profile.

It is important to keep separate the two different ways we will use the PIE: First to determine convergence, and secondly to generate a profile of candidate answers. We plan to automate the determination of the convergence values in the future, but for now, this must be performed manually as described below:

1. Make a new copy of the properties file for the convergence test:

```
>cp test-2atE4.properties conv-10at1e5.
properties
```

2. Modify this new properties file to do ten replicates of 100,000 steps:

   Edit the *conv-10at1e5.properties* file to the values shown in **boldface**:

   everyN=**10,000**
   consoleUpdate=**25,000**

```
runCount=10
maxSteps=100,000
```

3. Run the PIE, renaming the output directory when done

```
> java -jar /path/to/PIE/bin/pie.jar ./conv-10at1e5.properties
> mv 2009_12_17--18_25_14_391 conv-10at1e5
```

4. Examine the *pie.summary* output file. Determine the highest "bestScore" value and the number of times it is repeated.

   When done, the `pieSummary.txt` file (Listing 9) will contain the best scoring answers from each of ten runs. The `bestScore` column gives each run's total score. A spreadsheet program can be used to import, view, and manipulate this file (see Note 2). If the highest score is repeated in more than one of the runs, we are on our way to determining the convergence length to use in our full analysis (Fig. 4).

5. Execute a longer convergence run and check for repeated scores.

   We just repeat steps 1–4 but increasing the maximum number of steps by a factor of 5×.

```
>cp conv-10at1e5.properties conv-10at1e6.properties
Edit conv-10at1e6.properties:
    everyN=50,000
    maxSteps=500,000
>java-jar/path/to/PIE/bin/pie.jar./conv-10at1e6.properties
>mv2009_12_17--19_02_42_844 conv-10at1e6
```

6. Keep repeating with larger max steps until convergence has been obtained.

   Until we obtain two successive profiles where the same maximum score is repeated, we keep increasing the maximum number of steps and generating new profiles. The amount to increase the step size by matters only in the sense that we are trying to find the convergence length in a small number of guesses without wasting too much time on large guesses (Fig. 4).

7. Interpolate to find a max steps value giving approximately two runs with the same best score.

   Because our smallest run gave 4/10, we already obtained convergence. We therefore reduce the number of steps to check for the point at which we get closer to two runs with the same scoring best result. We try 50,000 steps (copying the parameters file, renaming it, and adjusting the same four run parameters as appropriate). This seems to be exactly the correct number of steps, resulting in two of the ten searches having the maximum value (Fig. 4).

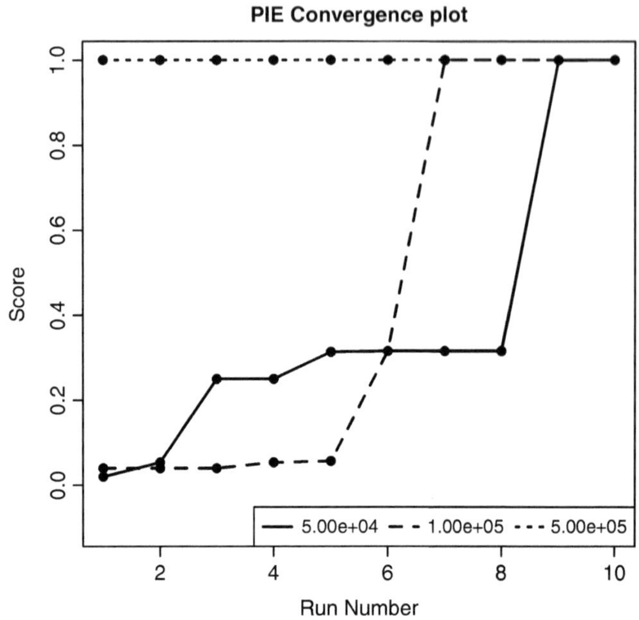

Fig. 4. Convergence. Three separate runs were performed for the L16 protein, each consisting of ten searches but using different search lengths: 50,000 steps (*solid*), 100,000 steps (*dashed*), and 500,000 steps (*dotted*). For each run, the ten search results are ordered from lowest scoring to highest scoring and plotted on the graph from *left* to *right*. The score axis has been normalized to the largest value found across all runs. The initial 100k step run appears to show convergence, with four of the ten runs having the same high score. As with any other stochastic process (such as measurement!), the results from the PIE will differ in specifics even for identical input, but on average are consistent. The second run at 500k steps, presents only the top scoring candidate, ten out of ten times. Together these provide strong evidence that this top scoring candidate really is the best candidate that can be found. A third run of 50k steps gets two out of ten values converging to the same high scoring value, making 50k our approximate optimal convergence length (see Note 1).

3.3. Obtaining a Complete PTM Answer Profile

Now that we have an estimate for the convergence length, 50,000 steps, we can move on to generating and interpreting the full answer profile, by generating many more replicate runs:

1. Repeat the steps used to generate the 50k convergence run, except use more replicates.

   ```
   >cp conv-10at5e4.properties profile-100at5e4.properties
   ```
   Edit `profile-100at5e4.properties` file:
   ```
       runCount = 100
   ```
   ```
   >java -jar /path/to/PIE/bin/pie.jar./profile-100at5e4.properties
   >mv 2009_12_17--20_43_03_709 profile-100at5e4
   ```

Although our goal is different, the process of generating a profile of answers is the same. Figure 5 is an example of the kind of result produced.

2. Examine the `pieResults.txt` and determine if it is good enough.

   (a) If the top candidate is represented too many or too few times, increase or decrease the run length and try again.

   Using a spreadsheet program or the provided R scripts, examine the number of identical best scoring results. There should be several replicates of the top scoring result. If the profile does not provide enough replicates, refine the convergence length estimate, and try again. Ideally, we would like a quantitative connection between the number of replicates of a given score and our confidence in the adequacy of coverage. Lacking that we choose five as a rule of thumb. For our example, 50k steps turns out to be an underestimate, providing only two replicates of the top-scoring answer. The larger number of runs provides a higher resolution result, indicating a larger number of steps are needed to increase separation between very nearly identical best-scoring answers.

   We refine our convergence length estimate using this additional data, and repeat step 1 using 100k steps, with results named `profile-100at1e5`. Figure 5 shows our results.

   We now have about ten of the answers in the top scoring set, and the following nearly best answers are well represented, each presented in its own five or more score wide, equally-scoring block.

   (b) If the top candidate is well represented, but there are other candidates in this high scoring group that have only one or two replicates, increase the number of searches (runCount) and try again.

   Creating a larger profile, say with 500 runs, increases the "resolution." Providing more searches is like adding pixels to a screen, by widening every scoring band. It may be possible to decrease the number of steps if running many extra searches (see Note 1). As discussed next in the interpretation section, we have one interesting candidate that occurs just once at a score 85% of the best result. If desired, we could increase the run count to look for other suboptimal solutions that could have been missed because of random chance. However, we have plenty of candidates to work with, so we will not do that here.

3.4. Interpretation

It has taken several steps to get to this answer profile, so it is probably a good idea to step back and check out what we have done.

We have collected several different kinds of experimental mass spectrometer data derived from a (theoretical) variant of L16 ribosomal protein. To this collection of experimental results, we have added some general prior knowledge about modified proteins, such as which modifications are more common. We also included information that applies to our specific domain, by specifying a lower likelihood of phosphorylation than average because of the protein being a ribosome component. We then dumped all that information into a directory and used the PIE to put it all together and tell us about the modified protein variant or variants described by this data.

This process of integration and interpretation took place in two phases, first finding a convergence length and then simply running pie with the correct length to find not only the best answer – the modification pattern that is most consistent with the PIE-evaluated data – but also the runner-up choices. Estimating the correct convergence length was not completely successful using our quick and short runs of ten searches each, but after one extra higher-resolution round of 100 searches, we obtained a useful value. By using longer searches in the second round an answer profile with enough resolution to be interpretable was generated (Fig. 5) How to interpret the answer profile is the subject for the rest of this section.

1. What modification is present in the highest scoring candidate?

   In our example, the highest scoring candidates presents a set of four modifications – one methylation, one oxidation, one acetylation and a single n-terminal amino acid loss (methionine). This accurately reflects the set of modifications expected.

2. Is there a consensus set of modifications present in the runner up (suboptimal scoring) candidates?

   In our example, almost all the highest scoring candidates present the same consensus set of four modifications – one methylation, one oxidation, one acetylation and a single n-terminal amino acid loss (methionine).

   The presence of these modifications are thus a highly supported predictions for our protein variant and have been identified despite there being no specific evidence for methylation or acetylation or the n-terminal methionine truncation anywhere in the experimental data. Essentially PIE extracts this information from the intact mass, solving the combinatorics puzzle of what pieces can be assembled to get this mass.

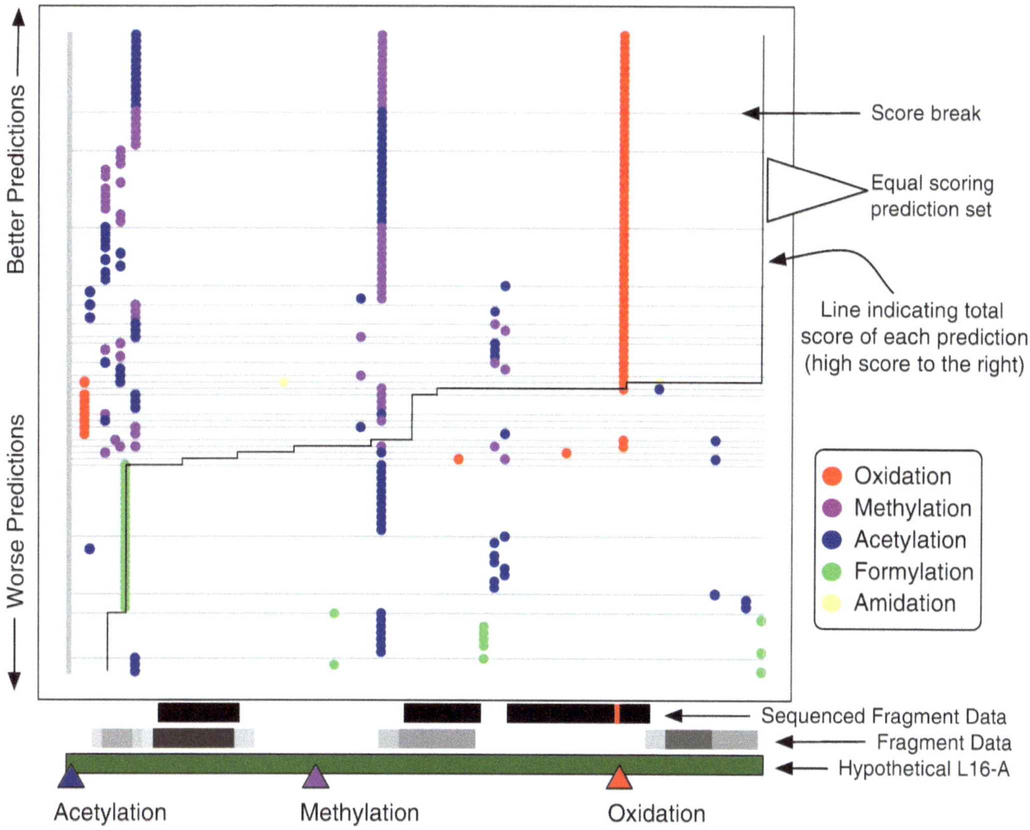

Fig. 5. Answer profile. This answer profile represents the combined results from 100 searches, each 100k steps long, for the L16-A target described in Fig. 3. The 100 candidate results are stacked vertically, ordered from the lowest scoring (*bottom*) to highest scoring (*top*). The horizontal axis represents the amino acid sequence of the protein from N terminus (*left*) to C terminus (*right*), with only the adduct modifications shown (*colored dots* positioned where predicted). Consensus modification positions are easily seen as overlapping vertical columns of dots. The *left* side of the graph shows a grey bar indicating the N-terminal truncation, predicted for every candidate. The relative score of each candidate is indicated by where it crosses the *dark-grey* score line. This line begins vertically on the *right* (at 100% relative score), transitions to a jagged horizontal line across the *middle*, and then ends at the *bottom left* (about 5% relative score). The *horizontal light-grey* lines delineate answer sets within which all answers have identical scores. Most of the high-scoring sets have correctly identified the three adducts and the N-terminal truncation, although there is one high scoring answer (about 85%) that suggests we have two amidations and an extra oxidation instead of the methylation, a surprising answer that cannot be ruled out given the available data. The position of the oxidation modification has also been correctly identified throughout most high scoring candidates, aligning with the oxidation modification presented in the sequenced fragment data. The positions for the methylation and acetylation modification are not correctly aligned with the target, but are generally placed where they do not conflict with the "unmodified position" data from the MS/MS and peptide data. More data is needed to pin down the exact location of these modifications.

3. If there are other high-scoring modification sets, compare the data-specific scoring columns from the pieSummary.txt file to determine the scoring cutoff.

   The first alternate candidate set, found only once, has two oxidations, two amidations, one acetylation, and a one

AA n-terminal truncation. It scores about 80% as the consensus modification set. To determine the "reason" for the lower candidate score, we compare each of the individual data module component scores to the component scores of the best candidate. The biggest drop is in the mod location prior score, indicating at least one of the modifications in this lower scoring answer is placed on an uncommon site.

The next suboptimal set of modifications is one formylation, one acetylation, and one N-terminal cleavage. It is only about 7% as good as the best one. Its main failing is that it does not contain any oxidation site, which conflicts with the explicit MS/MS information. This or worse problems present in every other lower scoring candidate, so we ignore those poor scoring results.

Multiple (2) high scoring candidate sets are present in our case. Because both sets are consistent with all experimental data, the data is not complete enough to pinpoint a single best answer.

One of the design goals for the PIE is to use it as a tool to determine what additional experiments or additional data may be needed to resolve cases such as this. New data can be acquired and PIE rerun. Data indicating a methylation would rule out the alternate candidate, giving only one high-scoring set. Likewise, data indicating a second oxidation or an amidation would raise the score of the alternate candidate, replacing and probably eliminating the top scoring candidate.

Although not the case in this example, the answer profile can be interpreted to detect multiple isoforms. If more than one high scoring candidate exists, but each conflicts with different parts of the bottom up data, this means the data that cannot be simultaneously satisfied by just one answer. Each of the conflicting answers are required, meaning multiple modification isoforms are present.

4. Given the set of modifications, do any have consensus positions?

In our example, all searches have the PIE placing the N terminus of the protein after the initial methionine, predicting one AA truncation, and the C terminus of the protein after the last AA, predicting no C-terminal truncation. These are both correct given the known L16 target. All top scoring answers also correctly localize the oxidation to 107-P by using the MS/MS data.

We only have one result including any amidations. If we believe this to be a viable candidate, we need to run more PIE searches (larger profiles) to add resolution to the bands in the middle of the graph. To keep this example simple, we will focus only on the best scoring modification profiles.

5. For modifications that aren't localized, is this because of conflicting information from MS/MS data, or missing information?

   There is no clear consensus in the output on where the methylation or acetylation modifications belong. Several different positions are suggested for each amongst the very high scoring candidates. If this lack of consensus had been caused by conflicting fragment information, that would be evidence for multiple position isoforms. This is not the case here, as the bottom up data contains information only on the oxidation site.

   In this case, we know that the acetylation should be on the N terminus (after truncation of the methionine), and that the methylation should be on 49-R. However, there is nothing in the input data to tell the PIE this. Intact mass data was no help here, as it only identifies *what* modifications there are, not *where* they go. The MS/MS data help PIE to localize the oxidation and, along with the matching precursor mass peptide data, also provide constraints to the PIE for *where not* to put modifications. However, after considering all this there are still many places were modifications could go, with no experimental evidence to choose any position preferentially.

   The PIE will try to fall back on prior data in this case. For example, if a selenocystine modification was present, and there was only one cystine, then prior data alone would pick the correct location. For acetylation and methylation, prior data is not as useful as there are many different amino acids that might be modified present throughout the protein. For methylation, our prior data gives highest priority to lysines (K), and only when it cannot place modifications there will it target arginines (R). For the acetylation, a similar background AA-based prior is supplemented with a prior rule that raises the score if it is placed on the N terminus However, after methionine loss, the N terminus of this protein presents leucine (L). Based on the background AA prior location data, this is nearly the last place the PIE will put an acetylation, first selecting K, then A then S, etc. One can see this background prior at work in Fig. 17.5 with different amino acids picked for methylation and acetylation. Free positions (not covered by bottom up data) with the same amino acid will score the same, and hence modification positions will vary within a same-scoring band.Enhancing the performance of priors by collecting more specific data on N- or C-terminal modifications is one possible way to improve the performance of the PIE in this kind of situation, and PIE's data scoring is modular to make this easy. However, priors will only help fill small gaps in knowledge. It is the necessary role of the experimental data to do the heavy lifting.

6. You are done!

   We now have a characterization of the posttranslational modification pattern for the L16-A protein. Extending this example to analyzing other data is easy if those data are clean and complete. If there are insufficient data, the PIE will suggest the most likely candidates, allowing determination of what experiments to run that will select efficiently between possible candidates. After collecting more data, you can add it to the PIE, and a reduced set of candidates will be generated. If data is really poor or complicated, a straightforward and simple interpretation is not possible in any event, but analyzing it with the PIE may provide additional insight that would not otherwise be possible. We discuss a variety of ways PIE might be used in more complex situations in Note 3.

4. Notes

1. **Determining the run length parameter.**

   We first find an upper bound on the number of steps parameter, the search length L that (probably) allows us to find the best scoring modification set. Because we do not know a priori what the best score is, we will not recognize it when we first see it. Only repeatedly finding the same best scoring answer several times and at several run lengths gives us confidence that there is no better answer left unreported. Specifically, in one profile reporting ten searches of length L, we find the same best scoring result at least two times, and, in a second profile of ten searches each at least $1.5 \times L$ or longer, we also find at least as many of the same best scoring result. This leads us to believe that the best scoring answer is likely to be contained in any profiles with ten or more runs of length L or longer.

   We next find the minimum search length L that will work. We know the best scoring answer has score B, and can be generated by a search length of L or smaller, we generate profiles using successively shorter search lengths (say by halves), until only one to two out of ten results of score B are found. This value for L is then a coarse estimate to use in generating a full answer profile, with about 100 runs.

   It is reasonably likely that the estimate for run length, L, will not be perfect. Larger convergence profiles would allow better estimates, but also require more time. There is really no need for better than a coarse estimate as the full answer profile itself can be used to "fine" tune the parameters, re-running with different run length and/or run count parameters. This is exactly what happens with the L16-A example (Subheading 3).

2. **R scripts.**

   Some "R"-based analysis scripts are available in the PIE distribution in the pie/R directory. These can be used with the R statistical analysis software – available from The R Foundation for Statistical Computing (http://www.R-project.org) – to generate convergence profile graphs similar to those in this chapter. See the user manual for more information.

3. **Using the PIE to solve more complex problems.**

   The example presented is a relatively straightforward application of PIE. Proteomic analysis not look like our example. We will continue to extend the example set distributed as the PIE grows, but a few quick tips are provided here.

   If the sample analyzed consists of a mixture of different proteins, not just different isoforms: A separate run for each possible protein can be done. Although scores are not comparable between runs with different data or protein targets, the number of replicates of the highest scoring value and the ease with which convergence is obtained can be used as a rough guide to the best targets, and bad choices for proteins will likely result in the uninformative prior.

   If there are many isoforms or a large number of identical modifications varying only by modification position: Nothing will help distinguish positions other than position-specific data. This could be top-down MS fragmentation data or bottom-up peptide MS/MS data indicating amino acid specific modification information, or a more accurate prior distribution, such as the variable weighting of phosphorylation sites generated by Netphos (23).

   If the protein has long truncations caused by signal peptide removal: Using a data module that scores cleavages based on information from a program like `SignalP` (25) would improve results. Such a module will be part of an upcoming release. You can always run such a program yourself and use the predicted truncated protein as a target.

   If the protein contains amino acid substitutions: these can be considered "modifications", although care must be used when cyclical modifications that result in a net zero mass are possible, i.e., if. `K→T`, `T→S`, and `S→K` are all allowed, many isobaric modification sets are possible (with 1, 2, 3, … *N* sets of these three modifications). If net-0 mass modifications are possible, only a narrow range for the number of modifications (less than the length of the cycle) will be efficient, and longer runs will likely be necessary.

A list of adduct modification names for the mass deltas is required. If bottom up MS/MS data contains an identified peptide, but with a mass shift caused by an unidentified PTM, this might be entered in the `modifications.txt` file as an "unknown adduct..." with appropriate mass data. However, it is hard to provide the requisite prior information for this modification relative to other modifications or target AAs.

Adduct modifications are variable, such as polysaccharides or lipids. If you cannot easily list each individual specific mass for all the possible modifications, the PIE cannot currently predict these modifications. We believe the PIE could be extended to allow for variable modifications, such as by adding or subtracting pieces of a modification instead of an entire modification at once, but that will have to wait for later versions.

The intact mass is an important piece of information.

Without this, the regions of the protein not covered by fragment data have only the prior distribution to describe where to put modification. Without an intact mass, it may be possible to run the PIE multiple times varying the numbers of expected modifications parameter of the modCount data module (see also Note 8).

With neither useful top down nor bottom up data: There is no experimental evidence, and the PIE has nothing to work with. The PIE can only integrate the data you give it. Without experimental evidence, only an uninformative prior result is obtained.

4. **Uninformative prior-only results.**

   If good data are available – data that provide a complete and consistent picture of a protein and its modifications – the PIE will be able to find unique high scoring answers for the modification state of the protein. If data significantly incomplete, convoluted, or contain contradictory information, the PIE will likely still provide useful knowledge by characterizing the modification scenarios that are supported, to what degree they are supported, and how each data type individually contributes to that support.

   However, if the experimental data are particularly uninformative, a "prior only" result will be obtained. For comparison, a true prior only result can be generated from the PIE by running with all experimental data modules turned "off."

5. **Isotopic type of mass measurements.**

   In the current implementation, all mass measurements across all data files are taken from the average isotope mass columns. To use a different mass measurement, such as the monoisotopic masses, you can simply specify those in place of the average mass column. Although the column label will not match the data in it, the calculations will be performed using the correct mass. It is important that the same mass type be used across all experimental data as well – there is yet no support for using more than one mass measurement type, or for converting between them. Picking the mass measurement type using a parameter.

6. **Specifying a modification set to search for.**

   The modification set considered by the PIE for a given run is provided in the file `modifications.txt`. *Not including a modification in the modification set in the* `modifications.txt` *file is one of the few actions that can cause correct answers to be excluded from the search space considered by the PIE.* To allow for novel solutions it is best to leave the list as long as possible, although if the list gets too long, run-times will be extended and the PIE will eventually have trouble finding the best answers. Even though the PIE does not suffer from exponential explosion in computing time, every modification specified will increases the time needed. Adding a reasonable number of modifications should not adversely affect performance; adding all the modifications listed in dbPTM (37) likely would. Currently, adding a modification to the *`modifications.txt`* file also requires adding associated data to the files used by the `ModType` and `ModLocation` prior data modules as described in Subheading 2.

7. **Evaluating peptide data sets.**

   For evaluating fragment data, several scoring models are implemented, each with their own effects on the outcome. These will be separated into different modules, but must currently be selected using the `fragmentScoringAlgorithm=parameter`. The default "deltaMass" algorithm used in the example is a center and spread model based on theoretical vs. actual peptide masses. This can be used when only precursor masses or "peptide mass fingerprint" style data are present. The "errorCounting" algorithm provides instead an inverse exponential model. This does not use the mass of the fragment, only its "interpreted" sequence and modification components from a program that can localize the PTM (e.g., `Findmod`, (22)).

8. **Setting the number of modifications.**

   One reasonable prior for the number of modifications is to set the expected number of modifications to zero, and the allowed range of modifications to the length of the protein. This guides selection of the fewest modifications that are consistent with remaining data.

   For the example, we only know of one possible modification, an oxidation. However, our intact mass does not reflect such a simple answer, so we set the expected number of modifications to 2. There is no need to be exact when setting this parameter as we have good intact data. Setting it to zero or one would still work, but as with any modeling process, the more accurate and consistent the data, the better. We have no reason to suspect a large number of modifications, so we set the spread about the expected number of modifications to 7. This allows zero to modifications with reasonable probability when supported by other data. The choice for seven as a range is a guess, based on the moderate coverage of the peptide without other modifications, the small number of modifications already seen, and a bias toward believing that if the protein has a large number of modifications, we would know.

Without experimental intact mass data, integration of results becomes harder, and the PIE is forced to rely on this prior as the guide for predicting modifications in regions of the protein not covered by peptide or fragment data. However, it is still possible to explore the space of candidates consistent with the peptide or fragment data by using a sequence of values (0, 1, 2, ...) for the number of expected modifications and setting the allowed range of modifications very low (like 0.25). The results then summarize the supported isoforms assuming one modification, then assuming two modifications, etc.

9. **Using background knowledge as prior information.**

One of the design goals for the PIE is to allow easy incorporation of various types of knowledge about a problem. Sometimes this knowledge is in the form of experience about what the answers should and should not look for a protein given its context. The PIE makes use of such information through explicitly stated prior distributions of expected results. For example, the distributions of expected modification types and locations is likely different between ribosomal proteins and signal transduction proteins. Existing scoring module parameters can be edited or new scoring modules can be written to reflect different prior expectations within different contexts. One module will apply a prior for phosphorylation based on the predictions of the program `Netphos` (23). Although not a substitute for good experimental data, prior data helps make better guesses where MS data is missing.

10. **PIE needs memory.**

Java assumes only a small amount of memory will be needed by any program by default, and the PIE may fail with an error message if it needs more. The quantity of raw data and properties such as the number of iterations increase the amount of memory needed.

To adjust the PIE to run with more memory, use the standard java memory setting parameter with the `java-jar` command

```
>java-Xmx256M-jar"/path/to/pie.jar""/path/to/run.properties"
```

This tells java it can use up to 256 MB of memory. If that is not enough, it may crash, and you can go larger.

References

1. Seo, J. and Lee, K. J. (2004) Post-translational modifications and their biological functions: proteomic analysis and systematic approaches. *J. Biochem. Mol. Biol.* **37**, 35–44.
2. Walsh, C. T., Garneau-Tsodikova, S., and Gatto, G. J. (2005) Protein posttranslational modifications: the chemistry of proteome diversifications. *Angew. Chem. Int. Ed. Engl.* **44**, 7342–7372.
3. Kollmann, M., Lovdok, L., Bartholome, K., Timmer, J., and Sourjik, V. (2005) Design principles of a bacterial signaling network. *Nature* **438**, 504–507.
4. Kentner, D. and Sourjik, V. (2006) Spatial organization of the bacterial chemotaxis system. *Curr. Opin. Microbiol.* **9**, 619–624.
5. Shi, Y. (2007) Histone lysine demethylases: emerging roles in development, physiology and disease. *Nat. Rev. Genet.* **8**, 829–833.
6. Minamoto, T., Buschmann, T., Habelhah, H., Matusevich, E., Tahara, H., Boerresen-Dale, A. L., et al. (2001) Distinct pattern of p53 phosphorylation in human tumors. *Oncogene.* **20**, 3341–3347.
7. Banerjee, A. and Gerondakis, S. (2007) Coordinating TLR-activated signaling pathways in cells of the immune system. *Immunol. Cell Biol.* **85**, 420–424.
8. Mann, M. and Jensen, O. N. (2003) Proteomic analysis of post-translational modifications. *Nat. Biotechnol.* **21**, 255–261.
9. Domon, B. and Aebersold, R. (2006) Mass spectrometry and protein analysis. *Science* **312**, 212–217.
10. Albrethsen, J. (2007) Reproducibility in protein profiling by MALDI-TOF mass spectrometry. *Clin. Chem.* **53**, 852–858.
11. Eng, J. K., McCormack, A. L., and Yates III, J. R. (1994) An approach to correlate tandem mass spectra data of peptides with amino acid sequences in a protein database. *J. Am. Soc. Mass Spectrom.* **5**, 976–989.
12. Perkins, D. N., Pappin, D. J., Creasy, D. M., and Cottrell, J. S. (1999) Probability-based protein identification by searching sequence databases using mass spectrometry data. *Electrophoresis* **20**, 3551–3567.
13. Craig, R. and Beavis, R. C. (2004) TANDEM: matching proteins with tandem mass spectra. *Bioinformatics* **20**, 1466–1467.
14. Little, D. P., Speir, J. P., Senko, M. W., O'Connor, P. B., and McLafferty, F. W. (1994). Infrared multiphoton dissociation of large multiply charged ions for biomolecule sequencing. *Anal. Chem.* **66**, 2809–2815.
15. Kelleher, N. L., Zubarev, R. A., Bush, K., Furie, B., Furie, B. C., McLafferty, F. W., and Walsh, C. T. (1999) Localization of labile posttranslational modifications by electron capture dissociation: the case of gamma-carboxyglutamic acid. *Anal. Chem.* **71**, 4250–4253.
16. Zubarev, R. A., Haselmann, K. F., Budnik, B., Kjeldsen, F., and Jensen, F. (2002) Toward and understanding of the mechanisms of electron-capture dissociation: a historical perspective and modern ideas. *Eur. J. Mass. Spectrom.* **8**, 337–349.
17. Siuti, N. and Kelleher, N. L. (2007) Decoding protein modifications using top-down mass spectrometry. *Nat. Methods.* **4**, 817–821.
18. VerBerkmoes, N. C., Bundy, J. L., Hauser, L., Asano, K. G., Razumovskaya, J., Larimer, F., et al. (2002) Integrating "top-down" and "bottom-up" mass spectrometric approaches for proteomic analysis of *Shewanella oneidensis*. *J. Proteome Res.* **1**, 239–252.
19. Strader, M. B., Verberkmoes, N. C., Tabb, D. L., Connelly, H. M., Barton, J. W., Bruce, B. D., et al. (2004) Characterization of the 70S ribosome from *Rhodopseudomonas palustris* using an integrated "top-down" and "bottom-up" mass spectrometric approach. *J. Proteome Res.* **3**, 965–978.
20. Yu, Y., Ji, H., Doudna, J. A., and Leary, J. A. (2005) Mass spectrometric analysis of the human 40S ribosomal subunit: native and HCV IRES-bound complexes. *Protein Sci.* **14**, 1438–1446.

21. Kertesz, V., Connelly, H. M., Erickson, B. K., and Hettich, R. L. (2009) PTMSearchPlus: software tool for automated protein identification and post-translational modification characterization by integrating accurate intact protein mass and bottom-up mass spectrometric data searches. *Anal. Chem.* **81**, 8387–8395.
22. Wilkins, M. R., Gasteiger, E., Gooley, A. A., Herbert, B. R., Molloy, M. P., Binz, P. A., et al. (1999) High-throughput mass spectrometric discovery of protein post-translational modifications. *J. Mol. Biol.* **289**, 645–657.
23. Blom, N., Gammeltoft, S., and Brunak, S. (1999) Sequence and structure-based prediction of eukaryotic protein phosphorylation sites. *J. Mol. Biol.* **294**, 1351–1362.
24. Monigatti, F., Gasteiger, E., Bairoch, A., and Jung, E. (2002) The sulfinator: predicting tyrosine sulfation sites in protein sequences. *Bioinformatics* **18**, 769–770.
25. Bendtsen, J. D., Nielsen, H., von Heijne, G., and Brunak, S. (2004) Improved prediction of signal peptides: SignalP 3.0. *J. Mol. Biol.* **340**, 783–795.
26. Frottin, F., Martinez, A., Peynot, P., Mitra, S., Holz, R. C., Giglione, C., and Meinnel, T. (2006) The proteomics of N-terminal methionine cleavage. *Mol. Cell. Proteomics* **51**, 2336–2349.
27. Metropolis, N., Rosenbluth, A. W, Rosenbluth, M. N., Teller, A. H., and Teller, E. (1953) Equation of state calculations by fast computing machines. *J. Chem. Phys.* **21**, 1087–1092.
28. Kirkpatrick, S., Gelatt, C. D., and Vecchi, M. P. (1983) Optimization by simulated annealing. *Science* **220**, 671–680.
29. Brocchieri, L. and Karlin, S. (2005). Protein length in eukaryotic and prokaryotic proteomes. *Nucleic Acids Res.* **33**, 3390–3400.
30. Hastings, W. K. (1970) Monte Carlo sampling methods using Markov chains and their applications. *Biometrika* **57**, 97–109.
31. Huelsenbeck, J. P., Ronquist, F., Nielsen, R., and Bollback, J. P. (2001) Bayesian inference of phylogeny and its impact on evolutionary biology. *Science* **294**, 2310–2314.
32. Holmes, M. R. and Giddings, M. C. (2004) Prediction of posttranslational modifications using intact-protein mass spectrometric data. *Anal. Chem.* **76**, 276–282.
33. Creasy, D. M. and Cottrell, J. S. (2004) Unimod: protein modifications for mass spectrometry. *Proteomics* **4**, 1534–1536.
34. Wisz, M. S, Suarez, M. K, Holmes, M. R, and Giddings, M. C. (2004) GFSWeb: a web tool for genome-based identification of proteins from mass spectrometric samples. *J. Proteome Res.* **3**, 1292–1295.
35. Searle, B. C, Dasari, S., Wilmarth, P. A., Turner, M., Reddy, A. P., David, L. L., and Nagalla, S. R. (2005) Identification of protein modifications using MS/MS de novo sequencing and the OpenSea alignment algorithm. *J. Proteome Res.* **4**, 546–554.
36. Gotoh, O. (1982) An improved algorithm for matching biological sequences. *J. Mol. Biol.* **162**, 705–708.
37. Lee, T. Y., Huang, H. D., Hung, J. H. Huang, H. Y., Yang, Y. S., and Wang, T. H. (2006) dbPTM: an information repository of protein posttranslational modification. *Nucleic Acids Res.* **34**, D622–D627.

Chapter 18

An Integrated Top-Down and Bottom-Up Strategy for Characterization of Protein Isoforms and Modifications

Si Wu, Nikola Tolić, Zhixin Tian, Errol W. Robinson, and Ljiljana Paša-Tolić

Abstract

Bottom-up and top-down strategies are two commonly used methods for mass spectrometry (MS) based protein identification; each method has its own advantages and disadvantages. In this chapter, we describe an integrated top-down and bottom-up approach facilitated by concurrent liquid chromatography-mass spectrometry (LC-MS) analysis and fraction collection for comprehensive high-throughput intact protein profiling. The approach employs a high resolution reversed phase (RP) LC separation coupled with LC eluent fraction collection and concurrent on-line MS with a high field (12 T) Fourier-transform ion cyclotron resonance (FTICR) mass spectrometer. Protein elusion profiles and tentative modified protein identification are made using detected intact protein mass in conjunction with bottom-up protein identifications from the enzymatic digestion and analysis of corresponding LC fractions. Specific proteins of biological interest are incorporated into a target ion list for subsequent off-line gas-phase fragmentation that uses an aliquot of the original collected LC fraction, an aliquot of which was also used for bottom-up analysis.

Key words: Protein, Peptide, Proteomics, Mass spectrometry, LC-MS, Top-down, Bottom-up, FT-ICR MS, FTMS, Post-translational modification, PTM

1. Introduction

Bottom-up and top-down strategies for mass spectrometry (MS) based protein characterization are complementary; each has its own strengths and weaknesses. In the bottom-up strategy, proteomic measurements at the peptide level offer a basis for comprehensive protein identification (1). However, important information, such as posttranslational modifications (PTMs), may be ultimately unobtainable as only a portion of the entire protein is generally detected. Also, information regarding proteolytic processing, an important biological process, is generally lost in

Cathy H. Wu and Chuming Chen (eds.), *Bioinformatics for Comparative Proteomics*, Methods in Molecular Biology, vol. 694, DOI 10.1007/978-1-60761-977-2_18, © Springer Science+Business Media, LLC 2011

peptide centric analysis. Similarly, even if a peptide with a PTM is detected and successfully identified, information regarding the coordination of PTMs is generally lacking and the same enzymatic peptide sequence can occur in multiple proteins (e.g., families of highly related genes) confounding the determination of which protein was actually modified.

In top-down proteomics, intact proteins (2, 3) instead of peptides are dissociated in the gas phase. A 100% protein sequence coverage has been demonstrated, which allows identification of protein isoforms, proteolytic processing events, and PTMs. However, this strategy presently suffers from limited sensitivity and throughput (4, 5). The most recent significant advancements in top-down proteomics are electron capture dissociation (ECD) (6, 7) and electron transfer dissociation (ETD) (8). These techniques typically provide more efficient dissociation than conventional collisionally induced dissociation (CID), while preserving the labile modifications. However, ECD/ETD based approaches have been demonstrated only occasionally for intact protein characterization and have not yet been effectively incorporated into a high-throughput comprehensive top-down proteomic analysis.

Here, we describe a method that combines intact protein separations with on-line mass spectral acquisition and fraction collection (Fig. 1). A key advantage offered by this approach in

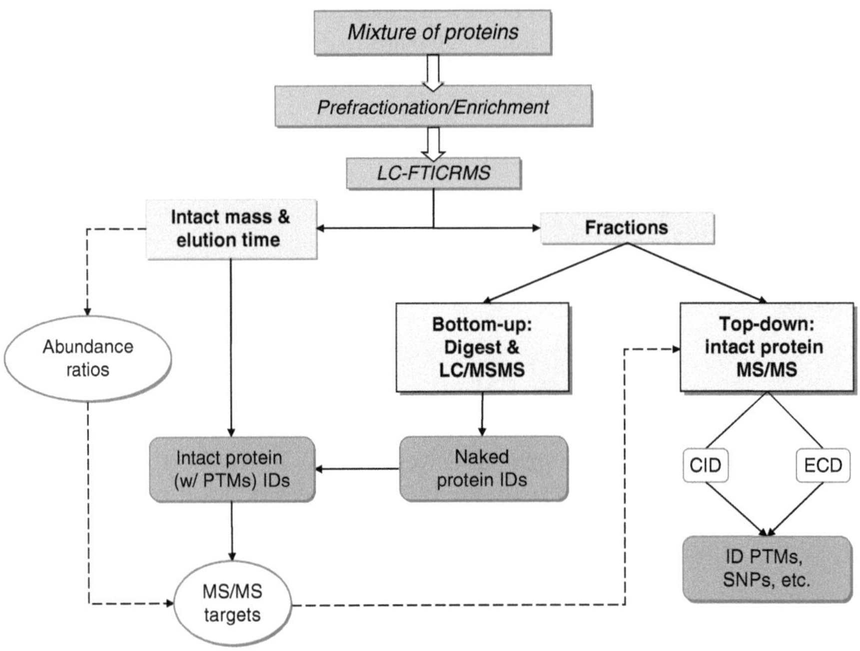

Fig. 1. Schematic diagram of the integrated top-down bottom-up proteomics strategy. Protein mixtures are fractionated to a level at which the reduced analyte complexity per fraction allows the inference of proteins from a peptide-level characterization of the same fraction. Accurate intact protein mass and time measurements facilitate protein abundance profiling that incorporates protein modifications, as well as high-throughput selection of biologically relevant targets for subsequent off-line gas-phase fragmentation using only an aliquot of the original collect RPLC fraction.

comparison to conventional top-down proteomics is that it minimizes the need for MS/MS analysis of the intact proteins that often cannot be effectively performed on the separation time scale, while still providing a protein fingerprint for more confident protein identification in the form of its corresponding peptides. More importantly, unlike the earlier strategy where complex intact protein mixtures (e.g., weak anion exchange fractions) were independently analyzed at the peptide and protein level (9, 10) this current approach preserves a direct link between the protein (mass) and its corresponding tryptic peptides (MS/MS data) because all tryptic peptides are confined to the chromatographic peak (i.e., on-line collected RPLC fraction) of the protein. Protein identification greatly benefits from this preserved linkage and time-consuming off-line intact protein MS/MS can be performed in a targeted fashion.

2. Materials

2.1. Standard Protein Mixture

1. Standard proteins were purchased from Sigma (St. Louis, MO): ubiquitin (U6253), cytochrome C (C2037), β-lactoglobulin A (L7880), β-lactoglobulin B (L8005), β-casein (C6905), carbonic anhydrase II (C3934), and myoglobin (M1882).
2. Calmodulin [accession number MCCH (PIR database) or P02593 (SWISS-PROT database)] was cloned, expressed, and purified under standard conditions, with polyhistidine tag (i.e., GHHHHHHGGGGGIL) on the C terminus for nickel affinity purification.

2.2. Top-Down: RPLC-FTICR Intact Protein MS Analysis and Fraction Collection

1. RPLC was used for on-line intact protein separation based on a custom in-house HPLC platform which was similar in principle to that developed by Shen et al. (11) (see Note 1).
2. Simultaneous nano-ESI and on-line fractionation was accomplished using the Triversa NanoMate 100 (Advion BioSciences, Ithaca, NY), on which both an auto-sampler and chip-based nanoESI device with 400 nozzles were available. Fractions were collected on a 96-well Eppendorf (Westbury, NY) twin-tec plate.
3. The composition of mobile phase A1 V/V was: 0.05% trifluoroacetic acid (TFA), 0.2% acetic acid, 5% isopropanol, 25% acetonitrile (ACN), and 69.75% water.
4. The composition of mobile phase B1 V/V was: 0.1% TFA, 9.9% water, 45% isopropanol, and 45% ACN.
5. The RPLC column (70 cm × 200 μm i.d.) was packed in-house with Phenomenex (Torrance, CA) Jupiter particles (C5 stationary phase, 5 μm particle diameter, 300 Å pore size).

6. Mass spectra of intact proteins were acquired using a modified Bruker 12 T APEX-Q Fourier-transform ion cyclotron resonance (FTICR) mass spectrometer (9) incorporating an electrodynamic ion funnel, quadrupoles for collisional focusing and ion pre-selection, a hexapole for external ion accumulation, and a quadrupole ion guide for transferring the ions to a novel compensated trapped-ion cell with improved DC potential harmonicity (12). A three-way pulsed leak valve was used to introduce N_2 gas during the external accumulation event to increase ion accumulation efficiency.

2.3. Bottom-Up: Trypsin Digestion and RPLC-Ion Trap Peptide MS/MS Analysis

1. Enzymatic digestion of an aliquot of the fraction collected RPLC eluent was performed using sequencing grade modified trypsin purchased from Promega (Madison, WI).
2. Digestion was performed in a buffer solution consisting of 50 mM ammonium bicarbonate in 30% (v/v) ACN and 70% water (pH 8.2).
3. A liquid chromatography separation of the digested peptide aliquot was performed using a custom in-house RPLC similar in principle to that developed by Shen et al. (11) (see Note 1).
4. The mobile phase A2 composition for peptide chromatography was: 0.05% TFA, 0.2% acetic acid, and 99.75% water.
5. The mobile phase B2 composition for peptide chromatography was: 0.1% TFA, 9.9% water, and 90% ACN.
6. The RPLC column (60 cm × 150 μm i.d.) was packed in-house with Phenomenex (Torrance, CA) Jupiter particles (C18 stationary phase, 5 μm particles, 300 Å pore size).
7. Mass spectra and MS/MS spectra of the peptide ions were acquired using an LTQ mass spectrometer (ThermoFisher Scientific, San Jose, CA).

3. Methods

3.1. Intact Protein LC-FTICR MS with On-Line Fractionation

1. The standard protein mixture was prepared in mobile phase A1, with protein concentrations as listed in Table 1. The total sample injection volume for the standard protein mixture was 20 μl (~20 μg total protein).
2. The RPLC system was equilibrated at 10,000 psi with 100% mobile phase A1. Next, a mobile phase selection valve was switched to create a near-exponential gradient as mobile phase B1 displaces A1 in a 2.5-ml mixer. A split was used to provide an initial flow rate through the column of ~5.5 μl/min and most proteins elute in less than 2 h.

Table 1
Standard protein mixture

Protein	M_r (Da)	Concentration (pmol/μl)
Ubiquitin	8564.6302	1.27
Carbonic anhydrase II	29024.7312	4.51
β-lactoglobulin A	18363.4538	7.92
β-lactoglobulin B	18277.4169	7.96
Calmodulin	18097.4710	9.04
β-casein	23983.1910	7.58
Cytochrome C	12229.2193	11.89
Myoglobin	16950.9920	5.36

3. The column eluent was split with ~300 nl/min of the flow directed to a modified Bruker 12 T APEX-Q FTICR mass spectrometer (9), and ~5.2 μl/min was collected into a 96-well Eppendorf Twin-tec plate using the NanoMate 100 system.
4. During the liquid chromatography-mass spectrometry (LC-MS) analysis, a single mass spectrum was recorded using 512 K data points, and the average of two mass spectra was used for data analysis.

3.2. Off-Line Protein MS/MS Analysis

1. One-third of the collected fraction volume (~5 μl) was transferred to a new 96-well plate for off-line MS/MS analysis.
2. CID-MS/MS analyses was accomplished by reducing the DC offset on the accumulation hexapole from 0 to −25 V. 10–50 mass spectra were averaged to obtain fragment ion information of desired quality (the S/N ratio was larger than three for more than 80% fragment peaks).
3. ECD-MS/MS analyses was performed using a heated hallow cathode located outside the ICR cell (i.e., the standard Bruker ECD arrangement). The cathode was heated with a current of 1.6–1.8 A. A 1–3-ms electron injection time was used with the potential on the solid cathode dispenser set at −7.5 to −15 V. Up to 50 mass spectra were averaged to obtain fragment ion information of sufficient quality.

3.3. On-Plate Fraction Digestion and RPLC-Ion Trap Peptide MS/MS Analysis

1. The remaining sample (~10 μl) was digested on-plate overnight at 37°C by adding 100 ng of trypsin and 10 μl of digestion buffer.
2. Samples were evaporated on the plate to remove organic solvent (to ~5 μl remaining volume) using a Savant SpeedVac

(ThermoFisher Scientific, San Jose, CA). The final volume was adjusted to 15 μl with mobile phase A2 for bottom-up analysis.

3.4. Data Analysis

3.4.1. Peptide Identification: Bottom-Up

1. Peptide RPLC MS/MS data were processed using SEQUEST (13) and a database that contained both genome derived possible *S. oneidensis* protein sequences and the standard protein sequences listed in Table 1.
2. No enzyme rules were applied, and identified peptides were filtered according to the criteria suggested by Washburn et al. (14) (see Note 2). Provisional databases that contained proteins supported by at least two distinct peptide identifications and in the protein mass range of 5–40 kDa were created for each fraction.
3. The elution profiles (Fig. 2a) for individual proteins were generated using relative protein abundances derived from the bottom-up data (see Note 3).

3.4.2. Intact Protein RPLC-FTICR Mass Spectra Processing

Intact protein RPLC-FTICR mass spectra were processed using in-house developed software (ICR-2LS and Viper) (15), available at http://ncrr.pnl.gov/software/) as previously described (9, 10) (see Note 4). Time-domain signals were Hanning apodized and twice zero-filled prior to FT. All spectra were externally calibrated, using myoglobin and ubiquitin spectra acquired in a separate LC-MS analysis.

3.4.3. Tentative Intact Protein Identification (Table 2)

1. The resulting mono-isotopic masses were clustered into features based on neutral mass, charge state, abundance, isotopic fit, and spectrum number (relating to RPLC retention time). Spectra that corresponded to a particular feature were summed, and the resulting spectra reprocessed as described above.
2. All charge states were collapsed into a zero charge state spectrum (i.e., neutral mass), which was then searched against the appropriate provisional protein database (assembled from bottom-up data) for tentative intact protein identifications. The procedure of collapsing to a zero charge state (sometimes referred as hyper-transform) and aligning different charge state conformers of the same molecular species was accomplished with in-house developed software (see Note 5).
3. Tentative identifications of proteins and modified proteins were accomplished by matching bottom-up data with the measured intact protein masses. Protein assignments were subsequently confirmed by protein MS/MS analyses using collected fractions (discussed in following section).
4. Mass measurements were manually inspected by matching the observed most abundant peak to the theoretical most

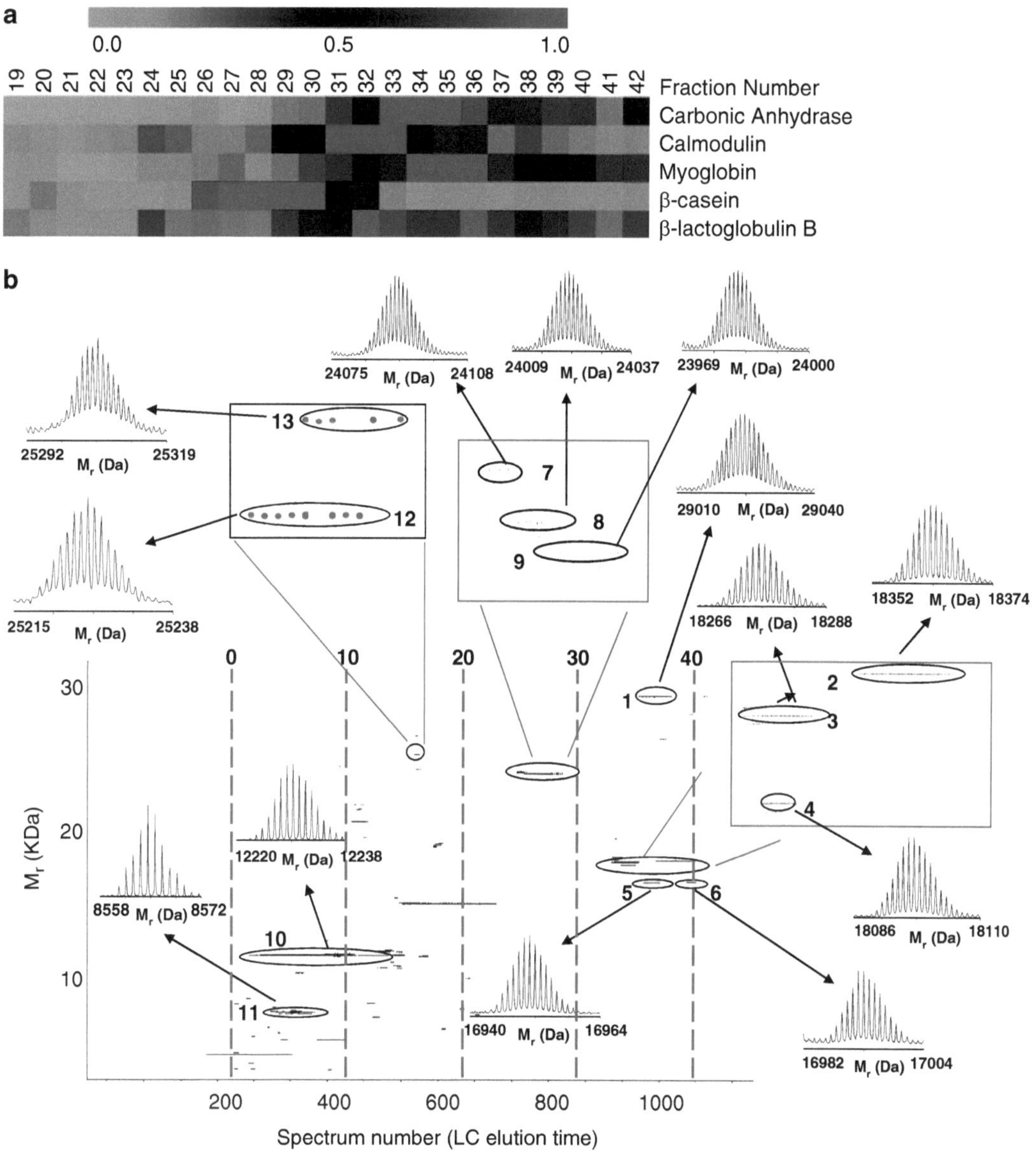

Fig. 2. Intact LC-MS analysis of a standard protein mixture: (**a**) A 2D display reconstructed from the LC-MS data. (**b**) A heat map representation of protein elution patterns generated for the later portion of the LC-MS analysis using tryptic peptides identified in each fraction.

abundant peak (generated using ICR-2LS). Discrepancies between the measured protein masses and the predicted masses for proteins in the provisional databases were used to search for a limited set of protein PTMs.

3.4.4. Intact Protein MS/MS Spectra Analysis

1. Target protein MS/MS spectra were analyzed using ICR-2LS and/or the online version of ProSight PTM 2.0 from the Kelleher group at Northwestern (16) (http://prosightptm2.northwestern.edu/).

Table 2
Summary of results obtained for the mixture of standard proteins using the integrated top-down bottom-up proteomics approach

Spot in Fig. 2b	Measured M_r (Da)	Theoretical M_r (Da)	Protein	Modifications	Eluted in Fractions	MMA (ppm)
1	29024.7313	29024.7312	Carbonic anhydrase II	Removal of N-term Met, N-acetylation	[33,36]	0.00
2	18363.4561	18363.4538	β-lactoglobulin A	2 Disulfide bonds	[35,37]	0.13
3	18276.4323	18276.4169	β-lactoglobulin B	2 Disulfide bonds	[32,34]	0.84
4	18097.4866	18097.471	Calmodulin	With His-tag	[31.33]	0.86
5	16950.9857	16950.992	Myoglobin	Intact protein	[34,36]	-0.37
6	16993.0104	16993.0025	Myoglobin	Lys-acetylation	42	0.46
7	24092.3436	24092.2661	β-casein, variant B	5 Phosphorylations, P67H(SNP) and S122R (SNP) to variant A2	26	3.22
8	24023.2204	24023.1971	β-casein, variant A1	5 Phosphorylations, P67H(SNP) to variant A2	27	0.97
9	23983.2346	23983.191	β-casein, variant A2	5 Phosphorylations	[28,30]	1.82
10	12229.2211	12229.2193	Cytochrome C	Removal of N-term Met, N-acetylation, with heme	[3,12]	0.15
11	8564.6471	8564.6302	Ubiquitin	Intact protein	[4,7]	1.97
12	25226.9975	25227.0194	α-casein2	11 Phosphorylations	14	0.87
13	25306.1167	25305.9875	α-casein2	12 Phosphorylations	14	–5.18

2. The THRASH algorithm was applied to deisotope the raw MS and MS/MS data.
3. Neutral monoisotopic masses were assigned for both precursor and fragment ions with a minimum S/N ratio of 3.
4. By allowing dynamic modifications (such as acetylation, oxidation and phosphorylation), ICR-2LS was used to identify b and y ion fragments within a 25-ppm mass tolerance.
5. ProSight software was used to further confirm the ICR-2LS findings and PTM localization (P-score less than 1E-2).
6. Figure 3 illustrates the MS/MS analysis of a protein with *m/z* of 1,161.95 (charge state of 25+), which was reconstituted from RPLC fraction 35. This protein was confirmed as carbonic anhydrase II acetylated at the N terminus.
7. For high throughput analysis of intact protein LC-MS/MS data can be processed using commercial version of ProSight software. The online version ProSight PTM 2.0 from the Kelleher group at Northwestern University was licensed by ThermoFisher Scientific as ProSightPC 2.0 (16, 17) (see Note 6).

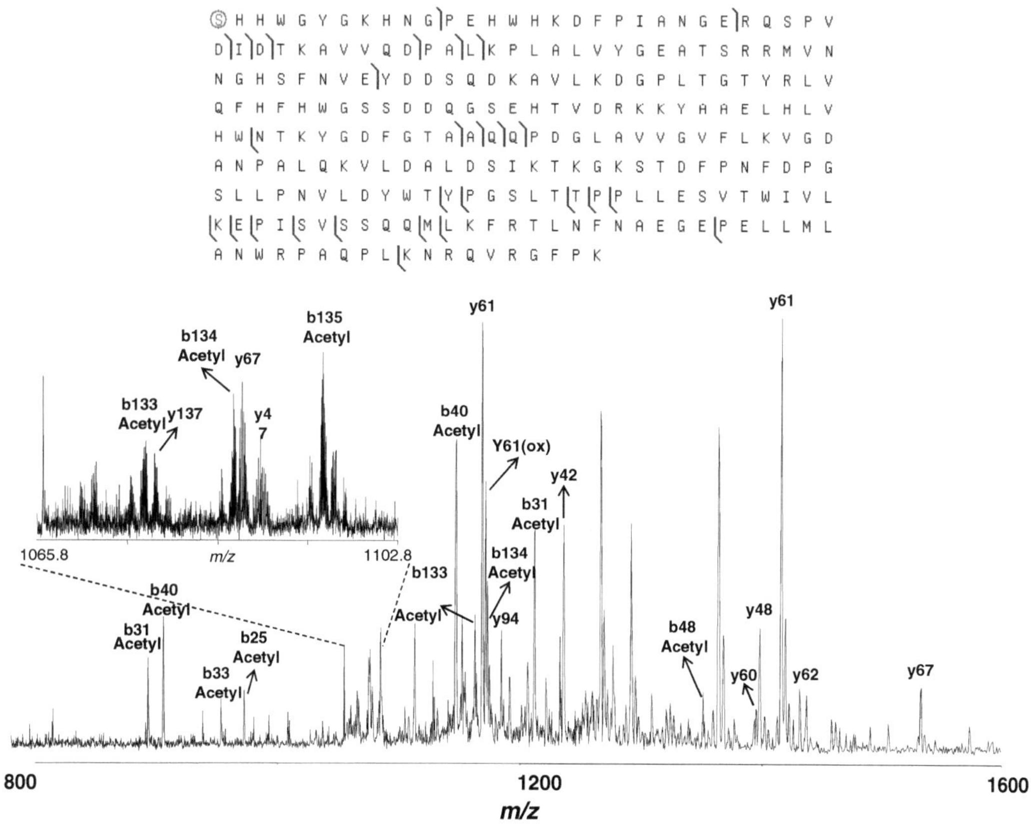

Fig. 3. Identification of carbonic anhydrase II in fraction 35 using CID MS/MS data acquired for *m/z* = 1,161.

Table 3
Summary of results obtained for yeast proteasome using the integrated approach[a]

ID	UniProt #	Description	Notes	M_{theo}	M_{exp}	MMA (ppm)
1	YLR421C	26S Proteasome regulatory subunit RPN13	1 Acetylation, 1 phosphorylation, loss of N-term 3 residues	17674.62	17674.58	−2.26
	YLR421C	26S Proteasome regulatory subunit RPN13	1 Acetylation, 2 phosphorylation, loss of N-term 3 residues	17754.58	17754.55	−1.69
2	YDR382W	60S Acidic ribosomal protein P2β	1 Phosphorylation	11129.36	11129.34	−1.80
3	YOL039W	60S Acidic ribosomal protein P2α	1 Phosphorylation	10825.22	10825.19	−2.77
4	YOR362C	Proteasome component C1	Loss of N-term Met, blocking of N-term Thr	N/A	31740.47	N/A
5	YBL041W	Proteasome component C5	Loss of N-term 12 residues	25628.92	25628.91	−0.39
6	YHR200W	26S Proteasome regulatory subunit RPN10	Loss of N-term Met	29615.84	29615.77	−2.36
7	YER012W	Proteasome component C11	1 Acetylation	22558.62	22558.60	−0.89
8	YGL011C	Proteasome component C7α	Loss of N-term Met, 1 acetylation	27911.20	27911.16	−1.43
9	YDR394W	26S Protease regulatory subunit 6B homolog	1 Acetylation	47935.21	47935.19	−0.42
10	YML092C	Proteasome component Y7	Loss of N-term Met, 1 acetylation	27072.08	27072.06	−0.74
11	YFR052W	26S Proteasome regulatory subunit RPN12	Loss of N-term Met	31787.19	31787.16	−0.94
12	YMR314W	Proteasome component PRE5	1 Acetylation	25645.24	25645.24	0.00
13	YGR135W	Proteasome component Y13	Loss of N-term Met, 1 acetylation	28624.59	28624.57	−0.70
14	YDR427W	26S Proteasome regulatory subunit RPN9	No Modification	45782.85	45782.87	0.44

[a]Tentative identifications of proteins and modified proteins were accomplished by matching bottom-up data with the measured intact protein masses

3.5. Yeast Proteasome Results

1. The described integrated top-down bottom-up method has been applied to characterize the yeast proteasome.
2. A small sample, ~15 μg, was analyzed by RPLC-MS analysis with concurrent fraction collection revealing the presence of several putative proteins with mass >10 kDa.
3. A subset of these proteins has been tentatively identified (Table 3) by matching bottom-up derived protein identifications and elution profiles with the intact protein accurate masses and elution profiles, as illustrated in Fig. 4.

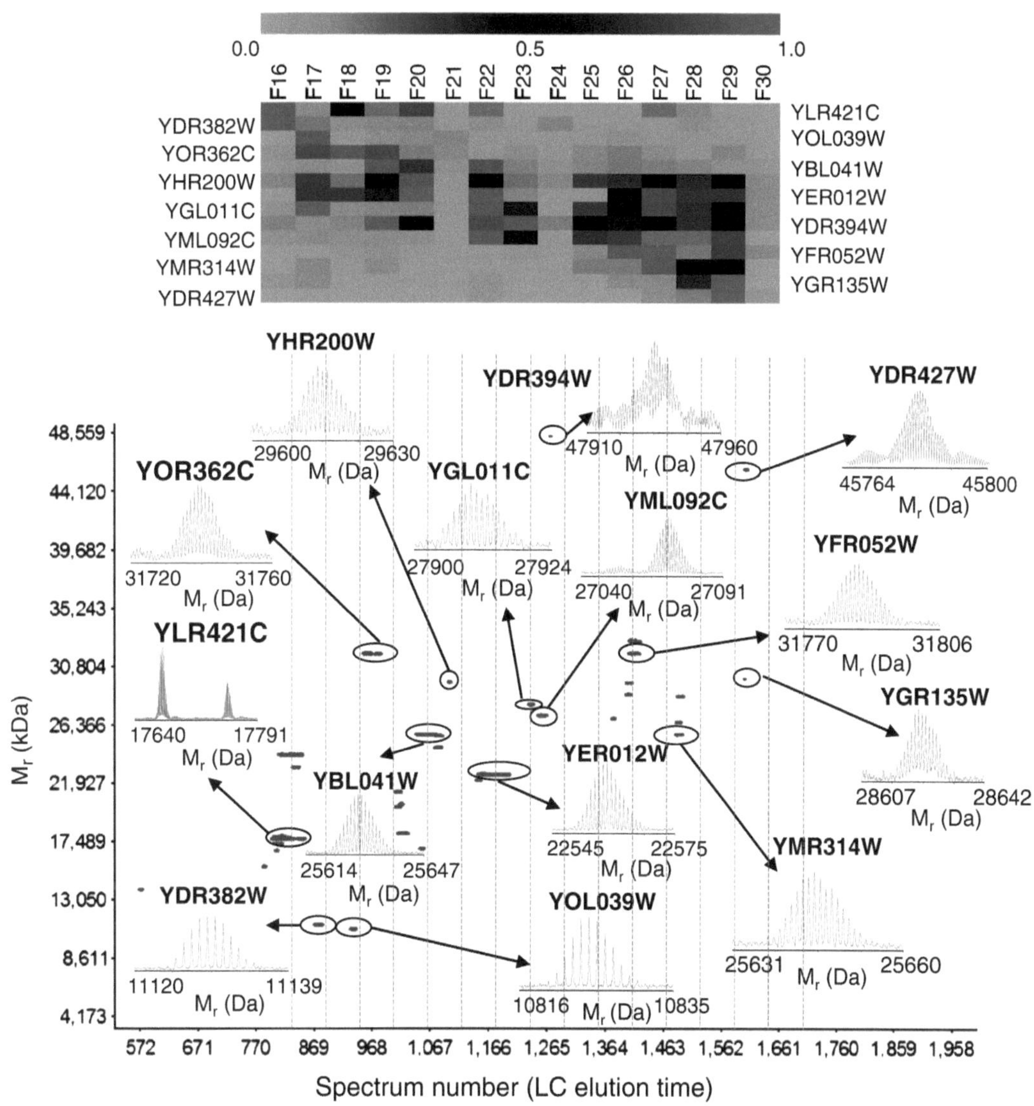

Fig. 4. *Top*: A heat map representation of protein elution patterns generated for the LC-MS analysis of the yeast proteasome using tryptic peptides identified in each fraction. *Bottom*: A 2D display (mass vs. spectrum number or RPLC elution time) reconstructed from the intact protein LC-MS data.

4. This hybrid analysis identified various classes of PTMs including oxidation, phosphorylation, methylation, as well as proteolytic processing events.

4. Notes

1. The home-built system used ISCO LC pumps (model 100DM, Isco, Lincoln, NE) to run LC separation at constant pressure of 10,000 psi. The mobile phases were delivered at a constant pressure of 10,000 psi by two ISCO pumps and were combined in a steel mixer (2.5 ml) containing a magnetic stirrer before entering the separation capillary. Fused-silica capillary flow restrictors (30-μm i.d. with various lengths) were used to control the concentration gradient of mobile phase B1 in the mixer.
2. The filtering criteria described by Washburn et al. were interpreted as follows: Xcorr ≥1.9 for the charge state 1+ and fully tryptic peptides, Xcorr ≥2.2 for the charge state 2+ fully and partially tryptic peptides, and Xcorr ≥3.75 for the charge state 3+ fully and partially tryptic peptides, all with a ΔCn value of ≥0.1.
3. The heat map representation of protein elution patterns was generated for the later portion of the RPLC-MS analysis using tryptic peptides identified in each fraction. Observed counts of the peptides for each protein were used to derive relative protein abundances. The observation counts were normalized by dividing the value obtained for each protein with the sum of the values for the protein (row), with the scale ranging from 0 (i.e., least abundant, green) to 1 (i.e., most abundant, red). The columns in the heat map represent the RPLC fraction number.
4. After peak picking, an autocorrelation calculation was applied to predict the charge state by looking at the frequency of the peaks surrounding the most intensive peak. This calculated charge state value as well as m/z value for the most abundant peak was then used to calculate an approximate molecular mass. This molecular formula was then used to calculate a theoretical isotopic distribution using the Mercury algorithm. Theoretical and experimental isotopic distributions were then compared to determine an isotopic fit value (i.e., the least-squares error between the theoretical and the experimental data). The charge state, monoisotopic, average, and most abundant molecular masses for the lowest (i.e., best) isotopic fit value were assumed to be correct and were reported. This process was repeated until every isotopic distribution in a

spectrum (above a given noise threshold) was processed and reduced to neutral mass.

5. Neutral mass spectra as well as charge state determination can also be achieved using some commercial available software, such as DataAnalysis from Bruker and Xtract from Thermo.
6. The stand-alone version ProSightPC represents the first tool on the market to address needs of high-throughput high-resolution top-down MS/MS analyses, although it is possible other vendors have similar proprietary software packages which we do not have access to evaluate. The ProSightPC tool could be utilized with the UStags platform (18) in combination with an in-house developed suite of functions. ProSightPC supports native Thermo "raw" files and XML based "puf" files of deisotoped neutral masses which together with a MySQL based repository and build-in THRASH algorithm presents an attractive all-in-one solution for both database and sequence tag based searches.

Acknowledgments

Portions of this work were supported by the National Center for Research Resources (RR 018522), the National Institute of Allergy and Infectious Diseases (NIH/DHHS through interagency agreement Y1-AI-4894-01), the National Institute of General Medical Sciences (NIGMS, R01 GM063883), and the U. S. Department of Energy (DOE) Office of Biological and Environmental Research. Work was performed in the Environmental Molecular Science Laboratory, a DOE national scientific user facility located on the campus of Pacific Northwest National Laboratory (PNNL) in Richland, Washington. PNNL is a multi-program national laboratory operated by Battelle for the DOE under Contract DE-AC05-76RLO 1830.

References

1. Aebersold, R., Mann, M. (2003) Mass spectromety-based proteomics. *Nature* **422**, 198–207.
2. Kelleher, N. L. (2004) Top-down proteomics. *Anal. Chem.* **76**, 197A–203A.
3. McLafferty, F. W., Breuker, K., Jin, M., Han, X., Infusini, G., Jiang, H., Kong, X., Begley, T. P. (2007) Top-down MS, a powerful complement to the high capabilities of proteolysis proteomics. *FEBS J.* **274**, 6256–6268.
4. Parks, B. A., Jiang, L., Thomas, P. M., Wenger, C. D., Roth, M. J., Ii, M. T., Burke, P. V., Kwast, K. E., Kelleher, N. L. (2007) Top-down proteomics on a chromatographic time scale using linear ion trap Fourier transform hybrid mass spectrometers. *Anal. Chem.* **79**, 7984–7991.
5. Siuti, N., Kelleher, N. L. (2007) Decoding protein modifications using top-down mass spectrometry. Nat. *Methods* **4**, 817–821.
6. Zabrouskov, V., Whitelegge, J. P. (2007) Increased coverage in the transmembrane domain with activated-ion electron capture dissociation for top-down Fourier-transform

mass spectrometry of integral membrane proteins. *J. Proteome Res.* **6**, 2205–2210.

7. Zubarev, R. A., Kelleher, N. L., McLafferty, F. W. (1998) Electron capture dissociation of multiply charged proteins cations. A nonergodic process. *J. Am. Chem. Soc.* **120**, 3265–3266.
8. Coon, J. J., Ueberheide, B., Syka, J. E. P., Dryhurst, D. D., Ausio, J., Shabanowitz, J., Hunt, D. F. (2005) Interpreting the protein language using proteomics. *Proc. Natl. Acad. Sci. USA* **102**, 9463–9468.
9. Sharma, S., Simpson, D. C., Tolic, N., Jaitly, N., Mayampurath, A. M., Smith, R. D., Pasa-Tolic, L. (2007) Probing proteomes using capillary isoelectric focusing-electrospray ionization Fourier transform ion cyclotron resonance mass spectrometry. *J. Proteome Res.* **6**, 602–610.
10. Wu, S., Lourette, N. M., Tolic, N., Zhao, R., Robinson, E. W., Tolmachev, A. V., Smith, R. D., Pasa-Tolic, L. (2009) An integrated top-down and bottom-up strategy for broadly characterizing protein isoforms and modifications. *J. Proteome Res.* **8**, 1347–1357.
11. Shen, Y., Tolic, N., Masselon, C., Pasa-Tolic, L., Camp, D. G., Hixson, K. K., Zhao, R., Anderson, G. A., Smith, R. D. (2004) Ultrasensitive proteomics using high-efficiency on-line micro-SPE-nanoLC-nanoESI MS and MS/MS. *Anal. Chem.* **76**, 144–154.
12. Tolmachev, A. V., Robinson, E. W., Wu, S., Kang, H., Pasa-Tolic, L., Smith, R. D. (2008) Trapped-ion cell with improved DC potential harmonicity for FT-ICR MS. *J. Am. Soc. Mass Spectrom.* **19**, 586–597.
13. Yates, J. R., Eng, J. K., McCormack, A. L., Schieltz, D. (1995) Method to correlate tandem mass spectra of modified peptides to amino acid sequences in the protein database. *Anal. Chem.* **67**, 1426–1436.
14. Washburn, M. P., Wolters, D., Yates, J. R. (2001) Large-scale analysis of the yeast proteome by multidimensional protein identification technology. *Nat. Biotechnol.* **19**, 242–247.
15. Monroe, M. E., Tolic, N., Jaitly, N., Shaw, J. L., Adkins, J. N., Smith, R. D. (2007) VIPER: an advanced software package to support high-throughput LC-MS peptide identification. *Bioinformatics* **23**, 2021–2023.
16. LeDuc, R. D., Taylor, G. K., Kim, Y. B., Januszyk, T. E., Bynum, L. H., Sola, J. V., Garavelli, J. S., Kelleher, N. L. (2004) ProSight PTM: an integrated environment for protein identification and characterization by top-down mass spectrometry. *Nucleic Acids Res.* **32**, W340–W345.
17. Zamdborg, L., LeDuc, R. D., Glowacz, K. J., Kim, Y. B., Viswanathan, V., Spaulding, I. T., Early, B. P., Bluhm, E. J., Babai, S., Kelleher, N. L. (2007) ProSight PTM 2.0: improved protein identification and characterization for top down mass spectrometry. *Nucleic Acids Res.* **35**, W701–W706.
18. Shen, Y., Tolic, N., Hixson, K. K., Purvine, S. O., Anderson, G. A., Smith, R. D. (2008) De novo sequencing of unique sequence tags for discovery of post-translational modifications of proteins. *Anal. Chem.* **80**, 7742–7754.

Part III

Comparative Proteomics in Systems Biology

Chapter 19

Phosphoproteome Resource for Systems Biology Research

Bernd Bodenmiller and Ruedi Aebersold

Abstract

PhospoPep version 2.0 is a project to support systems biology signaling research by providing interactive interrogation of MS-derived phosphorylation data from four different organisms. Currently the database hosts phosphorylation data from the fly (*Drosophila melanogaster*), human (*Homo sapiens*), worm (*Caenorhabditis elegans*), and yeast (*Saccharomyces cerevisiae*). The following will give an overview of the content and usage of the PhosphoPep database.

Key words: Systems biology, Protein phosphorylation, Database, Data integration, Signaling network

1. Introduction

In this chapter, we give an introduction to PhosphoPep, a database for phosphopeptides and phosphoproteins from model organisms and a suite of associated software tools as a resource for systems biology research. PhosphoPep currently contains FOR WORM over 5,444 unique high confidence phosphopeptides that could be assigned to 2,959 gene products, comprising 3,545 assigned unique phosphorylation sites. For *Saccharomyces cerevisiae* the database stores 9,554 high confidence phosphopeptides that could be assigned to 2,071 gene products, comprising 5,890 assigned unique phosphorylation sites. The contents of the *Drosophila melanogaster* data set include 16,875 phosphopeptides that could be assigned to 5,347 gene products, comprising 12,756 assigned phosphorylation sites. Finally, for human studies, 3,784 high confidence unique phosphopeptides that could be assigned to 5,160 gene products, comprising 2,810 assigned phosphorylation sites are included (see Table 1).

Cathy H. Wu and Chuming Chen (eds.), *Bioinformatics for Comparative Proteomics*, Methods in Molecular Biology, vol. 694, DOI 10.1007/978-1-60761-977-2_19, © Springer Science+Business Media, LLC 2011

Table 1
Data lodged in PhosphoPep v2.0 database (taken from (1))

Organism	Phosphopeptides with $P > 0.8$[a]	Total phosphorylation sites[b]	Phosphopeptides with assigned phosphorylation site(s)[b]
D. melanogaster	16,875	16,608	12,756
S. cerevisiae	9,554	8,901	5,890
C. elegans	5,444	4,986	3,545
H. sapiens	3,784	3,980	2,810

[a]PeptideProphet Score as computed by PeptideProphet (16)
[b]A phosphopeptide was considered to have an unassigned/assigned site if a dCn threshold was not reached and/or exceeded (see Supplementary Material and Methods in (1))

To support further experimentation and analysis of the phosphorylation data, different software tools were added to the PhosphoPep database. First, we implemented a search function to detect the sites of phosphorylation on individual proteins and to place phosphoproteins within cellular pathways as defined by the Kyoto Encyclopedia of Genes and Genomes (KEGG) pathway database (2). Such pathways, along with the identified phosphoproteins can be interrogated by a pathway viewer and exported to Cytoscape (3) a software tool, which supports the integration of the data from PhosphoPep and other databases. Second, we added utilities for the use of the phosphopeptide data for targeted proteomics experiments. In a typical experiment of this type, the known phosphorylation sites of a protein or set of proteins are detected and quantified in extracts representing different cellular conditions via targeted mass spectrometry experiments such as MRM (4–6). Third, we made the data in PhosphoPep searchable by spectral matching through SpectraST (7, 8). Specifically, for each distinct phosphopeptide ion identified in this study, all corresponding MS2 spectra were collapsed into a single consensus spectrum. Unknown query spectra can then be identified by spectral searching against the library of phosphopeptide consensus spectra. Collectively, PhosphoPep and the associated software tools and data mining utilities support the use of the data for diverse types of studies, from the analysis of the state of phosphorylation of a single protein to the detection of quantitative changes in the state of phosphorylation of whole signaling pathways at different cellular states and has been designed to enable the iterative cycles of experimentation and analysis that are typical for systems biology research (1, 9, 10).

2. Materials

PhosphoPep web interface is accessible at http://www.phosphopep.org/. After choosing the organism of interest on the starting page, the data can be queried and used in the following ways.

First, by using the "Search Peptides" the user can search for the protein of interest by using the gene/protein name, protein ID, or by pasting part of the protein sequence into the dedicated search interface.

Second, by using the "Identified Proteins" button, all proteins of a given organism with identified phosphoproteins will be shown and each of them can be selected.

Third, by using the "Bulk Search" function, a list of proteins can be queried by the identifiers as indicated under "Search Peptides."

All of the first three quries will direct the user to the protein information page. Besides showing general information about each of the proteins, additional information can be retrieved and functions useful for analyzing the phosphoprotein can be executed by using the tabs shown below.

- "View available KEGG pathways for this protein (http://www.genome.jp/kegg/) (2)". As shown later in this chapter, this button allows to place the protein within its (signaling) pathway(s).
- "Start cytoscape network with this protein (http://www.cytoscape.org/) (3)". Either the single phosphoprotein or the complete pathway (including all observed phoshoproteins) is exported into the Cytoscape environment. This software tool allows one to visualize networks and to integrate different data types and to perform network specific analyses. A similar analysis can also be performed using the next button.
- "Search for protein interaction information in String (http://string.embl.de/) (11)". In the String database protein–protein interactions of various kinds are stored and again can be used to built networks.
- ODB "View orthologs/homolog information (http://www.orthomcl.org/) (12)".
- PA "Look up protein information in PeptideAtlas (http://www.peptideatlas.org/) (13)".
- "Search protein sequence at Scansite (http://scansite.mit.edu/) (14)".

Furthermore, for the phosphopeptides identified using mass spectrometry, different information is given. These include

- PeptideProphet: When interpreting tandem mass spectrometry data, it is crucial to determine if an identification is correct. The PeptideProphet computes a probability of a given fragment ion spectrum to be correctly assigned to a peptide sequence by a given database search algorithm and assigns a score accordingly (15, 16). The range of the score is from 0 (worst) to 1 (best). Depending on the dataset or database the probabilities can slightly vary at a given threshold/score.
- Tryptic ends: As we analyze peptides in our tandem mass spectrometry experiments we have to digest the proteins using a protease. This is often done by using trypsin. Trypsin cleaves after arginine and lysine but exhibits also some non-specific cleavage (5). Two tryptic ends means that both ends were specifically cut by trypsin.
- Peptide mass: Molecular mass of the (phospho) peptide.
- deltaCn: The deltaCn score (dCn) is a score computed by the Sequest (17) algorithm, which we use to interpret tandem mass spectra. The dCn is the difference between the (normalized) cross-correlation parameter of the first- and second-ranked amino acid sequence assigned to a tandem mass spectrum. Simplified, the dCn tells you how much better the first (best) database search hit fits to a tandem mass spectrum than the second hit. In the case of phosphopeptides, the dCn also correlates to the correctness of the phosphorylation site assignment within the phosphopeptide sequence (18).
- # Obs: Number of times the phosphopeptide was identified in our experiments.
- # Mappings: Maps # of gene models/maps to # of transcripts.

Besides the functions to retrive information from singe proteins, several specialized functions are also provided by the PhosphoPep database. These include: first, "Pathway Search", which allows retrieval of complete signaling pathways as given by the KEGG database (2) in a graphical representation. Importantly, from each of the shown phosphoproteins the protein information page can be opened.

Second, "Spectral Search", underyling this web interface is SpectraST (7, 8), which annotates tandem mass spectra using a spectral library consensus search algorithm. By pasting the list of measured masses (as represented in a .dta file) a phosphopeptide tandem mass spectrum can be searched against the consenus phosphopeptide library of PhosphoPep. In case a high number of spectra should be searched, ideaLLy, SpectraST and the PhosphoPep libraries are downloaded (http://www.peptideatlas.org/speclib/) and used for that puropose.

Third, the last function "MRM Transitions" allows retrieval of the coordinates to perform targeted proteomics experiments using a triple quadrupole mass spectrometer (for more information see (6, 8)).

3. Methods

As for most users it is crucial to know whether the identified phoshopeptide and the site of phosphorylation are correct, two sections detailing these topics are given.

3.1. Is the Phosphopeptide Correctly Identified?

To understand the basic methods of peptide identification using tandem mass spectrometry, we strongly recommend studying the presentation that you can find under the link http://www.proteomesoftware.com/Proteome_software_pro_interpreting.html (Proteome software Inc.). The presentation is easy to understand and represents a nice introduction to proteomics.

Of note, as the following text was written for users without any experience in mass spectrometry, we attempted to describe each topic in a simplified manner, sometimes at the expense of accuracy. For users who wish to learn more about each topic we suggest reading the literature given at the end of this tutorial.

When phosphopeptides are analyzed using liquid chromatography – tandem mass spectrometry and phosphopeptide sequences are assigned to the resulting spectra using database search algorithms – primarily two types of error can occur. The first type of error is the misassignment of the fragment ion spectrum to a peptide sequence (15, 16). The second type of error is the misassignment of the site of phosphorylation in an otherwise correctly identified phosphopeptide (18).

Here we explain how each of the errors was assessed and how the users of PhosphoPep can use the computed scores and some rules to judge if a phosphopeptide was correctly identified and the site correctly assigned.

As mentioned above, one type of error in the automatic interpretation of tandem mass spectra is the misassignment of the fragment ion spectrum to a peptide sequence. This type of error can be estimated by applying statistical models such as the PeptideProphet (16) and/or by using decoy sequence databases (19).

All data loaded into PhosphoPep were assessed using both methods and we already applied a stringent cut off on all data. Therefore the false positive content in the case of the fly data is about 2.6% (for yeast, worm, and human this number is similar). This means that if you do not apply any further filter criteria about 1 out of 38 phosphopeptide entries are wrong. For bioinformatic large scale analyses of this false positive rate is in most cases very acceptable, but for a biologist who wants to perform follow-up experiments this can already be too high and therefore it is desirable to choose your own false positive rate. So how do you choose your own false positive rate?

One of the statistical tools to compute the false positive rate, the PeptideProphet (16), computes a score (ranging from 0 (worst) to 1 (best)). This score is displayed for every peptide in PhosphoPep (9) (see Fig. 1). As mentioned above, we have already pre-filtered the data, therefore the lowest PeptideProphet score you will find is 0.8. The closer the score is to 1.0 the lower is the chance that you pick a wrongly identified phosphopeptide. For example, at a Peptide Prophet cut off of 0.99 approximately 0.2% of all entries (equal or above this score) are estimated to be false positive assignments (1 out of 500 phosphopeptide entries) for the fly dataset.

With the button to the left you can choose the Prophet Score cut off on your own (Fig. 2).

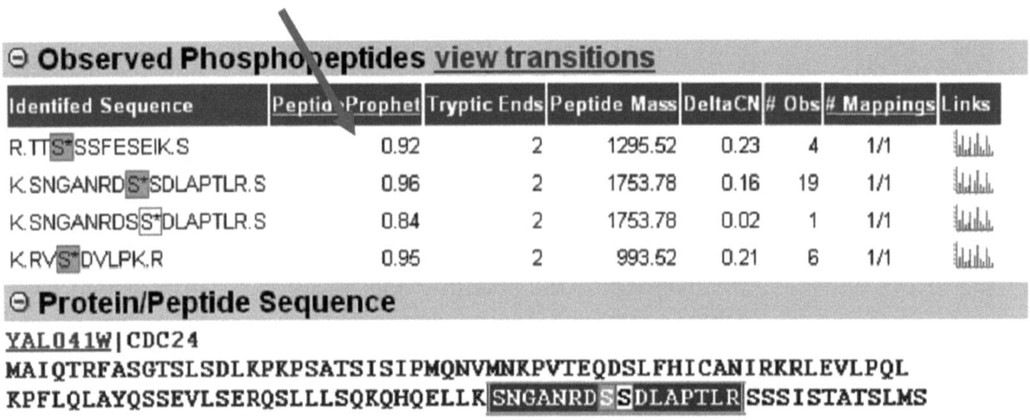

Fig. 1. The *arrow* points to the PeptideProphet score.

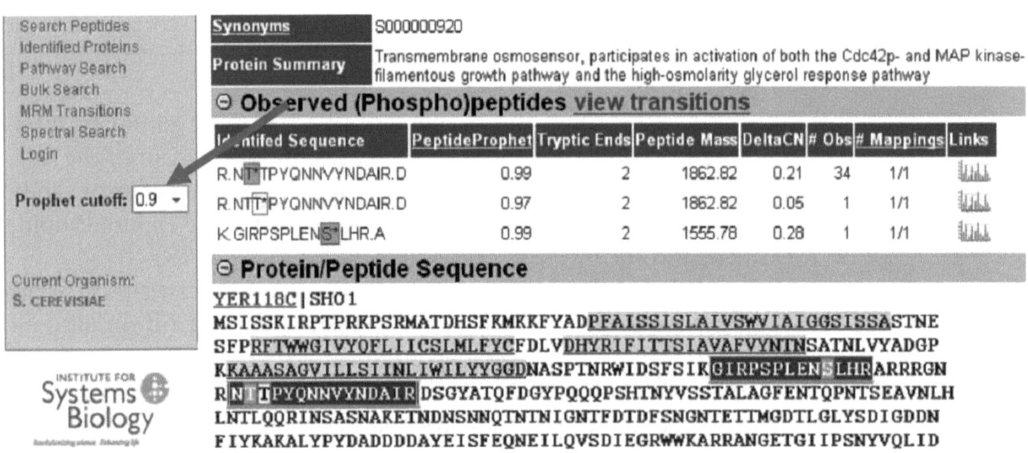

Fig. 2. The *arrow* points to the menu that allows users to choose a PeptideProphet score.

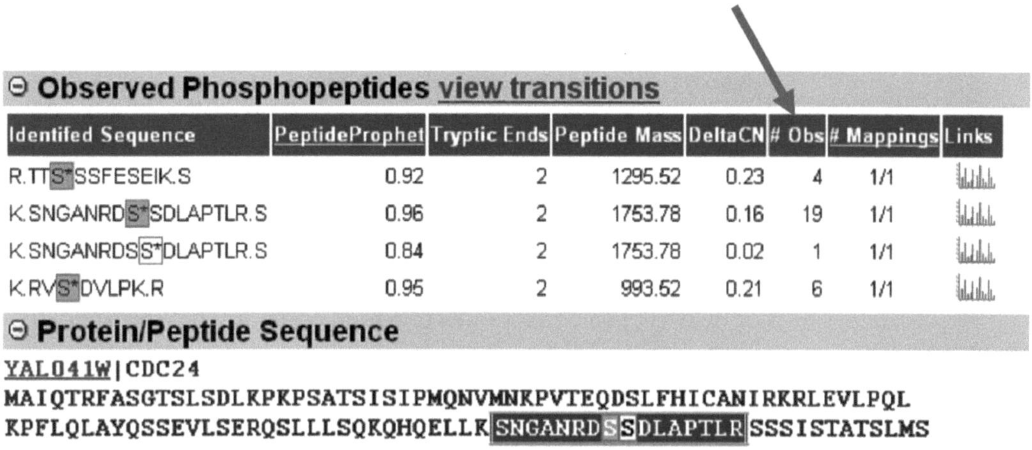

Observed Phosphopeptides view transitions

Identifed Sequence	PeptideProphet	Tryptic Ends	Peptide Mass	DeltaCN	# Obs	# Mappings	Links
R.TTS*SSFESEIK.S	0.92	2	1295.52	0.23	4	1/1	
K.SNGANRDS*SDLAPTLR.S	0.96	2	1753.78	0.16	19	1/1	
K.SNGANRDSS*DLAPTLR.S	0.84	2	1753.78	0.02	1	1/1	
K.RVS*DVLPK.R	0.95	2	993.52	0.21	6	1/1	

Protein/Peptide Sequence

YAL041W|CDC24
MAIQTRFASGTSLSDLKPKPSATSISIPMQNVMNKPVTEQDSLFHICANIRKRLEVLPQL
KPFLQLAYQSSEVLSERQSLLLSQKQHQELLKSNGANRDSSDLAPTLRSSSISTATSLMS

Fig. 3. The *arrow* points to column that indicates how often a phosphopeptide was identified in different samples.

One further criterion that increases the certainty that a phosphopeptide was correctly identified is the "# Obs" which tells you how often a phosphopeptide was identified in our experiments (Fig. 3).

The chance that a phosphopeptide, which was identified multiple times, is wrong is lower than that of a phosphopeptide that was just identified once (but keep in mind that this is only a rule of thumb and exceptions exist) (20).

So taken together, if you choose a phosphopeptide for follow up experiments make sure that it has a high PeptideProphet score and was observed multiple times.

3.2. Is the Site of Phosphorylation Correctly Assigned?

Often phosphopeptides are rich in serine and threonine residues, which can sometimes puzzle the algorithm for the automatic interpretation of tandem mass spectra in regards to which serine/threonine(/tyrosine) was phosphorylated (18). Therefore another type of error connected to phosphopeptides identified using tandem mass spectrometry is the misassignment of the site of phosphorylation in an otherwise correctly identified phosphopeptide (18).

This error was estimated by comparing the search engine output scores for the potential phosphorylated forms of a peptide, assuming that any hydroxy-amino acid in a phosphopeptide could be phosphorylated. Based on this estimation we highlighted the phosphopeptides either red (high probability of correct assignment) or yellow (low probability of correct assignment) (9, 17, 18).

As one typical approach to study protein phosphorylation is to mutate the site of phosphorylation to another amino acid residue, it is advisable to ascertain that you choose the correct amino acid. There are several steps you can take to ensure that the site of phosphorylation was correctly assigned.

3.2.1. Take a Look at the dCn Value

The first step to determine the certainty in the phosphorylation site assignment is to look at the dCn score. Simplified, the dCn tells you how much better the first (best) database search hit fits to a tandem mass spectrum than the second hit (Fig. 4). Now if the first and second hits are the same phosphopeptide but the Sequest algorithm has problems unequivocally assigning the phosphorylation site, the score will be very low, often close to zero.

Again as a rule of thumb: The higher the dCn score the more certain is the phosphorylation site assignment. Normally, a score of dCn > 0.125 corresponds to a high certainty that the site is correctly assigned (18).

In Fig. 5 a phosphopeptide is shown that was identified several times but the site of phosphorylation could never be assigned with high certainty. As a result the same phosphopeptide exists in several versions in PhosphoPep. Such agglomerations of the same peptide with many different phosphorylation sites are a hint that

⊖ Observed Phosphopeptides view transitions

Identifed Sequence	PeptideProphet	Tryptic Ends	Peptide Mass	DeltaCN	# Obs	# Mappings	Links
R.TTS*SSFESEIK.S	0.92	2	1295.52	0.23	4	1/1	
K.SNGANRDS*SDLAPTLR.S	0.96	2	1753.78	0.16	19	1/1	
K.SNGANRDSS*DLAPTLR.S	0.84	2	1753.78	0.02	1	1/1	
K.RVS*DVLPK.R	0.95	2	993.52	0.21	6	1/1	

⊖ Protein/Peptide Sequence

YAL041W | CDC24
MAIQTRFASGTSLSDLKPKPSATSISIPMQNVMNKPVTEQDSLFHICANIRKRLEVLPQL
KPFLQLAYQSSEVLSERQSLLLSQKQHQELLKSNGANRDSSDLAPTLRSSSISTATSLMS

Fig. 4. The *arrow* points towards the dCn value column.

R.RS*ST*PETENAFSATPR.A	0.93	2	1910.76	0.01	1	1/1	
R.RSST*PETENAFSAT*PR.A	0.84	2	1910.77	0.09	1	1/1	
R.RS*STPET*ENAFSATPR.A	0.81	2	1910.77	0.06	1	1/1	
R.RS*ST*PETENAFSAT*PR.A	0.90	2	1990.72	0.02	1	1/1	
R.RSST*PET*ENAFSATPR.A	0.84	2	1910.77	0.10	1	1/1	
R.RS*S*TPETENAFSATPR.A	0.94	2	1910.77	0.01	1	1/1	
R.RSST*PETENAFSATPR.A	0.92	2	1830.79	0.06	1	1/1	
R.RSST*PET*ENAFSAT*PR.A	0.95	2	1990.72	0.07	1	1/1	

Fig. 5. The *arrow* points towards the dCn value column.

the site is not well assigned (but keep in mind, some proteins are heavily phosphorylated and therefore the same peptide can exist in different phosphorylation forms).

3.2.2. Take a Look at the Kinase Phosphorylation Motif

An additional step to take to confirm a site of phosphorylation is to look at the possible kinase motifs surrounding the phosphorylation site (21). In the example below (Fig. 6) phosphorylation sites on the protein FUS3, a MAPK, are shown. Here it is not clear whether

R.IIDESAADNSEPTGQQ**S***GMTE**Y***VATR.W

or

R.IIDESAADNSEPTGQQSGM**T***E**Y***VATR.W

is correct. Knowing that the MAP kinases are activated by the phosphorylation in the **TXY** motif, we can assume that the R.IIDESAADNSEPTGQQSGM**T***E**Y***VATR.W is correct.

3.2.3. Predict the Motif Using Scansite

In case you do not have all the kinase motifs memorized you can use the Scansite algorithm (14) to search the protein sequence for possible kinase motifs. For this simply click on the button.

"Search protein sequence at Scansite" in the "Protein Info" section (Fig. 7).

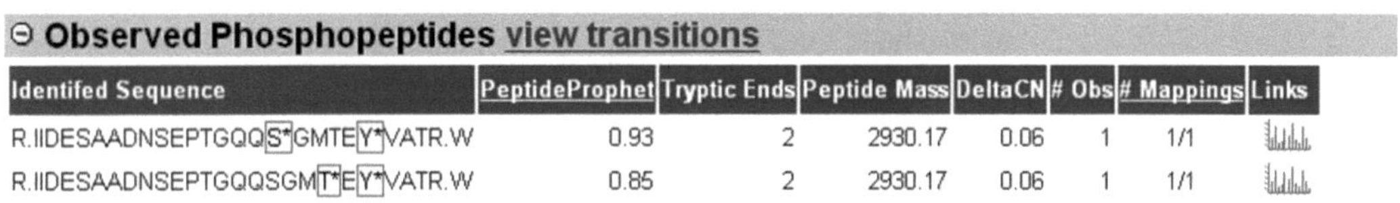

⊖ Observed Phosphopeptides view transitions

Identifed Sequence	PeptideProphet	Tryptic Ends	Peptide Mass	DeltaCN	# Obs	# Mappings	Links
R.IIDESAADNSEPTGQQS*GMTEY*VATR.W	0.93	2	2930.17	0.06	1	1/1	
R.IIDESAADNSEPTGQQSGMT*EY*VATR.W	0.85	2	2930.17	0.06	1	1/1	

Fig. 6. Two phosphopeptides with unassigned phosphorylation sites are highlighted. However, knowing the activation loop motif of MAP kinases the sites can be assigned.

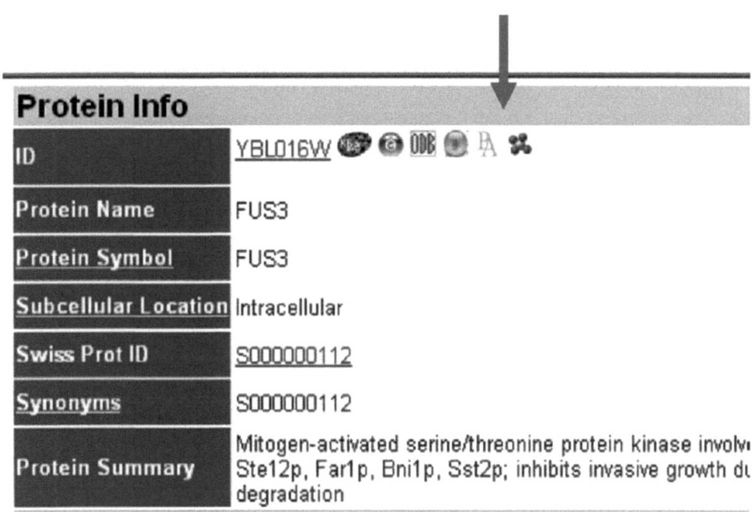

Protein Info

ID	YBL016W
Protein Name	FUS3
Protein Symbol	FUS3
Subcellular Location	Intracellular
Swiss Prot ID	S000000112
Synonyms	S000000112
Protein Summary	Mitogen-activated serine/threonine protein kinase involv Ste12p, Far1p, Bni1p, Sst2p; inhibits invasive growth du degradation

Fig. 7. The button that executes the Scansite algorithm is highlighted.

3.2.4. Check the Evolutionary Conservation of the Site

You can also check whether your phosphorylation site of interest is evolutionary conserved, which can be an additional indication for the correct assignment of a phosphorylation site. For this click on the button ODB "View orthologs/homolog information" (Fig. 8) and a new window will be opened, showing the alignment of the amino acid sequences with the identified phosphorylation sites between yeast, worm, fly, and human (Fig. 9).

Based on this alignment, we can conclude that the unassigned phosphothreonine is correctly assigned and that in the top amino acid sequence either the tyrosine or threonine in the **TXY** motif should be phosphorylated.

3.2.5. Take a Look at the Tandem Mass Spectrum

To assess whether the phosphorylation site was correctly assigned, it is always advisable to take a look at the tandem mass spectrum of the phosphopeptide. You can open it by clicking on the symbol (Fig. 10).

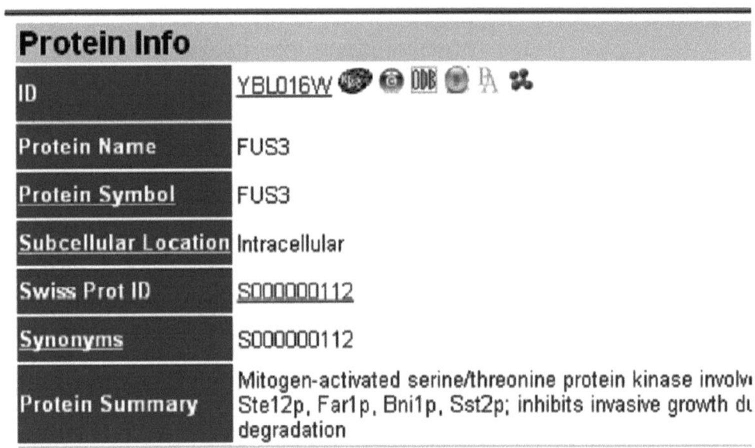

Fig. 8. The button that allows to show the orthologous proteins is indicated.

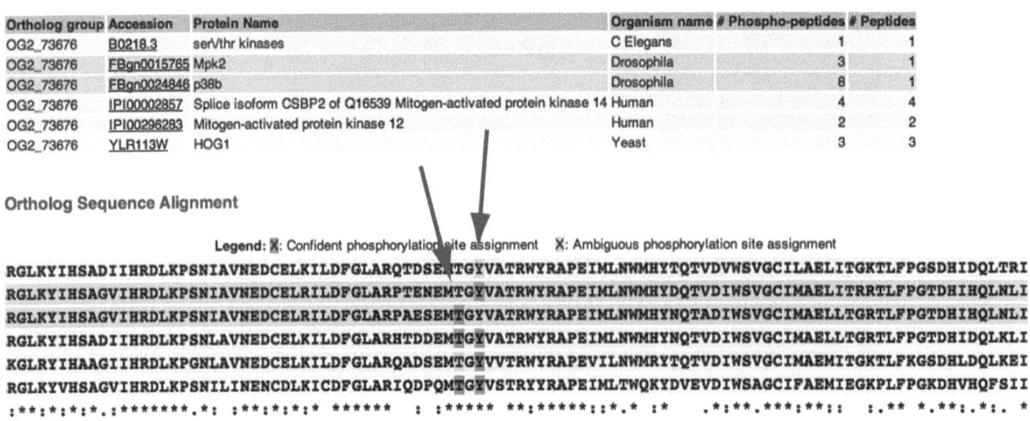

Fig. 9. Detailes are shown in the text.

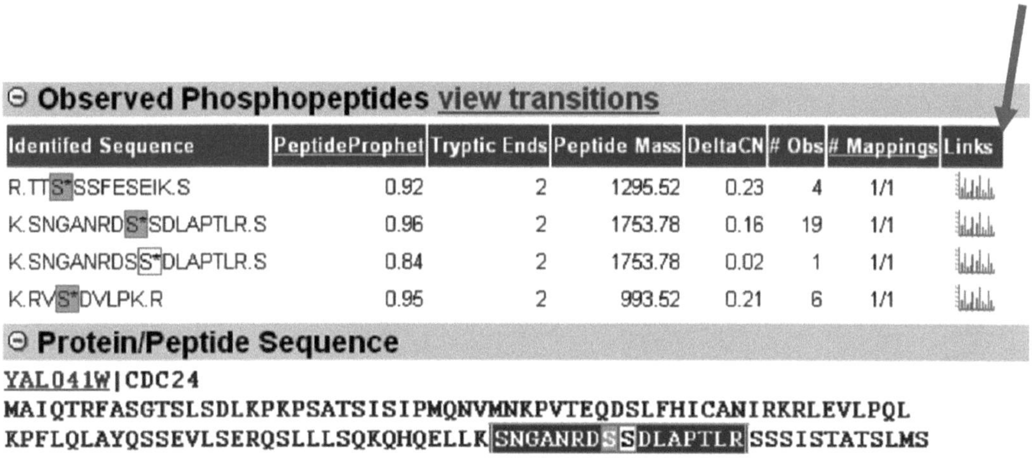

Observed Phosphopeptides view transitions

Identifed Sequence	PeptideProphet	Tryptic Ends	Peptide Mass	DeltaCN	# Obs	# Mappings	Links
R.TTS*SSFESEIK.S	0.92	2	1295.52	0.23	4	1/1	
K.SNGANRDS*SDLAPTLR.S	0.96	2	1753.78	0.16	19	1/1	
K.SNGANRDSS*DLAPTLR.S	0.84	2	1753.78	0.02	1	1/1	
K.RVS*DVLPK.R	0.95	2	993.52	0.21	6	1/1	

Protein/Peptide Sequence

YAL041W|CDC24
MAIQTRFASGTSLSDLKPKPSATSISIPMQNVMNKPVTEQDSLFHICANIRKRLEVLPQL
KPFLQLAYQSSEVLSERQSLLLSQKQHQELLKSNGANRDSSDLAPTLRSSSISTATSLMS

Fig. 10. The *arrow* indicates how to open the corresponding tandem mass spectrum of a given phosphopeptide entry.

The manual interpretation of tandem mass spectra can be, especially in the case of phosphopeptides, difficult. Therefore we recommend studying the following slides, which are a nice introduction to this topic. You can find them under the URL http://www.proteomesoftware.com/Proteome_software_pro_protein_id.html (Proteome Software Inc.)

This will open a new window in which the tandem mass spectrum is displayed (Fig. 11). In the upper window you see the tandem mass spectrum in which the fragment ion peaks are assigned with y-ion or b-ion together with a number (ion assignment nomenclature) as well as below the spectrum the amino acid sequence of the phosphopeptide is shown (a phosphoserine is indicated as "S[167]," a phosphothreonine as "T[181]" and a phosphotyrosine as "Y[243]"). Here you have to look for the following: left and right of the amino acid sequence the fragment ion signals that were found and could be assigned in the tandem mass spectrum are highlighted. In our example the question is, if the serine (at position 6) is phosphorylated LSLTDS$_{167}$TETIENNATVK or the adjacent threonine LSLTDST$_{167}$ETIENNATVK at position 7. Of note, most spectra loaded into the PhosphoPep database are consensus spectra (7), which means that only repeatedly observed peptide fragment ions are shown. Noise signals were removed.

The inspection of the highlighted ions shows that indeed all peptide fragment ions, including the one corresponding to the phosphorylated serine as well as the non-phosphorylated threonine, were identified and assigned, strengthening that the assigned serine phosphorylation is correct. In addition, take a look at the tandem mass spectrum in Fig. 11. Here you can see that both assigned fragment ions are rather intense. In addition, the y^{11+} fragment ion

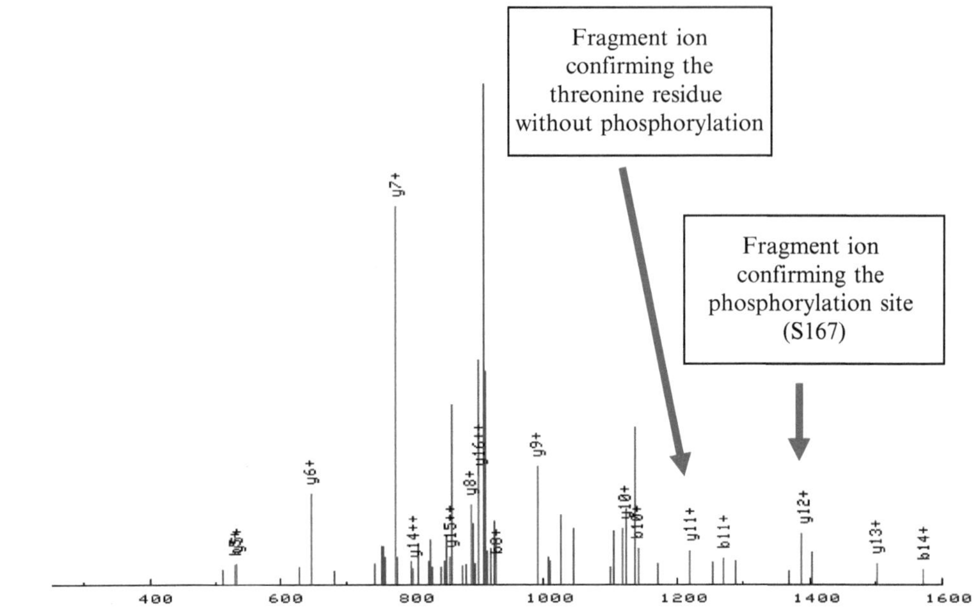

b^{1+}	b^{2+}	#	AA	#	y^{1+}	y^{2+}
114.1668	57.5871	1	L	17		
201.2450	101.1262	2	S	16	1803.8088	**902.4081**
314.4044	157.7059	3	L	15	1716.7306	**858.8690**
415.5095	208.2584	4	T	14	1603.5712	**802.2893**
530.5961	265.8017	5	D	13	**1502.4661**	751.7368
697.6542	349.3308	6	S[167]	12	**1387.3795**	694.1935
798.7593	399.8833	7	T	11	**1220.3214**	610.6644
927.8748	464.4411	8	E	10	**1119.2163**	560.1119
1028.9799	514.9936	9	T	9	**990.1008**	495.5541
1142.1393	571.5733	10	I	8	**888.9957**	445.0016
1271.2548	636.1311	11	E	7	**775.8363**	388.4219
1385.3586	693.1830	12	N	6	**646.7208**	323.8641
1499.4624	750.2349	13	N	5	**532.6170**	266.8122
1570.5412	785.7743	14	A	4	418.5132	209.7603
1671.6463	836.3268	15	T	3	347.4344	174.2209
1770.7789	885.8931	16	V	2	246.3293	123.6684
		17	K	1	147.1967	74.1021

Phosphorylation site

Fig. 11. On *top* the tandem mass spectrum and on *bottom* the assigned ions of the phosphopeptide "LSLTDS$_{167}$TETIENNATVK" are displayed.

at *m/z* 1,300.3 (1220.3+80) or the y^{11++} fragment ion at *m/z* 650.7, which would indicate that threonine 7 phosphorylated, are missing. The same is true for the b^{6+} ion at *m/z* 617.7, which would indicate that serine 6 is not phosphorylated. Both findings suggest the correct assignment of the phosphorylation site.

a

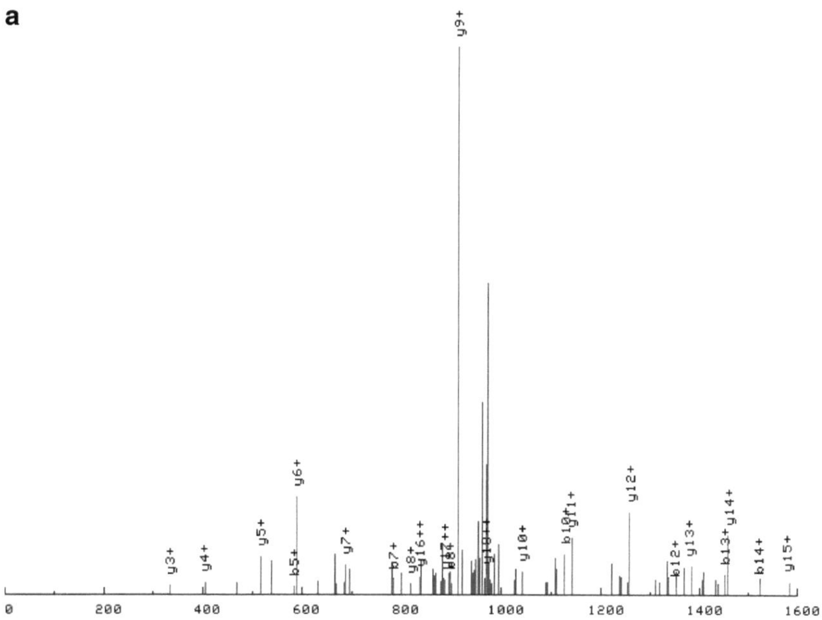

b^{1+}	b^{2+}	#	AA	#	y^{1+}	y^{2+}
102.1125	51.5599	1	T	19		
269.1706	135.0890	2	S[167]	18	1938.9370	**969.9722**
368.3032	184.6553	3	V	17	1771.8789	**886.4432**
455.3814	228.1944	4	S	16	1672.7463	**836.8769**
584.4969	292.7521	5	E	15	**1585.6681**	793.3378
655.5757	328.2915	6	A	14	**1456.5526**	728.7800
783.7064	392.3569	7	Q	13	**1385.4738**	693.2406
897.8102	449.4088	8	N	12	**1257.3431**	629.1753
998.9153	499.9613	9	T	11	**1143.2393**	572.1234
1127.0460	564.0267	10	Q	10	**1042.1342**	521.5708
1224.1627	612.5850	11	P	9	**914.0035**	457.5055
1352.2934	676.6504	12	Q	8	**816.8868**	408.9471
1451.4260	726.2167	13	V	7	**688.7561**	344.8818
1522.5048	761.7561	14	A	6	**589.6235**	295.3155
1636.6086	818.8080	15	N	5	**518.5447**	259.7761
1707.6874	854.3474	16	A	4	**404.4409**	202.7242
1822.7740	911.8907	17	D	3	**333.3621**	167.1848
1893.8528	947.4301	18	A	2	218.2755	109.6415
		19	K	1	147.1967	74.1021

Fig. 12. Two tandem mass spectra that fail to correctly assign the phosphorylation site in the peptide "T**S**VSEAQNTQPQVANADAK".

In Fig. 12a, b, an example is shown in which the assignment of the correct site of phosphorylation is difficult. It is not clear if the highlighted serine T**S***VSEAQNTQPQVANADAK or the highlighted threonine **T***SVSEAQNTQPQVANADAK is phosphorylated.

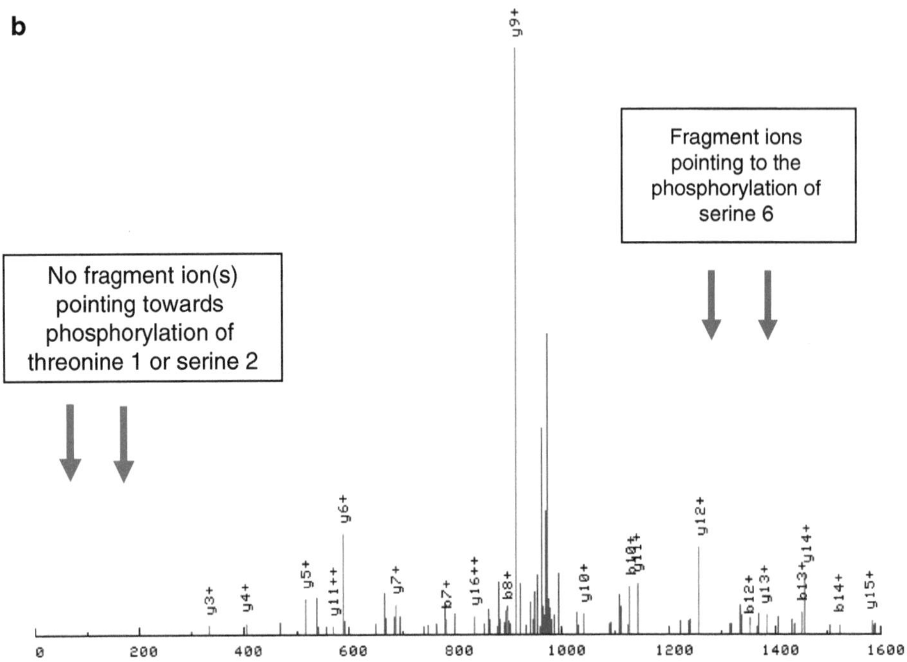

b^{1+}	b^{2+}	#	AA	#	y^{1+}	y^{2+}
182.0924	91.5499	1	T[181]	19		
269.1706	135.0890	2	S	18	1858.9571	929.9823
368.3032	184.6553	3	V	17	1771.8789	886.4432
455.3814	228.1944	4	S	16	1672.7463	**836.8769**
584.4969	292.7521	5	E	15	**1585.6681**	793.3378
655.5757	328.2915	6	A	14	**1456.5526**	728.7800
783.7064	392.3569	7	Q	13	**1385.4738**	693.2406
897.8102	449.4088	8	N	12	**1257.3431**	629.1753
998.9153	499.9613	9	T	11	**1143.2393**	**572.1234**
1127.0460	564.0267	10	Q	10	**1042.1342**	521.5708
1224.1627	612.5850	11	P	9	**914.0035**	457.5055
1352.2934	676.6504	12	Q	8	816.8868	408.9471
1451.4260	726.2167	13	V	7	**688.7561**	344.8818
1522.5048	761.7561	14	A	6	**589.6235**	295.3155
1636.6086	818.8080	15	N	5	**518.5447**	259.7761
1707.6874	854.3474	16	A	4	**404.4409**	202.7242
1822.7740	911.8907	17	D	3	**333.3621**	167.1848
1893.8528	947.4301	18	A	2	218.2755	109.6415
		19	K	1	147.1967	74.1021

Fig. 12. (continued).

First, most peptide fragment ions that could unequivocally distinguish the two phosphorylation sites are outside the recorded *m/z* range. Second, the y^{18++} fragment ion at *m/z* 969.97 that could indicate that the serine 2 is phosphorylated, but not threonine 1, is present at low relative intensity in a *m/z* region crowded with signals, therefore it is not sure whether this is a real fragment ion or noise.

References

1. Bodenmiller B., Campbell D., Gerrits B., Lam H., Jovanovic M., Picotti P., Schlapbach R., Aebersold R. (2008) PhosphoPep – a database of protein phosphorylation sites in model organisms. *Nat. Biotechnol.* **26**, 1339–1340.
2. Kanehisa M., Goto S., Hattori M., Aoki-Kinoshita K. F., Itoh M., Kawashima S., Katayama T., Araki M., Hirakawa M. (2006) From genomics to chemical genomics: new developments in KEGG. *Nucleic Acids Res.* **34**, D354–D357.
3. Shannon P., Markiel A., Ozier O., Baliga N. S., Wang J. T., Ramage D., Amin N., Schwikowski B., Ideker T. (2003) Cytoscape: a software environment for integrated models of biomolecular interaction networks. *Genome Res.* **13**, 2498–2504.
4. Lange V., Picotti P., Domon B., Aebersold R. (2008) Selected reaction monitoring for quantitative proteomics: a tutorial. *Mol. Syst. Biol.* **4**, 222.
5. Picotti P., Aebersold R., Domon B. (2007) The implications of proteolytic background for shotgun proteomics. *Mol. Cell. Proteomics* **6**, 1589–1598.
6. Picotti P., Lam H., Campbell D., Deutsch E. W., Mirzaei H., Ranish J., Domon B., Aebersold R. (2008) A database of mass spectrometric assays for the yeast proteome. *Nat. Methods* **5**, 913–914.
7. Lam H., Deutsch E. W., Eddes J. S., Eng J. K., King N., Stein S. E., Aebersold R. (2007) Development and validation of a spectral library searching method for peptide identification from MS/MS. *Proteomics* **7**, 655–667.
8. Lam H., Deutsch E. W., Eddes J. S., Eng J. K., Stein S. E., Aebersold R. (2008) Building consensus spectral libraries for peptide identification in proteomics. *Nat. Methods* **5**, 873–875.
9. Bodenmiller B., Malmstrom J., Gerrits B., Campbell D., Lam H., Schmidt A., Rinner O., Mueller L. N., Shannon P. T., Pedrioli P. G., Panse C., Lee H. K., Schlapbach R., Aebersold R. (2007) PhosphoPep – a phosphoproteome resource for systems biology research in Drosophila Kc167 cells. *Mol. Syst. Biol.* **3**, 139.
10. Bodenmiller B., Mueller L. N., Mueller M., Domon B., Aebersold R. (2007) Reproducible isolation of distinct, overlapping segments of the phosphoproteome. *Nat. Methods* **4**, 231–237.
11. von Mering C., Jensen L. J., Kuhn M., Chaffron S., Doerks T., Kruger B., Snel B., Bork P. (2007) STRING 7 – recent developments in the integration and prediction of protein interactions. *Nucleic Acids Res.* **35**, D358–D362.
12. Chen F., Mackey A. J., Stoeckert C. J., Jr., Roos D. S. (2006) OrthoMCL-DB: querying a comprehensive multi-species collection of ortholog groups. *Nucleic Acids Res.* **34**, D363–D368.
13. Desiere F., Deutsch E. W., Nesvizhskii A. I., Mallick P., King N. L., Eng J. K., Aderem A., Boyle R., Brunner E., Donohoe S., Fausto N., Hafen E., Hood L., Katze M. G., Kennedy K. A., Kregenow F., Lee H., Lin B., Martin D., Ranish J. A., Rawlings D. J., Samelson L. E., Shiio Y., Watts J. D., Wollscheid B., Wright M. E., Yan W., Yang L., Yi E. C., Zhang H., Aebersold R. (2005) Integration with the human genome of peptide sequences obtained by high-throughput mass spectrometry. *Genome Biol.* **6**, R9.
14. Obenauer J. C., Cantley L. C., Yaffe M. B. (2003) Scansite 2.0: proteome-wide prediction of cell signaling interactions using short sequence motifs. *Nucleic Acids Res.* **31**, 3635–3641.
15. Keller A., Eng J., Zhang N., Li X. J., Aebersold R. (2005) A uniform proteomics MS/MS analysis platform utilizing open XML file formats. *Mol. Syst. Biol.* **1**, 2005.0017.
16. Keller A., Nesvizhskii A. I., Kolker E., Aebersold R. (2002) Empirical statistical model to estimate the accuracy of peptide identifications made by MS/MS and database search. *Anal. Chem.* **74**, 5383–5392.

17. Eng J. K., McCormack A. L., Yates J. R. (1994) An approach to correlate tandem mass spectral data of peptides with amino acid sequences in a protein database. *J Am Soc Mass Spectrom.* **5**, 976–989.
18. Beausoleil S. A., Villen J., Gerber S. A., Rush J., Gygi S. P. (2006) A probability-based approach for high-throughput protein phosphorylation analysis and site localization. *Nat. Biotechnol.* **24**, 1285–1292.
19. Elias J. E., Gygi S. P. (2007) Target-decoy search strategy for increased confidence in large-scale protein identifications by mass spectrometry. *Nat. Methods* **4**, 207–214.
20. Reiter L., Claassen M., Schrimpf S. P., Buhmann J. M., Hengartner M. O., Aebersold R. (2009) Protein identification false discovery rates for very large proteomics datasets generated by tandem mass spectrometry. *Mol. Cell. Proteomics* **8**, 2405–2417.
21. Olsen J. V., Blagoev B., Gnad F., Macek B., Kumar C., Mortensen P., Mann M. (2006) Global, in vivo, and site-specific phosphorylation dynamics in signaling networks. *Cell* **127**, 635–648.

Chapter 20

Protein-Centric Data Integration for Functional Analysis of Comparative Proteomics Data

Peter B. McGarvey, Jian Zhang, Darren A. Natale, Cathy H. Wu, and Hongzhan Huang

Abstract

High-throughput proteomic, microarray, protein interaction and other experimental methods all generate long lists of proteins and/or genes that have been identified or have varied in accumulation under the experimental conditions studied. These lists can be difficult to sort through for Biologists to make sense of. Here we describe a next step in data analysis – a bottom-up approach at data integration – starting with protein sequence identifications, mapping them to a common representation of the protein and then bringing in a wide variety of structural, functional, genetic, and disease information related to proteins derived from annotated knowledge bases and then using this information to categorize the lists using Gene Ontology (GO) terms and mappings to biological pathway databases. We illustrate with examples how this can aid in identifying important processes from large complex lists.

Key words: Gene Ontology, Biological pathways, Protein database, UniProtKB, Proteomics, Bioinformatics

1. Introduction

High-throughput transcriptome and proteome projects have resulted in the rapid accumulation of large amounts of data comparing the expression of genes and proteins under many conditions. Proteomic, microarray, protein interaction, and other experimental methods all generate long lists of proteins and/or genes that have been identified or have varied in accumulation under the experimental conditions studied. Depending on the platform used, a variety of filtering tools and clustering algorithms are available to help identify and sort the significant results from the noise. The appropriate sorting and filtering methods are

Cathy H. Wu and Chuming Chen (eds.), *Bioinformatics for Comparative Proteomics*, Methods in Molecular Biology, vol. 694,
DOI 10.1007/978-1-60761-977-2_20, © Springer Science+Business Media, LLC 2011

dependent on the platform and experimental design. Many books, papers and software packages are available describing their application. However even the filtered and clustered lists often do not provide biological answers as to what genetic and cellular processes are significantly affected in a particular experiment. The long lists that result are difficult for biologists to sort through for insights. Further functional interpretation and knowledge discovery can be derived from the integration of protein sequence data with additional biomedical data. Here we describe the next step in data analysis – a bottom-up approach at data integration – starting with protein sequence information and then bringing in a wide variety of structural, functional, genetic, and disease information related to proteins and then using this information with various tools to assist biologists to functionally compare proteomic results and make inferences from complex lists. Figure 1 illustrates this functional approach using data integration. Here we illustrate, using tools available on the Protein Information Resource website and some real data use cases, how one can use Gene Ontology terms and pathway databases to assist in making sense out of large-scale proteomic data.

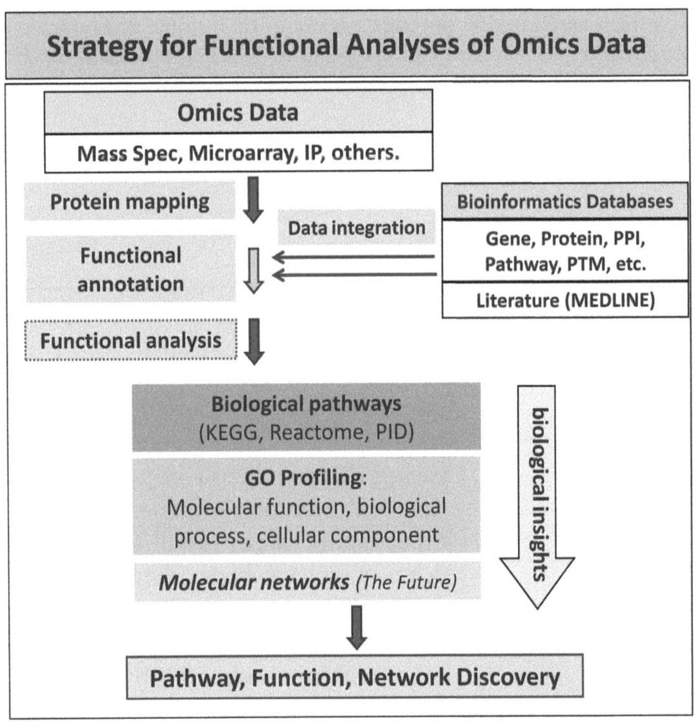

Fig. 1. Illustration of a bottom-up approach at omics data integration – starting with protein sequence information and then bringing in a variety of structural, functional, genetic, and disease information related to proteins and then using this information with tools to assist biologists to functionally compare proteomic results and make inferences from complex lists.

2. Materials

1. Lists of protein or gene identifiers from a proteomic or genomic experiment. Currently the PIR system supports retrieval of entries based on any 1 of 15 protein or DNA sequence identifiers, three gene identifiers (Entrez Gene, OMIM and UniProt gene names), and PDB ID as a protein structure identifier. Mixed groups of identifiers such as "Any Unique ID" or "Any Sequence ID" are also supported though this might retrieve more than you wish, as it is possible for one sequence ID such as a GI number to also match a taxonomy ID and an Entrez Gene ID or more and retrieve unrelated proteins. If you have a choice UniProt (1) accessions followed by GenBank/EMBL/DDBJ protein accessions are preferred as most all other identifiers are mapped to these databases.
2. Internet access to http://proteininformationresource.org/pirwww/search/batch.shtml.

The example identifier lists used in this chapter can be found here: ID mapping: ftp://ftp.pir.georgetown.edu/pub/MiMB/Comparative_Proteomics/ID_gi_list.txt

GO Analysis: ftp://ftp.pir.georgetown.edu/pub/MiMB/Comparative_Proteomics/Ba_infect_0hr.txt

ftp://ftp.pir.georgetown.edu/pub/MiMB/Comparative_Proteomics/Ba_infect_3+6hr.txt

ftp://ftp.pir.georgetown.edu/pub/MiMB/Comparative_Proteomics/Ba_infect_3+6hr_no-0hr.txt

Pathways Analysis: ftp://ftp.pir.georgetown.edu/pub/MiMB/Comparative_Proteomics/typhi_mgm_cluster_96.txt

3. Methods

3.1. Batch Retrieval, ID (Identifier) Mapping and Data Integration

The first step in this approach is to combine all available functional, genetic, structural, pathway, ontology, literature and disease information related to the genes or proteins on your list. This is done by mapping the lists of genes/proteins to a comprehensive data warehouse where this information is stored. Here we use the iProClass (2) warehouse that is composed of UniProtKB (1) proteins supplemented with selected UniParc (3) proteins and additional annotation, information and cross-references from over a 100 molecular databases. All iProClass accessions are UniProtKB or UniParc accessions.

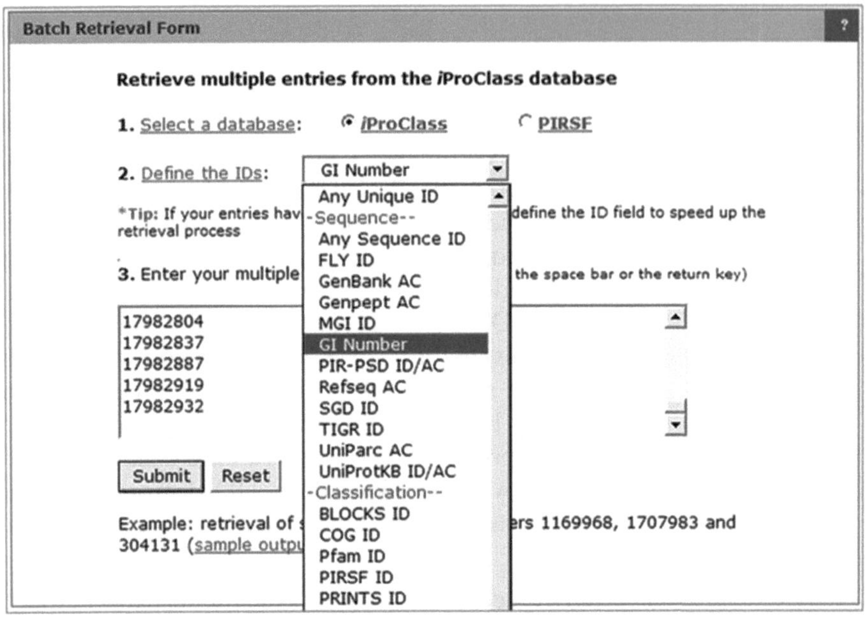

Fig. 2. Initiating batch retrieval with ID mapping. This example uses gi numbers to map to UniProtKB. The first step the analysis shown in Fig. 1.

1. Using the ID mapping example list, or your own list, open a browser window to http://proteininformationresource.org/pirwww/search/batch.shtml (see Fig. 2).
2. Copy/Paste the list into the window and set the appropriate input identifier. See Fig. 2 where we use a list of NCBI gi numbers.
3. Click "+" and give the list a reference name such as "Condition_1_strain_abc" so that you can use it to identify and retrieve the protein set later.
4. Click "retrieve" and wait for the result table. See Fig. 3 for following steps.
5. The resulting table contains all the iProClass protein entries that map to the identifiers you submitted. You can customize the result table altering the columns displayed and you can download the table. You can view detailed information about each protein in either the iProClass or UniProt website. For details on these and other analysis options, please see the PIR help pages at: http://proteininformationresource.org / pirwww/support/help.shtml#7
6. The reference name you entered is now a JOB identifier displayed next to the "retrieve" button. Click on the JOB identifier to bookmark your table to return to later and to distinguish this result from other results you may load for comparison. The JOB identifier will be displayed in some of the analysis windows we describe below (see Note 1).

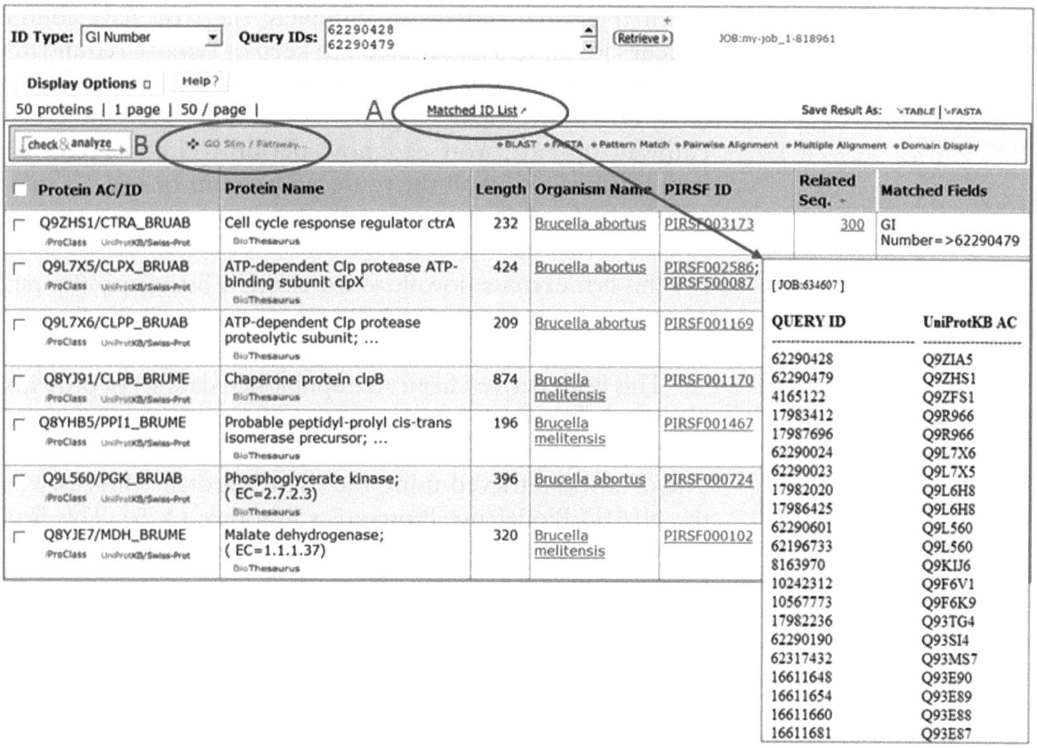

Fig. 3. The results of a batch retrieval using gi numbers. (**a**) Link to display the results of the ID mapping. (**b**) Link to initiate the GO and Pathway analysis options used in the methods.

7. Next review the ID mapping by clicking on the "matched ID List" link above the table. The resulting window shows the mapping from the ID submitted to iProClass/UniProt accessions. All matches are shown on one line, if multiple identifiers are seen in the UniProtKB AC column it means your identifier matched more than one sequence in the database. There are a number of reasons this may occur including multiple isoforms of the same gene/protein, redundancy in one of the mapped databases and others. You will observe "no match" if nothing is found. The results table will contain all protein accessions found.

   Checking this matched list is a critical step before proceeding with your analysis. If there are missing matches, you need to decide if you need to find them by other means now. If there are multiple (redundant) matches, you need to decide if you want to remove the extras to have a smaller and "cleaner" list before proceeding with the analysis. Keeping them is not necessarily a problem but the redundancy (common in high throughput experiments) is something you need to be aware of. Please refer to the Notes 5 and 6 for details on ID mapping and suggestions on the steps you can take to minimize redundancy and find missing protein matches.

8. When the batch retrieval is complete (i.e. you have found matches for all IDs and decided to keep or remove redundant matches), initiate the Pathway or GO analysis by clicking on the GO/Pathway icon (Fig. 3) which will modify the GO/Pathway options. You can now perform either a GO or Pathway analysis on all the retrieved proteins or select a subset of proteins first.

3.2. Gene Ontology Analysis

1. For this first exercise download the example list at ftp://ftp.pir.georgetown.edu/pub/MiMB/Comparative_Proteomics/Ba_infect_0hr.txt.

   This list is derived from an unpublished data set of proteins identified by mass spectrometry from experiments where mouse macrophage cells were infected with *Bacillus anthracis*. The data was retrieved using the Master Protein Directory in the NIAID Biodefense Proteomics Resource (4, 5). This first list is a control set containing all *Bacillus* proteins identified at zero time after inoculation.
2. Do batch retrieval at http://proteininformationresource.org/pirwww/search/batch.shtml using the UniProtKB accession option. After your batch retrieval, you should have a list of 831 proteins, at which point you should initiate the GO/Pathway analysis options (see step 8 in Subheading 3 and Fig. 3).
3. Click on the GO Slim "Proc." link to use GO Slims based on the Gene Ontology's (6) biological process ontology. This is a good place to start a characterization as it provides a high-level grouping of biological processes. A new analysis window will appear to display the results if your list of proteins is large (like this one) it may take a few minutes to display the results. See Fig. 4 for an example of the GO biological process display using this list of proteins. See Notes 2 & 3 for information on other GO ontology options and use of the GO analysis window.
4. The default GO display shown in Fig. 4 has four columns: (1) a check box where you can select to remove items from the display; (2) the GO ID which identifies the database and a unique ID that links to the Amigo database (http://amigo.geneontology.org) for additional information on the GO terms; (3) the GO Term or name; and (4) the frequency or number of proteins in the list that map to that Term. Some functions of the display are:
   (a) The GO ID, Term and Frequency columns can be sorted to help group similar IDs or Terms.
   (b) The frequency column has two sub-columns. (1) A number, which is the number of proteins that map to this Term. Clicking on this number opens a new window and does batch retrieval on these proteins only, allowing the user

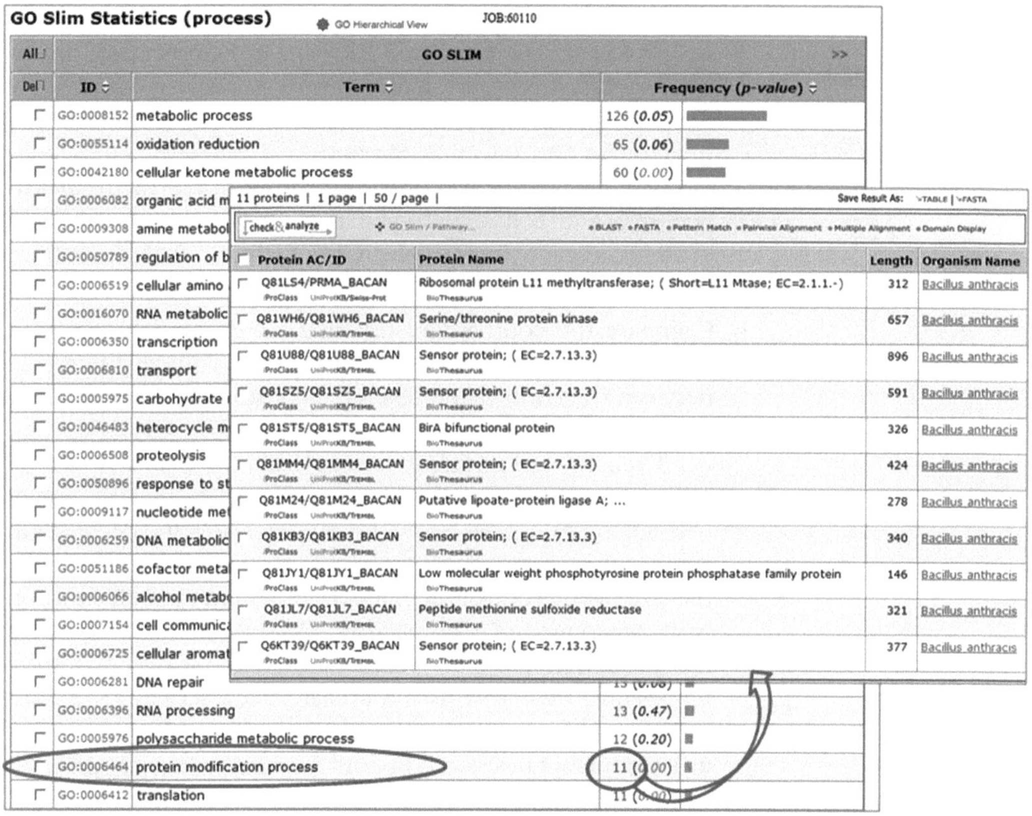

Fig. 4. Example of a GO process analysis discussed in the text. Shows the 11 "protein modification process" proteins that appeared in macrophages infected with *Bacillus anthracis*. These proteins are used for a GO function analysis shown in Fig. 5. See Note 5 for information on *p*-value.

to do an additional investigation and analysis on the selected set of proteins. Next to the number of proteins is a *p*-value whose calculation is based on the frequency of a particular term in the database. See Note 7 for more information on the use of this value. (2) A histogram bar, clicking on this bar opens a window with the GO term and iProClass accession numbers in plain text.

(c) The initial display shows only the PIR GO slim terms (http://www.geneontology.org/GO.slims.shtml). These terms are a subset of the complete GO ontology that groups the proteins into larger branches of the ontology tree but does not display all GO terms. To navigate to smaller branches and the leaves click on the ">>" icon. The first click opens a GO slim + 1 display that shows the original GO slim terms but includes the next terms in the ontology. Repeating the process shows ALL GO terms. Not all proteins have GO terms and not all proteins have additional terms at the more detailed levels.

5. After examining the GO display on the control set *open a second browser window or tab* and repeat the process of downloading, batch retrieval, and GO process analysis with the file ftp://ftp.pir.georgetown.edu/pub/MiMB/Comparative_Proteomics/Ba_infect_3+6hr.txt.

   This list is a test set containing all *Bacillus* proteins identified at 3 and 6 h after inoculation. It should return 1,525 proteins. Again, initiate the GO/Pathway analysis options and click on the GO Slim "Proc." link.

6. Compare the control and test lists on the screen (for such large lists it is often best to print out). Can you see differences between the control and test set? Probably not, though there are some differences.

   Even by categorizing a large list into a smaller list of biological processes, the result is still a relatively large list of overlapping categories and significant increases and decreases in composition are not apparent here as the test set is almost twice the size of the control set. This is not an uncommon situation with proteomic and genomic data. However, look down the list for "protein modification process" where the smaller "control" set has only two proteins and the larger test set has 13 (almost four times as large). This might be a significant change, but let us leave further investigation until later and refine our data.

7. Again, *open another browser window or tab* and repeat the download and batch retrieval process with the file ftp://ftp.pir.georgctown.edu/pub/MiMB/Comparative_Proteomics/Ba_infect_3+6hr_no-0hr.txt. This second list is a set of all *Bacillus* proteins identified at 3 and 6 h after inoculation but excludes all proteins also found in the control 0-time set. It should return 694 proteins. Again, initiate the GO/Pathway analysis options and click on the GO Slim "Proc." link.

8. Now compare the new list of GO biological process terms (see Fig. 4) to the previous two. All the proteins in this list appeared 3–6 h after infection and were not present at the time of infection, indicating they are likely to be important for the bacteria's adaptation to the macrophage environment and the process of infection. The list now gives a better indication of the biological processes affected, though it does not contain all the proteins in the processes as those also present at zero time were filtered out. We can now drill down further using other options.

9. Now let us investigate the "protein modification process" which moved up the list and showed significant changes in both test samples. Click on the number "11" in the frequency column next to the term (see Fig. 4). This pops up a new window and performs batch retrieval on only the proteins in this category. Initiate the GO/Pathway analysis options and click on the GO Slim "Func." link to view the GO molecular function ontology terms.

10. After the window opens click on “>>” to expand the GO Slim display to GO Slim + 1 and then click “>>” again to see all GO ontology terms. See figure z for the display. GO is an ontology where terms have parent-child relationships, the initial display shows a “GO slim” which represents early branches of the ontology tree, this can be expanded to smaller branches using “GO slim + 1” and to all available GO terms, the leaves, by showing all GO.

    The GO molecular function frequency display shown in Fig. 5 shows all the functional terms associated with the “protein modification process” proteins. A number of activities are shown and proteins can have multiple activities. Looking at the largest category “transferase activity,” we have seven proteins mostly involved in some sort of phosphate transfer characteristic of cellular signaling mechanisms (not shown). By examining each protein’s annotation individually in iProClass or UniProtKB or by using the pathway tool described in the next section, we can learn more about the role these proteins are known to play in the cell and possibly make some hypothesis to their roles related to infection here. Five of the seven proteins are sensor proteins involved in two-component sensory systems. Three of the five sensor proteins have pathways associated with them in the KEGG database as systems that a) sense oxygen levels for switching between aerobic and anaerobic respiration, b) sense environmental conditions that might induce sporulation, and c) signals to induce chemotaxis. The pathways for the other two sensor proteins are unknown.

3.3. Biological Pathways Analysis

1. For this exercise, use the Pathway example list at ftp://ftp.pir.georgetown.edu/pub/MiMB/Comparative_Proteomics/typhi_mgm_cluster_96.txt. This list of accessions is derived from a published data set of proteins identified by mass spectrometry from experiments where *Salmonella typhi* was

GO Slim Statistics (function) GO Hierarchical View JOB:755720

GO SLIM ID	GO SLIM Term	GO SLIM Frequency	GO SLIM + 1 ID	GO SLIM + 1 Term	GO ID	GO Term	GO Frequency
GO:0016740	transferase activity	7	GO:0016772	transferase activity, transferring phosphorus-containing groups	GO:0016301	kinase activity	6
					GO:0000155	two-component sensor activity	5
					GO:0004673	protein histidine kinase activity	5
					GO:0016772	transferase activity, transferring phosphorus-containing groups	5
					GO:0004672	protein kinase activity	1
					GO:0004674	protein serine/threonine kinase activity	1
			GO:0016741	transferase activity, transferring one-carbon groups	GO:0008168	methyltransferase activity	1
					GO:0008276	protein methyltransferase activity	1
GO:0001882	nucleoside binding	6	GO:0001883	purine nucleoside binding	GO:0005524	ATP binding	6
GO:0000166	nucleotide binding	6	GO:0017076	purine nucleotide binding	GO:0005524	ATP binding	6
			GO:0032553	ribonucleotide binding	GO:0005524	ATP binding	6
GO:0004871	signal transducer activity	5	GO:0004871	signal transducer activity	GO:0004871	signal transducer activity	4
GO:0016874	ligase activity	2	GO:0016879	ligase activity, forming carbon-nitrogen bonds	GO:0004077	biotin-[acetyl-CoA-carboxylase] ligase activity	1
GO:0030528	transcription regulator activity	2	GO:0016564	transcription repressor activity	GO:0016564	transcription repressor activity	1
GO:0003824	catalytic activity	2	GO:0016740	transferase activity	GO:0016740	transferase activity	7
			GO:0016874	ligase activity	GO:0016874	ligase activity	2
			GO:0016491	oxidoreductase activity	GO:0016491	oxidoreductase activity	1
GO:0031406	carboxylic acid binding	1	GO:0031406	carboxylic acid binding	GO:0031406	carboxylic acid binding	1
GO:0016787	hydrolase activity	1	GO:0016788	hydrolase activity, acting on ester bonds	GO:0004725	protein tyrosine phosphatase activity	1
GO:0016491	oxidoreductase activity	1	GO:0016667	oxidoreductase activity, acting on sulfur group of donors	GO:0016671	oxidoreductase activity, acting on sulfur group of donors, disulfide as acceptor	1
					GO:0008113	peptide-methionine-(S)-S-oxide reductase activity	1
GO:0008144	drug binding	1	GO:0008144	drug binding	GO:0008144	drug binding	1
	unclassified	0		unclassified		unclassified	0
Total UniProtKB ACCs in this JOB = 11							

Fig. 5. Example of a GO function analysis on the 11 GO process proteins identified in Fig. 4.

grown in different media (7) and compared with *Salmonella typhimurium* grown under the same conditions (8). This particular set represents a cluster of highly expressed proteins in *S. typhi* grown in magnesium-depleted minimal medium thought to mimic some of the intracellular conditions seen by the pathogen during infection.

2. Do batch retrieval at http://proteininformationresource.org/pirwww/search/batch.shtml using the UniProtKB accession option. After your batch retrieval, you should have a list of 96 proteins; initiate the GO/Pathway analysis options (see step 8 in Subheading 3 and Fig. 3).

3. Click on the Pathway link, a new analysis window will appear to display the result. See Fig. 6 for an example of the Pathway display and Fig. 7 for further detail. The Pathway analysis tool displays pathways from KEGG (9), Reactome (10) and PID (11). All proteins in the list that are members of a pathway are displayed by default. A protein may be a member in multiple pathways from multiple databases. Proteins that do not map to KEGG, Reactome or PID are displayed in an unclassified category at the bottom of the display. See Note 3 for additional information on using the pathway window. See Note 4 for an additional pathway analysis example.

4. The Pathway display has four columns similar to the GO display: (1) a check box where you can select to remove categories from the display; (2) the Pathway ID which identifies the database and a unique ID that links to the database for additional information on the pathway; (3) the pathway Term

PATHWAY Statistics JOB:typi_mgm_cluster_96-981631

All	PATHWAY		
Del	ID	Term	Frequency (*p-value*)
	(KEGG) sty01100	Metabolic pathways	18 (*0.07*)
	(KEGG) stt01100	Metabolic pathways	18 (*0.01*)
	(KEGG) sty02010	ABC transporters	8 (*0.02*)
	(KEGG) stt02010	ABC transporters	8 (*0.02*)
	(KEGG) sty02020	Two-component system	7 (*0.04*)
	(KEGG) stt02020	Two-component system	6 (*0.07*)
	(KEGG) stt00330	Arginine and proline metabolism	5 (*0.01*)
	(KEGG) sty00330	Arginine and proline metabolism	5 (*0.04*)
	(KEGG) stt00780	Biotin metabolism	4 (*0.01*)
	(KEGG) sty00780	Biotin metabolism	4 (*0.01*)
	(KEGG) stt00400	Phenylalanine, tyrosine and tryptophan biosynthesis	3 (*0.05*)
	(KEGG) sty00400	Phenylalanine, tyrosine and tryptophan biosynthesis	3 (*0.07*)
	(KEGG) stt00650	Butanoate metabolism	3 (*0.07*)
	(KEGG) stt03070	Bacterial secretion system	3 (*0.03*)
	(KEGG) sty03070	Bacterial secretion system	3 (*0.03*)
	(KEGG) sty00650	Butanoate metabolism	3 (*0.07*)

Fig. 6. Example of a pathway analysis on *S. typhi* proteins induced in Mg^{+2} depleted minimal media.

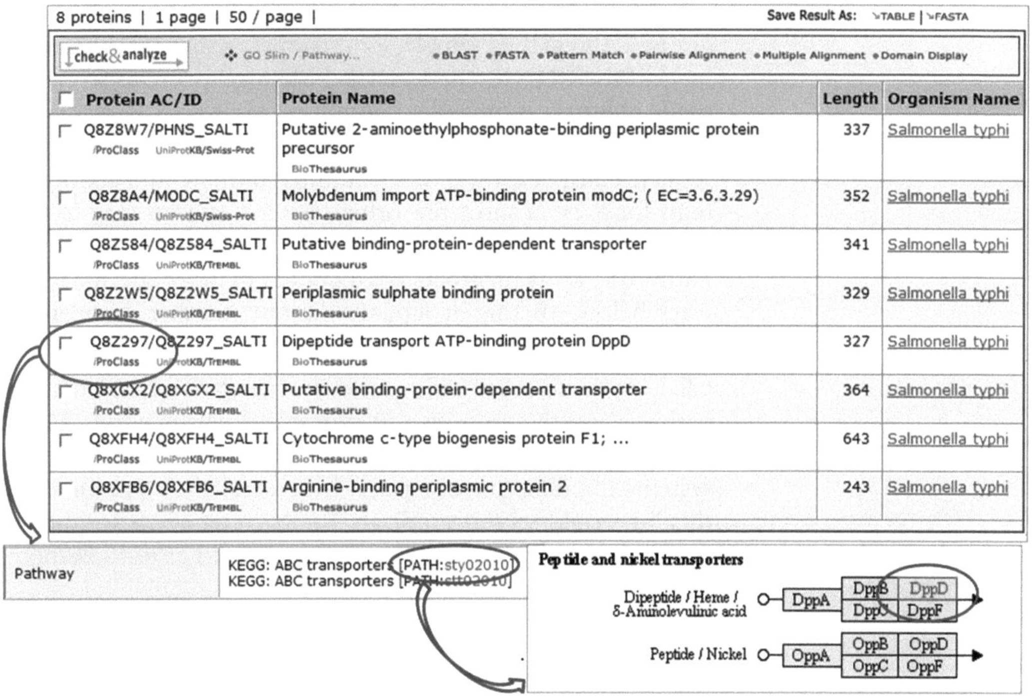

Fig. 7. Further examination of one of the eight "ABC Transporter" pathway proteins shown in Fig. 6. Following links to the iProClass report then to KEGG pathway database, we see that "Q8Z297-Dipeptide transport ATP-binding protein" DppD is highlighted as part of a complex involved in dipeptide/Heme/δ-aminolevulinic acid transport across the cell membrane.

(similar to name or description); and (4) the frequency or number of protein in the list that map to that pathway. Some functions of the display are:

(a) The Pathway ID, Name and Frequency columns can be sorted. For example, to group the displays by pathway database sort on ID that will put all KEGG, Reactome and PID data together. If you sort on Term pathways with similar descriptions are grouped. The frequency column has two sub-columns. (1) A number, which is the number of proteins that map to this pathway. Clicking on this number opens a new window and does batch retrieval on these proteins only, allowing the user to do additional analysis on the selected proteins. Next to the number of proteins is a *p*-value whose calculation is based on the frequency of a particular term in the database. See Note 7 for more information on the interpretation of this value. (2) A histogram bar. Clicking on this bar opens a window with the pathway term and iProClass accession numbers in plain text.

5. Examine the list of pathways on the display or in Fig. 6. In this example using bacterial proteins all the pathways are from KEGG as Reactome and PID pathways are mostly for Human

diseases. Also in this example, some pathway names are duplicated as UniProtKB in the past merged identical proteins from virtually identical strains of an organism while KEGG maintains a separate identical pathway display for every strain of a sequenced organism. As a result, most of the UniProtKB proteins for *S. typhi* and a few other bacterial species map to two pathways in KEGG. You can ignore the first "Metabolic Pathways" as its members are included in more specific pathways below. In this group, the largest number of proteins belongs to the first "ABC transporters" pathway.

6. Click on the number "8" next to the first "ABC transporters" pathway and the proteins will appear in a new window. These proteins are all part of bacterial membrane bound transport systems for different compounds or ions; their presence on this list would seem to indicate the bacteria needs to import their specific substrates in these conditions. Let us look more closely at one protein.
7. Click on the *iProClass* icon under accession Q8Z297-Dipeptide transport ATP-binding protein DppD, the fifth protein on the list. This will take you to the iProClass report with information on the protein and links to many additional resources. Scroll down to one of these resources, under Cross-References you will see Pathways and the links "ABC transporters [PATH: sty02010]" click on the link and a browser will open displaying the KEGG pathway with the gene name dppD for this protein highlighted in red. This protein is part of a complex that imports small peptides/Heme/δ-aminolevulinic acid. By inspection of the other proteins, this list of ABC transporters in a similar manner you should find that four of the nine proteins are involved in peptide/nickel transport complexes like this one. The other proteins are part of complexes that transport sulfate, molybdate, 2-aminoethylposphonate, arginine and heme. Let us look at some other pathways.
8. Back in the original Pathway Statistics window click on the number "7" next to the "Two component system" to investigate the proteins in the list. What signal and response systems seem to be induced? In the original published analysis of this data it was noted that Q8Z6B2 (PagC) and Q8Z4Y4 (Pâté) are both part of a system that responds to Mg^{+2} starvation (like in the media used here) and antimicrobial peptides. These proteins have been known to be involved in virulence.
9. Investigate the other pathways. The authors of the original paper noted that the four sequential enzymes in the biotin pathway shown in the display were induced and thought it significant but did not comment on the other biosynthetic pathways shown here.

Pathway or GO classification can aid in discovery but each has its limits. A combination of techniques is often required and in the end, detailed investigation of individual proteins and the related literature on the organism is needed. Let us now examine the proteins in this list not classified by pathways.

10. At the bottom of the Pathway Statistics window, click on the "55" unclassified proteins to retrieve a table of these proteins.
11. Click on the GO/Pathway icon to initiate the analysis options. Inspect the new columns that appear in the list. It seems some but not all these proteins have GO slim terms associated with them so let us do a GO analysis starting with biological process. Click on the GO "Proc." Icon.
12. Examine the GO biological process ontology results. The highest number of proteins in any category is seven and because this data set was designed to find proteins involved in infection and virulence the category called "interspecies interaction between organisms" seems most relevant. Click on the "7" and inspect the results. Six of the proteins are known or suspected to be toxins or virulence associated proteins that function to target and disrupt mammalian host cells including: Q8Z727-Cytolethal distending toxin subunit B; Q8Z727-Hemolysin E; Q8Z6A4-Putative pertussis-like toxin subunit; Q8Z2J2-mgtC protein (a virulence factor know previously to respond to low Mg^{+2}); and Q8Z550-Deubiquitinase sseL (a protein thought to disrupt the hosts ubiquitin pathway and a member of a known pathogenicity island in the *typhi* genome).

    Unfortunately, not all proteins are currently classified by GO terms or pathways. Some may have important annotation that can only be found by searches and inspection. As a final exercise examine the unclassified proteins at the bottom of our last GO biological process analysis by clicking on the number "27" next to the unclassified category.
13. Examine the list of protein names. Some are well annotated with a complete enzyme name and EC number but no GO biological process or pathway yet assigned. Many are named "Putative uncharacterized protein" or some variant, meaning little is known except they are in the genome and similar to other proteins of unknown function. A few names indicate they are known or predicted virulence proteins because of their presence on a pathogenicity island (7) see O84945-Putative pathogenicity island effector protein for an example. One intriguing protein is Q8Z4K1-Sigma-E factor negative regulatory protein, named for its similarity to an *E. coli* protein that inhibits transcription of some operons by cleaving a sigma factor.

4. Notes

1. *JOB identifiers* will be maintained and available for several weeks after first use but may be removed later, save some of your work locally.
2. *GO analysis options.* In addition to the GO process option, the "Func." link uses GO molecular function ontology (see example below), whereas the "Comp." link uses the GO cellular component ontology. The GO analysis tool displays GO terms annotated in UniProtKB. All proteins in the list with GO terms associated with it are displayed by default. A protein may have multiple GO terms associated with it. Proteins that do not map to GO terms are grouped in an "unclassified" box at the bottom of the display.
3. *GO and Pathway window.* The pop-up window for the GO or Pathway analysis can be overwritten if you do multiple pathway or GO analysis from the same batch retrieval. To prevent this do the following after the window appears. When you see the initial window text "To check status, click here." Right-click on "here" and select "Open in new Window". This will put the analysis in a new and independent browser window.
4. *Additional pathway example.* To see an example of the pathway analysis with PID and Reactome pathways use the list ftp://ftp.pir.georgetown.edu/pub/MiMB/Comparative_Proteomics/VACV_infect_Hela_no-control.txt which contains human proteins co-purified with Vaccinia virus IMV particles but excluding proteins seen in the uninfected control purifications (12).
5. *ID mapping background.* The identifier mapping in iProClass is the result of several automated processes including (a) extracting the rich cross-references in UniProtKB, (b) using these cross-references as a bridge to other database IDs (for example, a GenBank/EMBL/DDBJ accession can map a NCBI gi number to a UniProtKB accession), and (c) computationally mapping the sequences at 100% identity to establish a relationship as is done for UniRef100 (13) (used to map databases such as RefSeq to UniProtKB). Currently iProClass and ID mapping tables are updated every 3 weeks in sync with UniProtKB updates. The mappings are the most comprehensive protein-centric mappings available anywhere. They are available for download at ftp://ftp.pir.georgetown.edu/databases/idmapping/ and ftp://ftp.uniprot.org/pub/databases/uniprot/current_release/knowledgebase/idmapping. Web based mapping tools based on this data are available at http://proteininformationresource.org/pirwww/search/

idmapping.shtml and http://www.uniprot.org/?tab=mapping. And the source we use in this chapter to map IDs to iProClass the PIR batch retrieval at *Error! Hyperlink reference not valid.*

6. *ID mapping problems and solutions.* Difficulties usually fall into one of five categories.

   (a) *One-to-many mappings*: A common problem, especially when eukaryotic proteins derived from alternate splicing or viral polyproteins are involved. Some databases handle taxonomy names and identifiers differently causing one-to-many mappings. Possible solutions are to:

      i. Leave it as is, in which you should make note of the proteins affected or it may cause you confusion later where you could see multiple proteins listed in a category that is really only one protein.

      ii. Remove redundancy now, keeping in mind for this analysis you want to keep the protein with the most annotation especially GO and pathway annotation. This can involve some manual effort doing batch retrieval on the redundant item and manually checking and selecting the best one, however, there are some tricks you can use to do this quickly. If one of the UniProtKB accessions is from the Swiss-Prot section keep that one as it has the most complete annotation. If you use the PIR batch retrieval option in this manuscript any UniProtKB/Swiss-Prot accession will automatically be listed first in the "match list" and if there is no UniProtKB/Swiss-Prot accession all the accessions will be from the automatically annotated UniProtKB/TrEMBL section and are very unlikely to have significant differences in annotation. So we routinely copy/paste the match list into Excel do a find/replace looking for ";*" and take only the first UniProtKB match to do a second batch retrieval.

   (b) *No matches*: There are several common reasons for this:

      i. *Retired sequences*: some gene predictions and translations are retired with each new genome build. iProClass is not an archive and if an identifier is retired for the source database, it disappears from the mapping tables. This most often happens with RefSeq genome, IPI sequences from EBI and NCBI gi numbers. You can check gi numbers and RefSeq accession using Entrez tools at NCBI they will tell you is an identifier has been retired or replaced with a new sequence. IBI maintains tools to find retired sequences on their website. If a UniProtKB accession is not found search UniParc (14) on the uniprot.org website or UniSave at EBI (http://www.ebi.ac.uk/uniprot/unisave/).

If you recover the missing sequence and which to include it in your analysis, you can map it to the current version of iProClass using Blast or Peptide Match on the PIR website.

ii. *Protein sequences not available in iProClass or GenBank/EMBL/DDBJ*: This occurs most often for organisms whose genome sequencing was still in progress and stable builds and/or gene predictions were not yet available in GenBank/EMBL/DDBJ. There is no solution except to use blast to see if an acceptable alternative sequence is available.

iii. *Input identifier is not supported or identifier does not map to a protein*: If the identifier is not supported there is little you can do except try to find another supported identifier in the data. Contact PIR help at http://pir.georgetown.edu/pirwww/support/ if you have questions. iProClass supports DNA and gene identifiers that link to protein translations. Identifiers that point to RNA genes or pseudogenes are not supported.

(c) *Taxonomy issues*: Occasionally you may retrieve proteins with alternate species or strain names. This is most common when microbial pathogens are involved. This can occur for several reasons.

i. Not all molecular databases manage their taxonomy IDs and name the same and occasionally the names or IDs will change slightly even in the source databases. Usually this will not affect the analysis.

ii. Experimental data sets often report sequence identifiers for strains or variants other than the one used in the experimental sample. This is not an uncommon situation as the genetically most characterized variant is often an attenuated laboratory strain whereas the more virulent strains are either not yet fully sequenced or the sequence is of lower quality. Usually this will not affect the analysis, as the most annotated version of the protein is what you want.

7. *p-value use*: The *p*-value shown in the GO and Pathway statistics indicates if the proportion of an individual GO term or Pathway in your list deviates significantly from what would be the expected if a random sample of the same size was taken from the iProClass database. It is not calculated if the sample size is less than 30. For a large sample, it can give you some idea if a category deviates significantly from the database as a whole but as it does not take into account the distribution of terms in the database or the biology related to the term. Categories are not created or distributed through all organisms

equally; there are biases toward human proteins and disease pathways. Certain families of proteins are more abundant in some genomes than others depending on their environmental niche so some deviations might pre-exist in the database as a whole for these genomes.

References

1. The UniProt Consortium. (2009) The Universal Protein Resource (UniProt) 2009. *Nucleic Acids Res* **37**, D169–D174.
2. Wu CH, Huang H, Nikolskaya A, Hu Z, Barker WC. (2004) The iProClass integrated database for protein functional analysis. *Comput Biol Chem* **28**, 87–96.
3. Leinonen R, Diez FG, Binns D, Fleischmann W, Lopez R, et al. (2004) UniProt archive. *Bioinformatics* **20**, 3236–3237.
4. Zhang C, Crasta O, Cammer S, Will R, Kenyon R, et al. (2008) An emerging cyber infrastructure for biodefense pathogen and pathogen-host data. *Nucleic Acids Res* **36**, D884–D891.
5. McGarvey P, Huang H, Mazumder R, Zhang J, Chen Y, et al. (2009) Systems integration of biodefense omics data for analysis of pathogen-host interactions and identification of potential targets. *PLOS One* **4**, e7162.
6. Ashburner M, Ball CA, Blake JA, Botstein D, Butler H, et al. (2000) Gene ontology: tool for the unification of biology. The Gene Ontology Consortium. *Nat Genet* **25**, 25–29.
7. Ansong C, Yoon H, Norbeck AD, Gustin JK, McDermott JE, et al. (2008) Proteomics analysis of the causative agent of typhoid fever. *J Proteome Res* 7, 546–557.
8. Adkins JN, Mottaz HM, Norbeck AD, Gustin JK, Rue J, et al. (2006) Analysis of the *Salmonella typhimurium* proteome through environmental response toward infectious conditions. *Mol Cell Proteomics* **5**, 1450–1461.
9. Kanehisa M, Araki M, Goto S, Hattori M, Hirakawa M, et al. (2008) KEGG for linking genomes to life and the environment. *Nucleic Acids Res* **36**, D480–D484.
10. Matthews L, Gopinath G, Gillespie M, Caudy M, Croft D, et al. (2009) Reactome knowledgebase of human biological pathways and processes. *Nucleic Acids Res* **37**, D619–D622.
11. Schaefer CF, Anthony K, Krupa S, Buchoff J, Day M, et al. (2009) PID: the Pathway Interaction Database. *Nucleic Acids Res* **37**, D674–D679.
12. Manes NP, Estep RD, Mottaz HM, Moore RJ, Clauss TR, et al. (2008) Comparative proteomics of human monkeypox and vaccinia intracellular mature and extracellular enveloped virions. *J Proteome Res* 7, 960–968.
13. Suzek BE, Huang H, McGarvey P, Mazumder R, Wu CH. (2008) UniRef: comprehensive and non-redundant UniProt reference clusters. *Bioinformatics* **23**, 1282–1288.
14. Leinonen R, Diez FG, Binns D, Fleischmann W, Lopez R, Apweiler R. (2004) UniProt archive. *Bioinformatics* **20**, 3236–3237.

Chapter 21

Integration of Proteomic and Metabolomic Profiling as well as Metabolic Modeling for the Functional Analysis of Metabolic Networks

Patrick May, Nils Christian, Oliver Ebenhöh, Wolfram Weckwerth, and Dirk Walther

Abstract

The integrated analysis of different omics-level data sets is most naturally performed in the context of common process or pathway association. In this chapter, the two basic approaches for a metabolic pathway-centric integration of proteomics and metabolomics data are described: the knowledge-based approach relying on existing metabolic pathway information, and a data-driven approach that aims to deduce functional (pathway) associations directly from the data. Relevant algorithmic approaches for the generation of metabolic networks of model organisms, their functional analysis, database resources, visualization and analysis tools will be described. The use of proteomics data in the process of metabolic network reconstruction will be discussed.

Key words: Network reconstruction, Genome annotation, Metabolic modeling, Network expansion, Flux balance analysis, Expression analysis, Time-series data analysis, Granger causality, Systems biology

1. Introduction

Recent years have seen a rapid development of profiling technologies allowing to probe cellular systems across multiple levels of molecular organization, most importantly the metabolomic, transcriptomic, and proteomic systems levels. Although the degree of comprehensiveness still differs, available transcriptomics methods allow the near complete monitoring of the transcriptional activities of essentially all genes or genomic regions, whereas available proteomics, and even more so, metabolomics methods provide access to only a fraction of all proteins and metabolites, respectively, still,

Cathy H. Wu and Chuming Chen (eds.), *Bioinformatics for Comparative Proteomics*, Methods in Molecular Biology, vol. 694, DOI 10.1007/978-1-60761-977-2_21, © Springer Science+Business Media, LLC 2011

unseen opportunities for a holistic experimental approach creating an integral understanding of cellular systems upon applying these various profiling technologies have arisen.

Metabolic as well as signaling and regulatory pathways provide a natural framework for the integration of data from different molecular organizational levels. Pathways represent our accumulated scientific knowledge of molecular processes, structure the available data in a meaningful way, and allow the detection of coherent behaviors and, thus, a better separation of noise from real molecular signals. In particular, metabolic pathways can be expected to follow universal biochemical rules. Thus, metabolic pathways are expected to offer a suitable ordering framework even across different organisms. As a consequence, when studying system-wide responses of different organisms to external perturbations, the creation of this metabolic pathway reference framework, the metabolic network, frequently is among the first tasks when conducting systems biology experiments. Assuming that the underlying biochemical reactions are universal and catalyzed by similar enzymes, the task of assembling the metabolic network primarily means to detect all enzymes encoded in the organism's genomes – the so-called genome annotation. With this set of enzymes, all biochemically possible reactions can be derived, and thus the synthesizable set of metabolites can be determined. Comparison with actual experimental data then leads to the validation and refinement of the network, the detection of obvious gaps (missing enzymes), and a targeted search for filling these gaps (identification of enzymes in the genome).

At the same time, the available molecular profiling data sets also allow the reverse approach. Profile data, especially when followed over time, are frequently interpreted as results of as yet unknown pathways and other types of cause–effect relationships. To detect these pathways, various statistical data analyses techniques have been applied.

In this chapter, we describe the major steps involved in creating an integrated view of proteomics and metabolomics organizational domains. We will describe how the inventory of all enzymes encoded in a genome can be established, and how proteomics data can be used to obtain an improved view of the genomic complement. Flux balance analysis (FBA) as an approach to functionally characterize the resulting network is described in more detail. Furthermore, very basic statistical methods for the data-driven investigations to infer pathway associations between different molecules are introduced and relevant resources, software packages, and visualization means described.

2 Methods

2.1. Reconstruction of Genome-Scale Metabolic Networks Using Proteomics and Metabolomics Data

The basic steps involved in creating an integrated and network-based view of different molecule types in a given organism can be summarized as follows.

1. Functional gene annotation.
2. Automated genome-scale reconstruction.
3. Determination of discrepancies between the predicted network and measured data.
4. Expansion of the network to fill in the gaps and reconcile inconsistencies.

The reconstruction steps will be described in the subsequent paragraphs in more detail. Detailed description of the reconstruction process can also be found in (1, 2).

2.1.1. Functional Gene Annotation

As metabolites are processed by enzymes that in turn are encoded in the genome, the knowledge of the complete set of enzymes in a given organism is pivotal. The metabolic network reconstruction is normally done using all sequence and functional annotation data that is available in public databases combined with manual curation using literature and experimental data (see Note 1).

If available, all genomic, transcript (Unigenes or EST data), and protein sequences of the organism of interest should be downloaded from the webpage of the corresponding genome project [e.g., for *Chlamydomonas reinhardtii* from the JGI webpage (http://genome.jgi-psf.org/Chlre3/Chlre3.home.html) or the NCBI webpage (http://www.ncbi.nlm.nih.gov)). Functional annotations of genes and proteins can be retrieved from public databases or literature (see Table 1).

Enzyme functions can be obtained by transferring functional annotations like EC numbers, GO terms (3) or MapMan (4) bins across organisms using comparative analysis (see Note 1). Typically, proteins are then annotated using BLAST (5) against annotated transcripts or proteins. Instead of BLAST, more sensitive methods like PSI-BLAST (6) or HHpred (7) can be used. The annotation is transferred if a certain hit identity and score threshold hold (standard values are 40% sequence identity and a blast score of at least 50 to ensure a sufficient alignment length). Another, more reliable, method to functionally annotate a set of genes is using orthology information of an annotated genome. The Inparanoid (8) software, the OrthoMCL-DB database (9), or the KEGG (10) Orthology (KO) can be used to obtain evolutionary relationships. An automated method to map sequences to KO groups and KEGG pathways and reactions is KAAS (11) (KEGG Automatic Annotation Server), which is based on

Table 1
Public resources for functional genome annotation

Database	Data	URL
Entrez Gene	Gene annotation	http://www.ncbi.nlm.nih.gov/sites/entrez?db=gene
Entrez Genomes	Genomes	http://www.ncbi.nlm.nih.gov/genomes/lproks.cgi
Uniprot	Protein annotation	http://www.uniprot.org/
Interpro	Domain annotation	http://www.ebi.ac.uk/interpro/
TransportDB	Transporter annotation	http://www.membranetransport.org/
Brenda	EC numbers	http://www.brenda-enzymes.org/
KEGG	Pathways	http://www.genome.jp/kegg/
MetaCyc	Pathways	http://metacyc.org/
MapMan	Plant pathways	http://www.gabipd.org/projects/MapMan/
GabiPD	Plant annotation	http://www.gabipd.org/
PSORTdb	Subcellular localizations	http://db.psort.org/
Pubmed	Literature references	http://www.ncbi.nlm.nih.gov/pubmed

reciprocally best BLAST hits against all KO groups of functionally related genes assigned in the KEGG GENES database. To assign functional motifs and domains, InterproScan (12) can be used. The genome annotation can provide additional information such as subcellular localization, protein subunits, and protein complexes. If no experimental subcellular localization data is available, subcellular localization of proteins can be predicted using bioinformatics tools. A comprehensive list of methods is available at: http://en.wikipedia.org/wiki/Protein_subcellular_localization_prediction.

2.1.2. Automated Genome-Based Reconstruction

The genome annotation provides lists of metabolic enzymes that are present in the organism of interest catalyzing metabolic reactions (see Note 2). The next step in the reconstruction process is to determine which biochemical reactions are carried out by these enzymes. This can be determined manually or by using automated tools. Starting from the functional annotation of a genome given as EC number, KO group, GO term, or MapMan bin, there are a number of methods (see Table 2) that can be used to produce an initial draft metabolic network or to refine an existing metabolic network filling the missing reactions (see Subheading 2.1.3). Transport reactions have to be defined to connect the separated networks of the single compartments (see Note 3).

Table 2
Automated network reconstruction methods

Network reconstruction system	URL/reference
PathwayTools	http://www.biocyc.org/
GEM System	http://www.biomedcentral.com/1471-2105/7/168
metaShark	http://bioinformatics.leeds.ac.uk/shark/
SEED	(DeJongh 2007)
AUTOGRAPH	(Notebaart 2006)
KAAS	http://www.genome.jp/tools/kaas/

2.1.3. Determine Discrepancies Between the Predicted Network and Measured Data

A metabolic draft network that has been derived from sequence homologies to known enzymes may be incomplete. First not for all enzymes, the protein sequences are known and, second, homology matches may fail because of low sequence but high structural similarities (see Notes 5, 6). Metabolic profiles determined experimentally under well-characterized conditions can efficiently be exploited to identify metabolic capabilities missing in the derived draft network. Clearly, all observed metabolites must have been produced by the organism from the provided nutrients (see Note 10). To identify discrepancies between the predicted network and measured data, the draft network derived above is analyzed by structural modeling techniques to determine whether it is capable of carrying fluxes that allow for the synthesis of the observed metabolites from the applied nutrients. Evidently, the more growth conditions have been experimentally tested and the more metabolites could unambiguously be identified, the more discrepancies may be discovered. Furthermore, it is possible to exploit proteomics data to define even more synthesis routes that the network must be able to synthesize. If, for example, observed amino acid sequences strongly indicate a gene model for which a function is clearly assigned, then it is highly plausible that this reaction takes place and thus the participating substrates and compounds must be producible from the nutrients.

The underlying test for determining whether the draft network can carry the necessary flux can in principle be performed by the method of FBA (see Subheading 2.4.1). However, to avoid the tedious step in generating a manually curated stoichiometrically balanced model that is necessary for FBA, we propose the more robust method of network expansion (13). Although with this methodology, it is not possible to quantify flux ratios, the principle capability to produce metabolites can be tested very efficiently. In comparison to FBA, it is less mathematically stringent than relying on heuristics. However, for well-curated networks, it could be shown that almost identical results for the producibility of compounds can be expected (14), using only a fraction of the computing time.

For a given set of nutrients (the *seed*) and a given set of observed metabolites (the *target*), the identification of discrepancies involves the following steps:

1. Define a suitable set of cofactors that are assumed to be present (e.g., ADP/ATP, NAD(P)/NADH(P) and Co-A).
2. Denote by S the seed set of all nutrients.
3. Determine all reactions for which all substrates are either contained in S or belong to the cofactors defined in step 1.
4. Expand the set S by all products of the identified reactions.
5. Repeat the iteration with step 3 until no new products can be added.
6. Identify all those observed target compounds that are not contained in S. The draft network cannot produce these metabolites and therefore disagrees with the metabolite profile.

A web-based front end to the network expansion algorithm is available at http://scopes.biologie.hu-berlin.de (15).

2.1.4. Expand the Network to Fill in the Gaps and Reconcile Inconsistencies

There exist several attempts to fill gaps in metabolic networks (16–18) (see Note 2). Some methods are based on analyzing the local context of the reactions, e.g., by adding reactions that belong to predefined pathways if a certain number of reactions within this pathway have already been annotated. This bears the danger of missing possible solutions that are not contained in manually and rather arbitrarily defined pathways. Other approaches are based on FBA and apply mixed-integer linear programming techniques to identify minimal sets of reactions that are needed to allow for the network to carry a flux to synthesize a given set of products. This implies the disadvantage that a stoichiometrically balanced model has first to be built and embedded in a larger network derived from databases. In Christian et al. (19) these approaches are discussed and an alternative is presented that employs the method of network expansion, which was described above to identify discrepancies between the network draft and experimental observations. The presented method has the advantage that it can directly operate on networks derived from databases and thus the integration of the draft network into a larger reference network is greatly facilitated. In general, the identification of candidate reactions that should be added to the network relies on a *draft network* (see Subheading 2.1.2) and a *reference network* (derived from a database comprising known biochemical reactions from a large number of species, e.g., KEGG or MetaCyc (20)). The algorithm involves the following steps (see Notes 7–9):

1. All reactions from the reference network, which are not part of the draft network, are written to a list of possible candidate reactions (candidate list) (see Note 11).
2. The draft network is extended by this list.
3. The network expansion algorithm is used to test whether the extended draft network is in agreement with experimentally observed metabolite profiles. If this is not the case for some target metabolites, then our complete knowledge of biochemical reactions is not sufficient to explain their presence and for these metabolites, no extension can be predicted. In the following steps, we will therefore focus only on those target metabolites that may be produced from the fully extended network.
4. Remove the reaction from the top of the candidate list.
5. Test (with the method of network expansion) whether all targets can still be produced.
6. If this is the case, permanently remove the reaction. If not, add the reaction to the network and store it as a predicted extension.
7. Continue steps 4–6 until the complete candidate list is traversed.

This greedy algorithm will result in one particular minimal extension that is sufficient to reconcile inconsistencies. To sample various possible minimal extensions, this algorithm is repeated a large number of times for different list orderings of the candidate reaction list. Comparison of the solutions can give hints about the plausibility of the occurrence of a reaction. Those reactions, for example, which are found in all solutions, are very strong candidates that indeed have to be included in the metabolic network.

The quality of the predicted extensions can be considerably improved by including genomic sequence information. By a systematic comparison of the amino acid sequences predicted by the gene models to protein sequences from other organisms, a likelihood score can be defined representing the probability that some gene in fact encodes a protein catalyzing a particular candidate reaction. This information can be used to randomize the candidate reaction list in such a way that there is a tendency for those reactions for which a strong signal is detected for a catalyzing enzyme which is encoded in the genome, placed at a later position and is thus more likely to be retained in the predicted list of reactions.

The sequence information is also useful to assign reactions in the predicted extensions to a particular gene. In this way, hypotheses are generated such as which particular genes code for which enzymes. These hypotheses are in principle testable either directly by isolation of the gene product and in vitro studies or indirectly by knockout experiments. Further hints whether predicted sequences are in fact translated are obtained by proteomics measurements as described in the following section.

During experimental validation of the predicted genes and proteins and their functions, new evidence is likely to arrive about the existence of so far unobserved proteins and metabolites. This information can then be used to reiterate the reconciliation process such that a repeated cycle of experiments and theoretical predictions result in an increasingly accurate description of the genome-scale metabolic network.

2.2. Using Proteomics Information to Improve Genome Annotation and the Metabolic Network

Often, not all genes encoding enzymes in an organism are known, or many different gene models from different gene prediction tools are available, or EST data are incomplete. The problem is even more evident in genomes of organisms that are not fully sequenced. Then, shotgun proteomics as well as transcriptomics methods (21) can be used to generate new or validate hypothetical gene models. Such a strategy also helps to eliminate metabolic reactions with experimentally unverified transcripts. Moreover, network gaps can be filled by alternative isoforms and better functional annotation of new or changed gene models can be generated.

Like EST sequencing, high-throughput, high-mass-accuracy proteomics profiling methods provide actual evidence for the presence of gene products and thus can serve as validation of gene models (22). In proteomics, peptides, and proteins are normally identified using annotated protein sequence databases. Besides applying de novo sequencing, alternative gene model predictions, exon splice graphs (23), or EST and genomic sequence translated in all six reading frames can be used to identify as of yet unannotated peptides and proteins. Exon-splice graphs compactly encode putative splicing events. In proteomics, it is standard to require at least two peptides per protein for identification. For new genes, only cases in which two or more previously unannotated peptides are mapping within a 1 kb of the genomic sequence are accepted. Identified peptides can then be used to predict new gene-models using software such as Augustus (24). The new gene models and their products can then be functionally annotated using the methods described in Subheading 2.1.1.

Figure 1 provides a schematic overview of the integrative approach using proteomics and metabolomics data and mathematical modeling to improve the quality of metabolic network.

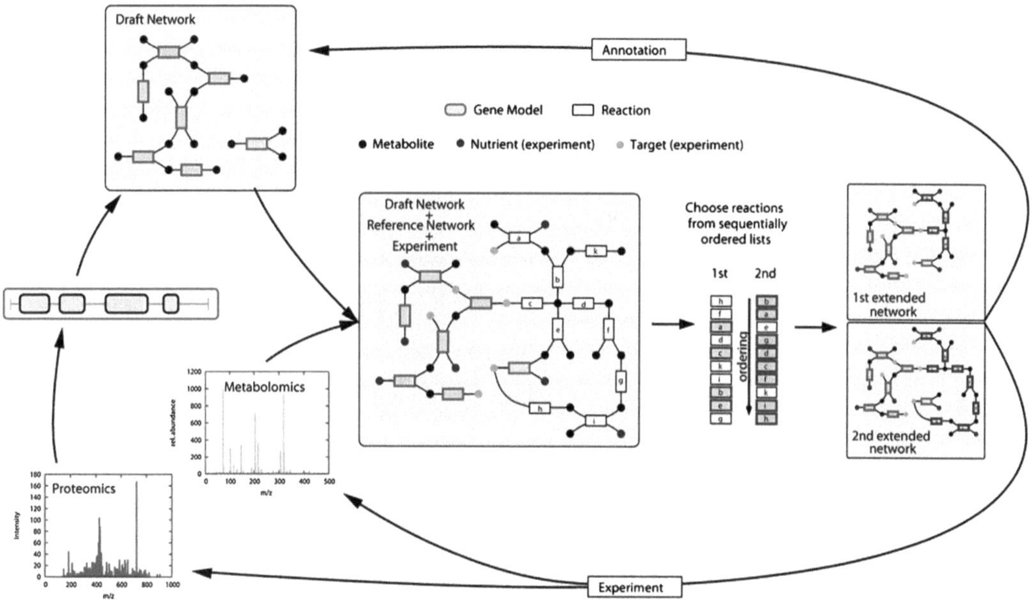

Fig. 1. Integrative approach using proteomics and metabolomics data and mathematical modeling to improve the completeness of the metabolic network. The initial network is derived from genomic data. The draft network may not be sufficient to explain the presence of all metabolites observed in metabolomics measurements or part of isolated reactions for new gene models predicted from proteomics experiments. The draft network is then embedded into a reference network consisting of reactions collected in databases such as MetaCyc or KEGG. A greedy algorithm calculates minimal sets of reactions (*extensions*) that have to be added to the draft network to make it compliant with all experimental data. A network is in agreement with observations if it is able to carry fluxes producing the measured metabolites from the applied nutrient medium. The calculation of a large number of extensions is achieved by initializing the algorithm with many differently ordered lists of reactions (*see* Subheadings 2.1.3 and 2.1.4). The solutions are compared and used to derive hypotheses about the existence of biochemical reactions and genes encoding the respective enzymes. These hypotheses can be tested experimentally or by bioinformatics methods. With this strategy, modeling, bioinformatics and experiment are combined in an iterative process to improve gene annotations and arrive at more complete genome-scale metabolic networks.

2.3. Visualization Tools for the Integrated Metabolic Pathway Analysis of Profiling Data

Various functional genomics data from gene expression, protein expression, and metabolic profiling experiments can be visualized in the context of the reconstructed metabolic network using various visualization tools (see Table 3). These visualization tools enable the visualization of the user's own experimental data in the context of the reconstructed metabolic network.

2.3.1. PathwayTools Omics Viewer

The PathwayTools Omics Viewer (25) as part of the PathwayTools software is a user data visualization and analysis tool allowing lists of genes, enzymes, or metabolites with experimental values to be drawn on a diagram of the full pathway map for an organism for which Pathway Genome Databases (PGDBs) have been developed. Examples are EcoCyc (26), AraCyc (27), YeastCyc (28), or ChlamyCyc (29).

Table 3
Network visualization tools

Visualization tool	URL
PathwayTools Omics Viewer	http://www.biocyc.org/
Cytoscape	http://www.cytoscape.org/
MapMan	http://www.gabipd.org/projects/MapMan/
Vanted	http://vanted.ipk-gatersleben.de/
Pajek	http://pajek.imfm.si/doku.php
MetaViz	http://www.labri.fr/perso/bourqui/software.html
SimPheny™	http://www.genomatica.com/technology/technologySuite.html

2.3.2. Cytoscape

Cytoscape is an open source bioinformatics software platform for visualizing molecular interaction networks and biological pathways and integrating these networks with annotations, gene expression profiles and other state data. Cytoscape supports the standard network and annotation file formats used in systems biology: GML, BioPAX, SBML, and OBO.

2.3.3. MapMan

MapMan (30) is a visualization platform that has been developed for the display of metabolite, transcript, and proteomics data onto metabolic pathways of *Arabidopsis* and other plant genomes and thus features a special emphasis on plant-specific pathways (31).

2.3.4. VANTED

VANTED (Visualization and Analysis of Networks containing Experimental Data) (32) is a platform independent tool for analyzing biological networks. VANTED combines the following features: dynamic network editing and layout, mapping of medium- to large-scale experimental data sets from different time points or conditions on networks, statistical tests, generation of correlation networks, and clustering of similarly behaving substances.

2.3.5. Pajek

Pajek (Slovene word for Spider) (33) is a program, for the analysis and visualization of large networks providing efficient algorithms for network analysis, e.g., partitions, paths, components, flow, decompositions, reduction, etc..

2.4. Functional Metabolic Network Analysis

Once a metabolic network model has been created, several approaches have been developed to investigate their quantitative and qualitative behavior. For example, it is possible to predict the flux distribution; i.e., the metabolic throughput per unit time across all reactions in the network that optimizes growth of an organism. Quantitative network analysis includes methods such as kinetic modeling using differential equations (34, 35), Elementary Mode Analysis (36) to identify subpathways that can operate at steady state thus providing an objective criterion for the definition of pathways, and Flux Balance Analysis (FBA) (35). In this chapter, we will describe FBA as it directly relates to the reconstructed metabolic network (see Subheading 2.1.1) and uses additional proteomics data such as subcellular localization of enzymes.

2.4.1. Flux Balance Analysis

Ultimately, quantitative results are sought from an integrated network analysis, in particular in the context of metabolic engineering, where optimized reaction kinetics and fluxes through the metabolic network are determined that increase yields of certain desired product metabolites. FBA has become a popular quantitative network analysis approach (35, 37, 38). Unlike complete deterministic modeling using differential equations that require the determination of (prohibitively) many reaction parameters (rate constants etc.), FBA operates under the most basic assumption of conservation of mass as reflected in the stoichiometric matrix dictated by the chemical pathways. Thus, FBA explores the possible steady-state operating modes of a given network, modes that are consistent with the conservation of mass. All interconverting processes [including transport processes (see Notes 3 and 4)] are treated as fluxes (V) with reversible reactions split into two separate fluxes, a forward and reverse flux. The change of the level of particular compound, X_i, then is the integrative effect of all fluxes, V, acting on it:

$$\frac{\mathrm{d}X_i}{\mathrm{d}t} = V_{\mathrm{synthesis}} - V_{\mathrm{degradation}} - V_{\mathrm{growth/use}} \pm V_{\mathrm{transport}} \quad (1)$$

The time dependent change of all metabolites, X (vector notation), in the system can then be computed from the product of the stoichiometric matrix, S, and all fluxes, V:

$$\frac{\mathrm{d}X}{\mathrm{d}t} = SV \quad (2)$$

At steady state, the net change of all metabolites is zero. Thus,

$$0 = SV \quad (3)$$

Assuming additional constraints – most importantly non-negative and bounded values for fluxes and concentrations, the solution of this equation can be determined that optimizes the yield of preselected target metabolites via linear programming techniques, such that

$$T = \sum_{i=1}^{N} c_i v_i \tag{4}$$

is maximized for T, where T represents the desired optimization parameter, c_i are the coefficients (weights) to be determined for all N fluxes.

As a result, a numeric solution for all fluxes in the system is found that are consistent with an optimal yield of a particular metabolite or target parameter that can be expressed as a result of fluxes.

In general, the FBA workflow includes the following steps:

1. *Determine the metabolic network for the organism under study* (see Subheading 2.1). Of particular importance are the correct assignments of subcellular compartment in which the reactions are occurring. The reconstruction of the network also includes the generation of the stoichiometric matrix, which follows basic biochemical principles of conservation of mass (see Notes 7, 8, and 12).
2. *Define constraints.* The steady-state condition is already a limiting constraint. Other constraints on maximally possible flux values can be derived from consideration rate kinetics of particular enzymes determining the maximally possible conversion rate. Physical constraints such as thermodynamic considerations based on Gibbs free energy have recently been proposed (39) to avoid implausible solutions. The biomass composition that needs to be maintained adds additional constraints on elemental and compound composition.
3. *Specify the optimization criteria.* Define the objective function; i.e., the parameter that is to be maximized (T in Eq. 4). Examples are yield of a specific metabolite (ATP, for example), maximized growth rate and others.
4. *Solve the linear equation systems under constraints to maximize objective function.* Apply linear programming as a mathematical means to find the solutions with optimized results specified under step 3 and constraints defined under step 2. Several linear programming optimizers are available such as the ILOG CPLEX solver (ILOG, Inc. Mountain View, CA, http://www.ilog.com/products/cplex/) or the optimization routines available under the Matlab mathematical programming environment (for additional software tools, see http://en.wikipedia.org/wiki/Linear_programming).
5. *Analyze results.* FBA provides information on the possible operating modes of metabolic networks at steady-state and helps identify suitable sites for metabolic engineering efforts that aim at boosting the yield of a particular compound or rendering processes more efficient (reduced nutrient uptake).

The FBA framework also allows studying hypothetical flux distributions for knockout mutants by deleting the knocked-out gene (enzyme) from the metabolic network. Furthermore, questions of robustness (sensitivity analysis) can be addressed as well. Thus, FBA allows integrating the proteomics level (presence or absence of enzymes) with the functional consequences on metabolism. An approach to integrate gene expression levels into the FBA formalism has been described recently (40).

A visual inspection of resulting flux distributions from FBA mapped onto the metabolic network and additional analyses such as knockout studies and robustness as well as flux variability analysis can be conveniently performed using the FBA-SimVis plug-in (41) for the VANTED software. An illustration of the resulting flux distribution and their visualization in the FBA-SimVis tool is shown in Fig. 2.

Recently, the concept of FBA with focus on metabolic reactions has been expanded to also include time-dependent regulatory steps (42).

An example for the successful application of FBA to the study of primary metabolism of *C. reinhardtii* under three growth conditions (autotrophic, heterotrophic, and mixotrophic) based on a reconstructed network generated under consideration of subcellular compartmentalization was presented recently by Boyle and

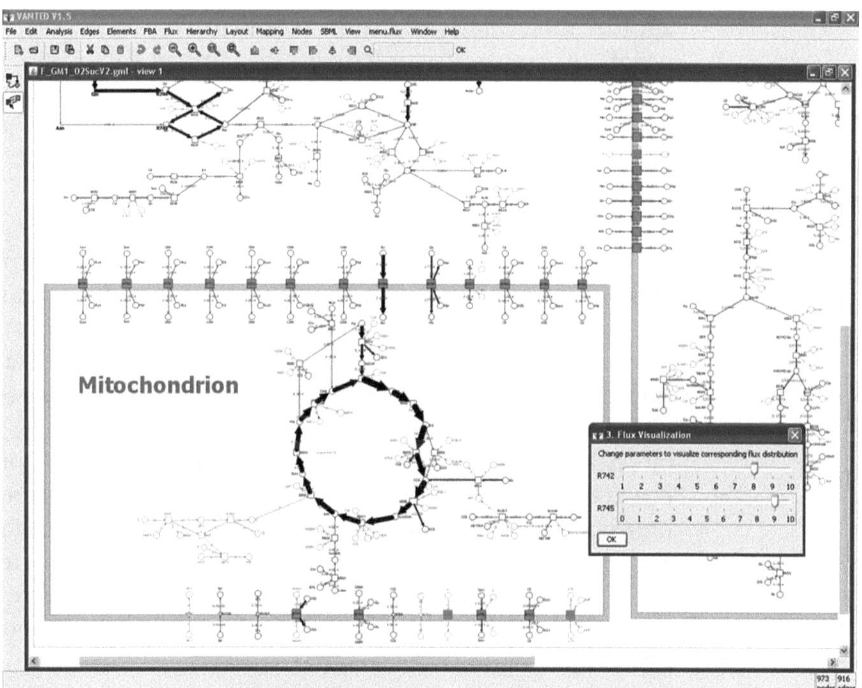

Fig. 2. Visualization of flux distribution in a model of barley seed metabolism with FBA-SimViz (41). Width of *reaction arrows* reflect flux values. Image courtesy Falk Schreiber.

Morgen (43). For the various conditions examined, the dominating metabolic routes were identified. Furthermore, FBA revealed the conditions associated with carbon efficiency. Thus, the study identified intervention sites for rational engineering of *Chlamydomonas* with the objective to modify its economically relevant or environmental properties (CO_2 fixation, for example).

2.4.2. FBA Software

Software programs for FBA computations include CellNetAnalyzer (44), the COBRA Toolbox (45), FBA (http://gcrg.ucsd.edu/Downloads/Flux_Balance_Analysis), and TinkerCell (http://www.tinkercell.com/).

2.5. Statistical Methods for the Integrated Analysis of Profile Data

To detect and understand relationships between molecules is a central goal of systems biology experiments that involve the parallel profiling of different molecule types (transcripts, proteins, metabolites). In the general sense, the interest is to determine, which molecules are involved in the same molecular processes. These associations can be inferred from profiling data derived from different samples taken at different steady states applying correlation followed by clustering techniques or from time series data monitoring the molecular response to external perturbations. Beyond functional associations, time series data also offer the potential to infer cause–effect relationships between molecules. The basic logic is that causes must precede effects. Thus, correlations of the time profile associated with one molecule with another molecule at later time points may be indicative, but not proof, of cause–effect relationships. In the following, we will focus our discussion of methods on the integrated analysis of protein with metabolite data. Evidently, the same concepts apply to other data types as well.

2.5.1. Correlation Analysis

As a hallmark of their association, molecules participating in the same process can be expected to follow a similar pattern of up- and down-regulation, in essence, to be correlated. Quantitatively, this is most frequently measured by their Pearson correlation. As profile data representing different molecule types (metabolites, transcripts, and proteins) can fall onto very different scales, other distance measures, such as Euclidean distance cannot be applied directly. Instead, all data sets need to be standardized beforehand, but transforming all values to a new range with zero mean and unit standard deviation. Correlation measures, on the other hand, are insensitive to absolute values, but identify similar patterns.

The linear correlation coefficient between two vectors (columns of data, e.g., level data for proteins, x, and metabolites, y, across n common samples) is defined as:

$$r_{xy} = \frac{\sum_i (x_i - \overline{x})(y_i - \overline{y})}{(n-1)s_x s_y}, \tag{5}$$

where $\overline{x}$ and $\overline{y}$ are the sample mean of x and y, s_x and s_y are the sample standard deviation of x and y and the sum is from $i = 1$ to n, the length of the data vector (see Notes 13, 14).

Based on the pairwise correlation coefficient as defined in Eq. 5, all variables (e.g., metabolites and proteins) can be clustered to identify subgroups of compounds and proteins that behave similarly. A number of different clustering techniques can be applied and depend to some degree on the question at hand (see Subheading 2.6.1, the Multiexperiment Viewer).

2.5.2. Time-Lagged Correlation Analysis of Time Series Data

Correlation analysis can be applied to identify groups of proteins, genes, metabolites that behave coherently and may thus be associated with similar processes. If time series data are available, the concept of correlation can also be used to identify potential cause–effect relationships with the ultimate goal to elucidate pathways from the data. For example, one could ask, whether a change of a particular metabolite is caused by a preceding change of enzyme levels. The basic assumption is that any cause resulting in an effect must precede the effect in time. Thus, time shifted (or time-lagged/time-delayed) correlation is performed to identify such shifted cause–effect patterns (see Fig. 3). Its use has been demonstrated for the detection of gene interaction networks (46). The conceptual expansion to correlate different molecule types is straightforward.

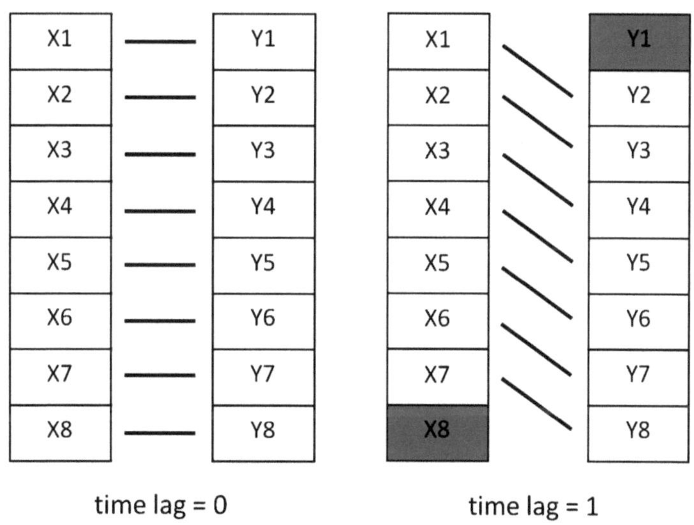

Fig. 3. Illustration of the concept of time-lagged correlation between two data vectors (e.g., protein and metabolite levels). The indexes denote the different time points in sequential order. The *lines* connecting the cells denote the pairwise association for which the correlation is computed. With every time lag increment, data points are lost (one for every variable in the example, shaded cells). Thus, the extent of possible time-delays is determined by the number of available time points. In the example, the scenario that X precedes Y is tested.

High time-lagged correlation levels are no proof of causal relationship as the observed correlations can also be caused by unconnected processes. However, absence of correlation is a strong indicator of independence.

2.5.3. Granger-Causality

As an alternative to time-lagged correlation analysis, Granger causality testing (47) can be applied to detect significant and directed (cause–effect) associations between metabolites and proteins (or any other combinations of molecule types). Granger causality tests whether past values of a time series associated with a variable (e.g., a particular metabolite) contain information that significantly improve the prediction of a future value of another variable (e.g., protein level) above and beyond the past values for this variable alone. Significance is established by applying a series of *F*-tests on the cross-term-coefficients for a linear regression model (see Eq. 6) for time dependent values of protein, $P(t)$, and metabolite data, $M(t)$, (or any other combination of variables) and computing associated *p*-values, with

$$P(t) = \sum_{i=1}^{d} A_{P,i} P(t-i) + \sum_{i=1}^{d} A_{MP,i} M(t-i) + E_P(t)$$
$$M(t) = \sum_{i=1}^{d} A_{PM,i} P(t-i) + \sum_{i=1}^{d} A_{M,i} M(t-i) + E_M(t) \qquad (6)$$

where $P(t)/M(t)$ denote protein/metabolite levels at time point t, the matrix A contains the linear regression coefficients, E the resulting residual error, and d is the maximal time lag (number of considered past values in the time series). In the model, if either one of the cross-term-coefficients (A_{MP} or A_{PM}) is significantly different from zero as tested by the *F*-test, past values of this variable improve the prediction of future values of the respective other variable. The variable is said to be Granger-caused by the respective other.

Granger causality was shown in the past to yield meaningful directed relationships between transcripts when applied to gene expression time series (48, 49).

Compared to correlation measures, Granger causality assigns very low mutual predictive values to variables showing monotonic behavior. Although in such cases, any time lag – forward or backward – will yield significant Pearson correlations, the Granger causality will be low, as the future values of a variable can be predicted from the variable itself. Thus, these trivial time-lagged correlations (that can, nonetheless, indicate true causal relationships) are eliminated under the concept of Granger causality.

Granger causality assumes covariance stationality, which in cases of perturbed systems is or may not be fulfilled. Nonetheless, Granger causality was shown to yield meaningful results even if this assumption is violated (49).

Granger causality computations can be performed using the MSBVAR-R package (http://cran.r-project.org/web/packages/MSBVAR/index.html; Method description: http://rss.acs.unt.edu/Rdoc/library/MSBVAR/html/granger.test.html). As discussed above for time-lagged correlation analysis, possible time lag values *d* will depend on the length of the available time series data. Obviously, the more time points available, the better.

2.6. Software for the Statistical Analysis of Multilevel Profiling Data

For the computation of pairwise correlations and clustering of data, many different software solutions and packages are available. At the most generic level, statistical computing environments, such as the freely available R or commercial solutions such as Matlab or Statistica, can be used to compute quantitative measures of interest. Because they essentially represent programming environments, they offer the greatest flexibility. By contrast, customized software packages that operate via a graphical application interface are more easily usable. Stand-alone applications [MultiExperiment Viewer (MeV)] are available as well as web-based software solutions (Metagenealyse) (see Table 4).

Table 4
Selected software packages for integrated, multivariate data analysis

Software	Commercial/free	Source	Features
Multiexperiment Viewer	Free	http://www.tm4.org/mev.html	Menu-driven statistical analyses options including biclustering, principal component analysis, correlation network generation
Statistica	Commercial	Statsoft, http://www.statsoft.com	Implementation of most clustering and many multivariate data analysis techniques.
R	Free	http://www.r-project.org/	Statistical computing environment with implementations of essentially all known statistical procedures
Matlab	Commercial	http://www.mathworks.com	Mathematical and statistical programming environment
Metagenealyse	Free	http://metagenealyse.mpimp-golm.mpg.de	Web-based suite of statistical analyses including imputation of missing values, clustering, principal component analysis, independent component analysis

2.6.1. The MultiExperiment Viewer

Designed for the analysis of microarray gene expression datasets, the freely available MeV offers a broad spectrum of standard and advanced statistical data analysis methods that can also be applied to other data types. Most noteworthy is the very intuitive graphical user interface, in which the various applied methods remain neatly organized such that the results of the various approaches are easily comparable. For the purpose of integrated analysis, the various clustering methods [hierarchical, K-means performed optionally as biclustering; i.e., simultaneously clustering rows (e.g., representing protein levels) and columns (e.g., representing samples)] can be applied. The program allows choosing between different distance measures. Without data standardization, correlation measures are most appropriate as different variables can fall into different value ranges. From the computed correlations, network views can be generated (so called Relevance Networks), thereby quickly revealing any significant associations between the different molecules. The program also allows investigating whether particular functions are overrepresented in user selected clusters. For several organisms and gene expression platforms, built-in annotation files are available. For other custom data, custom annotation files need to be generated and uploaded. Note, as the program assumes to process gene expression data, some options and predefined labels may not apply. For data import,if treatment minus control datasets are to be analyzed, choose the

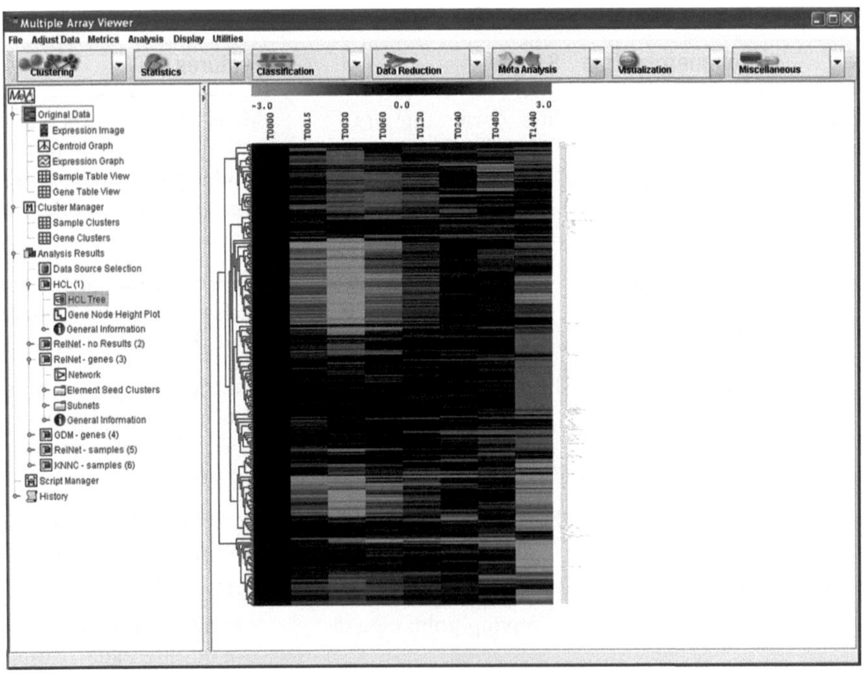

Fig. 4. Hierarchical clustering of molecular level data based on Pearson correlation distances using the Multiexperiment Viewer. Columns refer to different samples, rows correspond to the different gene transcript levels and metabolite levels. Evidently, any other data types can be treated similarly (protein levels, for example). Similar correlation patterns will result in a clustering of the corresponding molecules that may be indicative of functional association.

"Two-color Array" option as then also negative values will be colored appropriately, otherwise choose "Single-color Array" if only positive values are contained in the data matrix. Despite these idiosyncrasies, we find the program very valuable to easily perform sophisticated statistical analyses on the data at hand. Figure 4 shows an example of hierarchical clustering of molecular level data across different samples using the MeV.

A comprehensive review of integrated data analysis methods can be found in (50).

3. Notes

Here, we list some common problems encountered during automated genome-scale metabolic network reconstruction that can be used as a guide for the use of such methods (more details can be found in Feist et al. (2)) and add notes on the application of statistical concepts for the inference of pathways from data.

1. Functional annotations can change very quickly, but annotations are not continuously updated. Try to use regularly updated databases and automated methods for updating.
2. Manually check your reconstructed network for incorrect annotations or use methods that can test for inconsistencies (51).
3. Transporter reactions often have to be added manually, because annotation of transporters is still very insufficient.
4. The correct assignment of the enzymes to their respective subcellular compartment is of particular relevance for the analysis of metabolic pathways.
5. Protein–enzyme relationships are often not clearly defined. Problems can arise from the incorrect or missing annotation of isozymes, subunits, and protein complexes.
6. Reactions are often unspecifically defined. They can be associated with general classes of compounds, which can result in ambiguous connections in networks. Examples include electron carriers (NAD and NADP) or D-glucose (α-D-glucose and β-D-glucose).
7. Reactions are often unbalanced for H, C, P, N, O, or S in public databases (52).
8. Reactions are most often defined as reversible throughfully are not. Automated methods have been developed to address this problem (53, 54).
9. The protonation state of metabolites within reactions are often wrongly annotated.
10. Enzymes often need cofactors to be functional. The network must be able to produce them.

11. Often network and pathway annotation is derived by homology, but not all pathways are general across species, e.g., the photorespiration pathway between algae and higher plants.
12. In FBA, it is essential that the network is stoichiometrically fully balanced as otherwise the conservation of mass criterion is not fulfilled. Networks should also be balanced with regard to charge.
13. In correlation analysis, fewer data points will lead to increasing proportions of high correlation levels. In the extreme case of only two data points, the correlation will always be perfect, but trivial. Especially for the interpretation of time series data, where typically only few data points are available, this effect needs to be taken into account. Every additional time point significantly improves the statistical power. With six time points, there are 720 random orderings possible. By adding one more time point, this number goes up to 5,040. As a consequence, establishing statistical significance via randomized (shuffled) data set will yield much improved results in the latter case.
14. The Pearson correlation coefficient is sensitive to outliers and assumes Gaussian distributions. Thus, an apparent high degree of correlation can also result data points that are far removed from the majority of data points. To circumvent this problem, the rank-based Spearman correlation coefficient should be used. Instead of correlating the original values, the correlation is computed using the respective ranks associated with original level data in the different samples. Thus, the impact of outliers on the overall correlation is reduced significantly as the maximal rank difference can only be one unit. In practice, both measures should be used and/or the observed data points be examined beforehand for occurrences of outliers.

Acknowledgements

This work was supported by BMBF-funded GoFORSYS systems biology project. We wish to thank Falk Schreiber for generously providing us with Fig. 2.

References

1. Reed, J. L., Famili, I., Thiele, I., and Palsson, B. O. (2006) Towards multidimensional genome annotation. *Nat Rev Genet* 7, 130–141.
2. Feist, A. M., Herrgard, M. J., Thiele, I., Reed, J. L., and Palsson, B. O. (2009) Reconstruction of biochemical networks in microorganisms. *Nat Rev Microbiol* 7, 129–143.
3. Moxon, S., Schwach, F., Dalmay, T., Maclean, D., Studholme, D. J., and Moulton, V. (2008) A toolkit for analysing large-scale plant small RNA datasets. *Bioinformatics* 24, 2252–2253.

4. Thimm, O., Blasing, O., Gibon, Y., Nagel, A., Meyer, S., Kruger, P., Selbig, J., Muller, L. A., Rhee, S. Y., and Stitt, M. (2004) MAPMAN: a user-driven tool to display genomics data sets onto diagrams of metabolic pathways and other biological processes. *Plant J* 37, 914–939.
5. Altschul, S. F., Gish, W., Miller, W., Myers, E. W., and Lipman, D. J. (1990) Basic local alignment search tool. *J Mol Biol* 215, 403–410.
6. Altschul, S. F., and Koonin, E. V. (1998) Iterated profile searches with PSI-BLAST – a tool for discovery in protein databases. *Trends Biochem Sci* 23, 444–447.
7. Soding, J., Biegert, A., and Lupas, A. N. (2005) The HHpred interactive server for protein homology detection and structure prediction. *Nucleic Acids Res* 33, W244–W248.
8. Remm, M., Storm, C. E., and Sonnhammer, E. L. (2001) Automatic clustering of orthologs and in-paralogs from pairwise species comparisons. *J Mol Biol* 314, 1041–1052.
9. Chen, F., Mackey, A. J., Stoeckert, C. J., Jr., and Roos, D. S. (2006) OrthoMCL-DB: querying a comprehensive multi-species collection of ortholog groups. *Nucleic Acids Res* 34, D363–D368.
10. Kanehisa, M., Araki, M., Goto, S., Hattori, M., Hirakawa, M., Itoh, M., Katayama, T., Kawashima, S., Okuda, S., Tokimatsu, T., and Yamanishi, Y. (2008) KEGG for linking genomes to life and the environment. *Nucleic Acids Res* 36, D480–D484.
11. Moriya, Y., Itoh, M., Okuda, S., Yoshizawa, A. C., and Kanehisa, M. (2007) KAAS: an automatic genome annotation and pathway reconstruction server. *Nucleic Acids Res* 35, W182–W185.
12. Quevillon, E., Silventoinen, V., Pillai, S., Harte, N., Mulder, N., Apweiler, R., and Lopez, R. (2005) InterProScan: protein domains identifier. *Nucleic Acids Res* 33, W116–W120.
13. Handorf, T., Ebenhoh, O., and Heinrich, R. (2005) Expanding metabolic networks: scopes of compounds, robustness, and evolution. *J Mol Evol* 61, 498–512.
14. Kruse, K., and Ebenhoh, O. (2008) Comparing flux balance analysis to network expansion: producibility, sustainability and the scope of compounds. *Genome Inform* 20, 91–101.
15. Handorf, T., and Ebenhoh, O. (2007) MetaPath Online: a web server implementation of the network expansion algorithm. *Nucleic Acids Res* 35, W613–W618.
16. Reed, J. L., Patel, T. R., Chen, K. H., Joyce, A. R., Applebee, M. K., Herring, C. D., Bui, O. T., Knight, E. M., Fong, S. S., and Palsson, B. O. (2006) Systems approach to refining genome annotation. *Proc Natl Acad Sci U S A* 103, 17480–17484.
17. Satish Kumar, V., Dasika, M. S., and Maranas, C. D. (2007) Optimization based automated curation of metabolic reconstructions. *BMC Bioinformatics* 8, 212.
18. Green, M. L., and Karp, P. D. (2004) A Bayesian method for identifying missing enzymes in predicted metabolic pathway databases. *BMC Bioinformatics* 5, 76.
19. Christian, N., May, P., Kempa, S., Handorf, T., and Ebenhoh, O. (2009) An integrative approach towards completing genome-scale metabolic networks. *Mol Biosyst* 5, 1889–1903. DOI: 10.1039/b915913b.
20. Caspi, R., Foerster, H., Fulcher, C. A., Kaipa, P., Krummenacker, M., Latendresse, M., Paley, S., Rhee, S. Y., Shearer, A. G., Tissier, C., Walk, T. C., Zhang, P., and Karp, P. D. (2008) The MetaCyc Database of metabolic pathways and enzymes and the BioCyc collection of Pathway/Genome Databases. *Nucleic Acids Res* 36, D623–D631.
21. Manichaikul, A., Ghamsari, L., Hom, E. F., Lin, C., Murray, R. R., Chang, R. L., Balaji, S., Hao, T., Shen, Y., Chavali, A. K., Thiele, I., Yang, X., Fan, C., Mello, E., Hill, D. E., Vidal, M., Salehi-Ashtiani, K., and Papin, J. A. (2009) Metabolic network analysis integrated with transcript verification for sequenced genomes. *Nat Methods* 6, 589–592.
22. May, P., Wienkoop, S., Kempa, S., Usadel, B., Christian, N., Rupprecht, J., Weiss, J., Recuenco-Munoz, L., Ebenhoh, O., Weckwerth, W., and Walther, D. (2008) Metabolomics- and proteomics-assisted genome annotation and analysis of the draft metabolic network of Chlamydomonas reinhardtii. *Genetics* 179, 157–166.
23. Castellana, N. E., Payne, S. H., Shen, Z., Stanke, M., Bafna, V., and Briggs, S. P. (2008) Discovery and revision of Arabidopsis genes by proteogenomics. *Proc Natl Acad Sci U S A* 105, 21034–21038.
24. Stanke, M., and Morgenstern, B. (2005) AUGUSTUS: a web server for gene prediction in eukaryotes that allows user-defined constraints. *Nucleic Acids Res* 33, W465–W467.
25. Zhang, P., Foerster, H., Tissier, C. P., Mueller, L., Paley, S., Karp, P. D., and Rhee, S. Y. (2005) MetaCyc and AraCyc. Metabolic pathway databases for plant research. *Plant Physiol* 138, 27–37.

26. Keseler, I. M., Collado-Vides, J., Gama-Castro, S., Ingraham, J., Paley, S., Paulsen, I. T., Peralta-Gil, M., and Karp, P. D. (2005) EcoCyc: a comprehensive database resource for Escherichia coli. *Nucleic Acids Res* 33, D334–D337.
27. Mueller, L. A., Zhang, P., and Rhee, S. Y. (2003) AraCyc: a biochemical pathway database for Arabidopsis. *Plant Physiol* 132, 453–460.
28. Christie, K. R., Weng, S., Balakrishnan, R., Costanzo, M. C., Dolinski, K., Dwight, S. S., Engel, S. R., Feierbach, B., Fisk, D. G., Hirschman, J. E., Hong, E. L., Issel-Tarver, L., Nash, R., Sethuraman, A., Starr, B., Theesfeld, C. L., Andrada, R., Binkley, G., Dong, Q., Lane, C., Schroeder, M., Botstein, D., and Cherry, J. M. (2004) Saccharomyces Genome Database (SGD) provides tools to identify and analyze sequences from Saccharomyces cerevisiae and related sequences from other organisms. *Nucleic Acids Res* 32, D311–D314.
29. May, P., Christian, J. O., Kempa, S., and Walther, D. (2009) ChlamyCyc: an integrative systems biology database and web-portal for Chlamydomonas reinhardtii. *BMC Genomics* 10, 209.
30. Usadel, B., Nagel, A., Thimm, O., Redestig, H., Blaesing, O. E., Palacios-Rojas, N., Selbig, J., Hannemann, J., Piques, M. C., Steinhauser, D., Scheible, W. R., Gibon, Y., Morcuende, R., Weicht, D., Meyer, S., and Stitt, M. (2005) Extension of the visualization tool MapMan to allow statistical analysis of arrays, display of corresponding genes, and comparison with known responses. *Plant Physiol* 138, 1195–1204.
31. Goffard, N., and Weiller, G. (2006) Extending MapMan: application to legume genome arrays. *Bioinformatics* 22, 2958–2959.
32. Junker, B. H., Klukas, C., and Schreiber, F. (2006) VANTED: a system for advanced data analysis and visualization in the context of biological networks. *BMC Bioinformatics* 7, 109.
33. Batagelj, V., Mrvar, A. (1998) Program for large scale network analysis. *Connections* 21, 47–57.
34. Fell, D. A. (1992) Metabolic control analysis: a survey of its theoretical and experimental development. *Biochem J* 286 (Pt 2), 313–330.
35. Varma, A., and Palsson, B. O. (1994) Stoichiometric flux balance models quantitatively predict growth and metabolic by-product secretion in wild-type Escherichia coli W3110. *Appl Environ Microbiol* 60, 3724–3731.
36. Schuster, S., Fell, D. A., and Dandekar, T. (2000) A general definition of metabolic pathways useful for systematic organization and analysis of complex metabolic networks. *Nat Biotechnol* 18, 326–332.
37. Kauffman, K. J., Prakash, P., and Edwards, J. S. (2003) Advances in flux balance analysis. *Curr Opin Biotechnol* 14, 491–496.
38. Lee, J. M., Gianchandani, E. P., and Papin, J. A. (2006) Flux balance analysis in the era of metabolomics. *Brief Bioinform* 7, 140–150.
39. Hoppe, A., Hoffmann, S., and Holzhutter, H. G. (2007) Including metabolite concentrations into flux balance analysis: thermodynamic realizability as a constraint on flux distributions in metabolic networks. *BMC Syst Biol* 1, 23.
40. Colijn, C., Brandes, A., Zucker, J., Lun, D. S., Weiner, B., Farhat, M. R., Cheng, T. Y., Moody, D. B., Murray, M., and Galagan, J. E. (2009) Interpreting expression data with metabolic flux models: predicting Mycobacterium tuberculosis mycolic acid production. *PLoS Comput Biol* 5, e1000489.
41. Grafahrend-Belau, E., Klukas, C., Junker, B. H., and Schreiber, F. (2009) FBA-SimVis: interactive visualisation of constraint-based metabolic models. *Bioinformatics* 25, 2755–2757.
42. Lee, J. M., Gianchandani, E. P., Eddy, J. A., and Papin, J. A. (2008) Dynamic analysis of integrated signaling, metabolic, and regulatory networks. *PLoS Comput Biol* 4, e1000086.
43. Boyle, N. R., and Morgan, J. A. (2009) Flux balance analysis of primary metabolism in Chlamydomonas reinhardtii. *BMC Syst Biol* 3, 4.
44. Klamt, S., Saez-Rodriguez, J., and Gilles, E. D. (2007) Structural and functional analysis of cellular networks with CellNetAnalyzer. *BMC Syst Biol* 1, 2.
45. Becker, S. A., Feist, A. M., Mo, M. L., Hannum, G., Palsson, B. O., and Herrgard, M. J. (2007) Quantitative prediction of cellular metabolism with constraint-based models: the COBRA Toolbox. *Nat Protoc* 2, 727–738.
46. Schmitt, W. A., Jr., Raab, R. M., and Stephanopoulos, G. (2004) Elucidation of gene interaction networks through time-lagged correlation analysis of transcriptional data. *Genome Res* 14, 1654–1663.
47. Granger, C. W. J. (1980) Testing for causality: a personal viewpoint. *J Econ Dyn and Contr* 2, 329–352.
48. Lozano, A. C., Abe, N., Liu, Y., and Rosset, S. (2009) Grouped graphical Granger modeling for gene expression regulatory networks discovery. *Bioinformatics* 25, i110–i118.

49. Mukhopadhyay, N. D., and Chatterjee, S. (2007) Causality and pathway search in microarray time series experiment. *Bioinformatics* 23, 442–449.
50. Steinfath, M., Repsilber, D., Scholz, M., Walther, D., and Selbig, J. (2007) Integrated data analysis for genome-wide research, *EXS* 97, 309–329.
51. Sauro, H. M. and Lugalls, B. (2004) Conservation analysis in biochemical networks: Computational issues for software writes. *Biophys Chem* 109, 1–15.
52. Gevorgyan, A., Poolman, M. G., and Fell, D. A. (2008) Detection of stoichiometric inconsistencies in biomolecular models. *Bioinformatics* 24, 2245–2251.
53. Henry, C. S., Jankowski, M. D., Broadbelt, L. J., and Hatzimanikatis, V. (2006) Genome-scale thermodynamic analysis of Escherichia coli metabolism. *Biophys J* 90, 1453–1461.
54. Kummel, A., Panke, S., and Heinemann, M. (2006) Systematic assignment of thermodynamic constraints in metabolic network models. *BMC Bioinformatics* 7, 512.

Chapter 22

Time Series Proteome Profiling

Catherine A. Formolo, Michelle Mintz, Asako Takanohashi, Kristy J. Brown, Adeline Vanderver, Brian Halligan, and Yetrib Hathout

Abstract

This chapter provides a detailed description of a method used to study temporal changes in the endoplasmic reticulum (ER) proteome of fibroblast cells exposed to ER stress agents (tunicamycin and thapsigargin). Differential stable isotope labeling by amino acids in cell culture (SILAC) is used in combination with crude ER fractionation, SDS–PAGE and LC-MS/MS to define altered protein expression in tunicamycin or thapsigargin treated cells versus untreated cells. Treated and untreated cells are harvested at different time points, mixed at a 1:1 ratio and processed for ER fractionation. Samples containing labeled and unlabeled proteins are separated by SDS–PAGE, bands are digested with trypsin and the resulting peptides analyzed by LC-MS/MS. Proteins are identified using Bioworks software and the Swiss-Prot database, whereas ratios of protein expression between treated and untreated cells are quantified using ZoomQuant software. Data visualization is facilitated by GeneSpring software.

Key words: Time series, Proteome profiling, SILAC, LC-MS/MS, ER stress response, Subcellular proteomics

1. Introduction

Time series proteome profiling is a powerful approach for deciphering the molecular mechanisms of biological processes because this method allows for the tracking of both the quantitative and the dynamic aspects of complex protein networks. Although the changes in protein expression and trafficking that occur over time can be assessed via a proteomic approach, it is almost impossible to increase throughput and proteome coverage without losing quantitative accuracy. For instance, the extensive subcellular fractionation and separation techniques that must be used to increase proteome coverage for an organelle can introduce large variations in results from sample to sample during preparation. To circumvent such obstacles,

Cathy H. Wu and Chuming Chen (eds.), *Bioinformatics for Comparative Proteomics*, Methods in Molecular Biology, vol. 694, DOI 10.1007/978-1-60761-977-2_22, © Springer Science+Business Media, LLC 2011

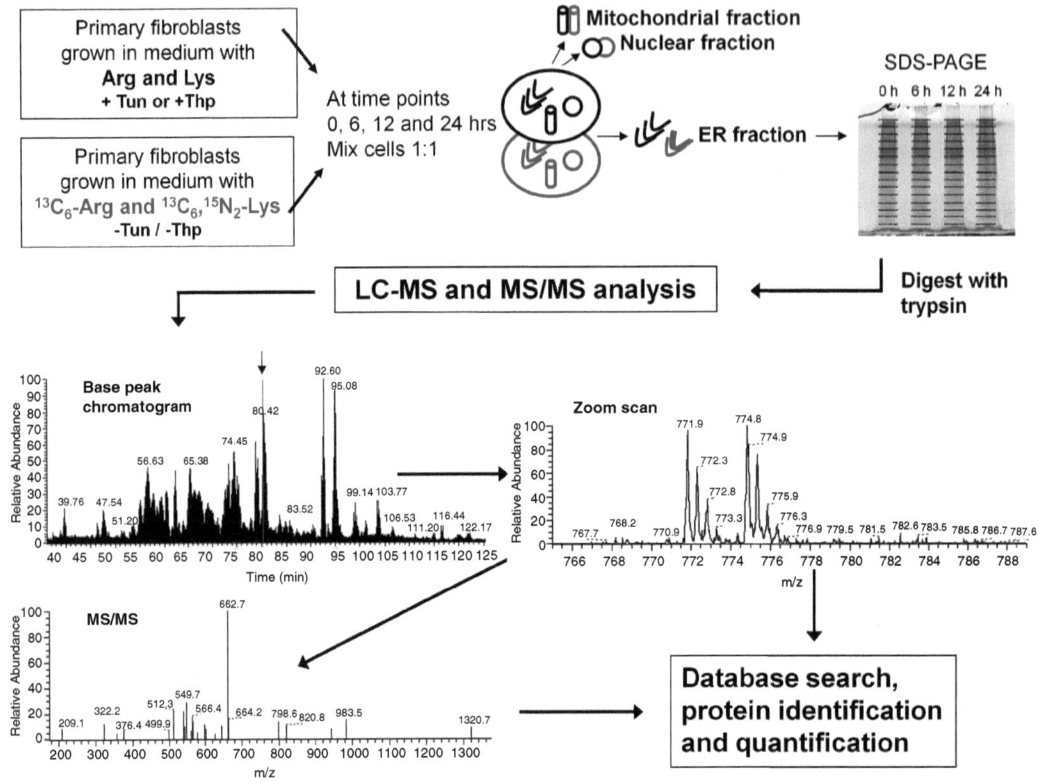

Fig. 1. Overview of the experimental design used to study temporal changes in ER stress response following treatment with tunicamycin (Tun) or thapsigargin (Thp). Control human primary fibroblasts are grown in medium in which Lys and Arg are replaced by $^{13}C_6$, $^{15}N_2$-Lys and $^{13}C_6$-Arg. The cells fully incorporate these amino acids after about five cell doublings. Labeled control cells remain untreated (–Tun/–Thp) while unlabeled cells are treated with an ER stress agent (+Tun or +Thp). At the indicated times, treated and untreated cells are mixed at a 1:1 ratio, then processed for subcellular fractionation. In this case, the ER fraction is prepared and proteins are extracted and further separated by SDS–PAGE. Each lane is sliced into 30–40 bands, digested by trypsin and the resulting peptides analyzed by LC-MS/MS. The base peak chromatogram is representative of the labeled and unlabeled peptide mixture obtained from one single gel band. The zoom scan image shows the mass spectrum of a pair of labeled and unlabeled peptides eluting at the retention time indicated by the *arrow* in the base peak chromatogram. The MS/MS window depicts the fragment ions generated from one peptide. Proteins are identified from the MS/MS data of their tryptic peptides using Bioworks software and ratios between treated and untreated samples are determined from the peak areas of labeled and unlabeled peptide pairs using the zoom scan and ZoomQuant software.

samples to be compared can be paired for analysis and processed under the same conditions using differential stable isotope labeling techniques. In this approach, proteins or peptides in control and experimental pools are labeled with light and heavy stable isotope tags and then mixed together for a single liquid chromatography tandem mass spectrometry (LC-MS/MS) run. The light and heavy peptide pairs coelute from the chromatographic column while their masses are resolved by the mass spectrometer. Therefore, their respective intensities allow for relative quantitation between the control and experimental samples. Though a variety of differential labeling techniques are available (1–3) we and others have found that stable isotope labeling by amino acids in cell culture (SILAC) is ideal for subcellular proteome profiling (4–6) because:

1. Cells to be analyzed are mixed before subcellular fractionation and protein extraction, greatly reducing any variation caused by experimental handling and sample processing.
2. It is the most comprehensive way to uniformly label all cellular proteins, thereby ensuring more accurate quantitative analysis.
3. Relative quantities are obtained for each tryptic peptide pair allowing for better assessment of differential protein expression.
4. It allows accurate temporal proteome profiling and monitoring of protein translocation.

Time series proteome profiling using the SILAC strategy can be implemented for any subcellular organelle (Fig. 1). Because the samples to be compared are mixed and processed in parallel, any organelle cross contamination will affect both samples equally, thus distinguishing true biological variations from technical variations. We recently implemented this strategy to examine temporal changes in the endoplasmic reticulum (ER) proteome of human fibroblast cells exposed to the ER stress inducers tunicamycin and thapsigargin (5). Our ability to quantify expression changes at six time points was made possible by pairing each time point with the same control. This control then acted as a reference point against which all data could easily be cross-correlated. Quantitative data was obtained with the use of ZoomQuant software and visualization was facilitated using the GeneSpring GX analysis platform, originally designed to process Affymetrix microarray data.

2. Materials

Unless otherwise noted, all reagents are made using distilled, deionized water (ddH_2O).

2.1. Cell Culture and Reagents

1. *Human primary fibroblasts* established from a punch skin biopsy explant from a 5-year-old donor (gift from Dr. Raphael Schiffmann, NINDS/NIH).
2. T-25 and T-75 tissue culture flasks.
3. *Low glucose Dulbecco's Modified Eagle Medium (DMEM)* containing 1 g/L D-glucose, 110 mg/L sodium pyruvate, 0.4 mg/mL pyridoxine HCl without Arg and Lys (Atlanta Biologicals, Lawrenceville, GA).
4. *Fetal bovine serum* (Invitrogen Corporation, Carlsbad, CA).

5. *Penicillin* (10,000 U/mL)/*streptomycin* (10,000 μg/mL) (100×) (Invitrogen Corporation, Carlsbad, CA).
6. *Stable isotope labeled (heavy) amino acids*: $^{13}C_6$-L-Arginine:HCl ($^{13}C_6$-Arg) and $^{13}C_6$, $^{15}N_2$-L-Lysine:2HCl ($^{13}C_6$, $^{15}N_2$-Lys) (Cambridge Isotopes Laboratories, Inc., Andover, MA).
7. *Unlabeled (light) amino acids*: L-Arginine:HCl (Arg) and L-Lysine:2HCl (Lys) (Sigma-Aldrich Corp., St. Louis, MO).
8. *SILAC "labeled" medium*: Dissolve 84 mg of $^{13}C_6$-Arg and 146 mg of $^{13}C_6$, $^{15}N_2$-Lys in 890 mL of DMEM. Add 10 mL of penicillin/streptomycin and 100 mL of FBS. Sterilize by passing through a 0.22-μm filter.
9. *SILAC "unlabeled" medium*: Prepare as above, but using "unlabeled" Arg and Lys.
10. *ER stress stock reagents*: 5 mg/mL tunicamycin in DMSO (1,000×), 1 mM thapsigargin in DMSO (1,000×) (Sigma-Aldrich Corp., St. Louis, MO).
11. *Phosphate buffered saline (PBS)*: 1 mM KH_2PO_4, 155 mM NaCl, 3 mM Na_2HPO_4. Adjust to pH 7.4.
12. *Cell lysis buffer*: 10 mM Tris–HCl, pH 7.4, 1 mM ethylenediaminetetraacetic acid (EDTA) and 2.5 M sucrose. One complete, Mini protease inhibitor cocktail tablet is added fresh for every 10 mL of buffer used (Roche Pharmaceuticals, Nutley, NJ).
13. *Protein extraction buffer*: 7 M urea, 2 M thiourea, 2% CHAPS (w/v) and fresh 50 mM DTT.

2.2. SDS–PAGE

1. *Protein concentration assay*: Bio-Rad protein assay kit II (Bio-Rad Laboratories, Inc., Hercules, CA).
2. *Sample desalting and clean up*: Bio-Spin 6 columns with Bio-Gel P-6 in Tris buffer (Bio-Rad Laboratories, Inc., Hercules, CA); vacuum centrifuge.
3. *Pre-cast polyacrylamide gel*: 10–20% Criterion Tris–HCl gel (Bio-Rad Laboratories, Inc., Hercules, CA).
4. *Laemmli sample buffer*: 2% SDS, 25% glycerol, 0.01% bromophenol blue, 62.5 mM Tris–HCl, pH 6.8 and 50 mM DL-dithiothreitol (DTT) added just before use.
5. *Tris/Glycine/SDS (TGS) running buffer (10×)*: 25 mM Tris-Base, 192 mM glycine, 0.1% SDS, pH 8.3 (Bio-Rad Laboratories, Inc., Hercules, CA).
6. *Gel fixing solution*: 45% methanol, 5% acetic acid (Prepare one liter and store at room temperature).
7. *Gel staining solution*: Ready to use Bio-Safe Coomassie stain (Bio-Rad Laboratories, Inc., Hercules, CA).

2.3. In-Gel Digestion and Peptide Extraction

Except for digestion buffer, prepare 100 mL of each solution and store at room temperature. Solutions are stable at room temperature for up to 2 months.)

1. 100% acetonitrile (ACN).
2. 50% ACN.
3. 50% ACN, 5% formic acid (FA) (v/v).
4. 100 mM NH_4HCO_3.
5. 50 mM NH_4HCO_3.
6. 25 mM NH_4HCO_3.
7. 0.1% trifluoroacetic acid (TFA).
8. *Digestion buffer*: 12.5 ng/μL of mass spectrometry grade Trypsin Gold (Promega Corp, Madison, WI) in 50 mM NH_4HCO_3. Dissolve one vial containing 100 μg of lyophilized trypsin in 8 mL of ice cold 50 mM NH_4HCO_3 solution. Prepare 50–100 μL aliquots in ice chilled Eppendorf tubes and store immediately at −80°C. The solution is stable at this temperature for up to a year.

2.4. Mass Spectrometry Instruments and Bioinformatics Tools

2.4.1. Buffers

1. *Aqueous mobile phase*: 0.1% formic acid (A).
2. *Organic mobile phase*: 95% acetonitrile with 0.1% formic acid (B).

2.4.2. Instrumentation

1. *Sample loading:* Autosampler (Dionex LC Packings, Sunnyvale, CA).
2. *Reverse-phase high pressure liquid chromatography (HPLC) system*: Dionex LC Packings nano-HPLC (Dionex-LC Packings, Sunnyvale, CA).
3. *Mass spectrometer*: LTQ (Thermo Fisher Scientific, Inc., Waltham, MA).
4. *Sample washing*: C18 trap column (5 μm, 300 μm i.d. × 5 mm), (LC Packings, Sunnyvale, CA).
5. *Sample fractionation (stationary phase)*: Zorbax C18 (3.5 μm, 100 μm × 15 cm) reverse-phase nanocolumn (Agilent Technologies, Palo Alto, CA).
6. *Sample injection*: 10-μm silica tip (New Objective Inc., Ringoes, NJ).

2.4.3. Bioinformatics

1. *Raw data collection*: Xcalibur 2.0.7 (Thermo Fisher Scientific, Inc., Waltham, MA).
2. *Protein identification*: Bioworks 3.1 (Thermo Fisher Scientific, Inc., Waltham, MA); UniProt/Swiss-Prot database (ftp://ftp.ncbi.nih.gov/blast/db/FASTA/) (see Note 1).

3. *Protein quantification*: ZoomQuant software (http://proteomics.mcw.edu/ZoomQuant).
4. *Data normalization and visualization*: GeneSpring software (Agilent Technologies, Palo Alto, CA).

3. Methods

3.1. Stable Isotope Labeling by Amino Acids in Cell Culture

1. Thaw and seed one vial of cells into a T-25 tissue culture flask with SILAC labeled medium. Similarly thaw and seed one vial of cells into a T-25 tissue culture flask with unlabeled medium (see Note 2).
2. Culture cells at 37°C, 5% CO_2, and replace with corresponding labeled or unlabeled medium every 2–3 days, until they have reached 70–80% confluence.
3. Passage cells into a T-75 flask using their respective labeled or unlabeled medium. Continue splitting cells 1:3 each time they reach 70–80% confluence until the cells have been fully labeled with the stable isotopes (see Notes 3 and 4). Labeled and unlabeled cells are cultured in parallel with the same number of passages and subcultures.
4. Continue to culture the cells until they have reached 100% confluence, then proceed with ER stress experiment as follows.

3.2. ER Stress Induction

1. Add 100 μL of tunicamycin stock solution to 100 mL of the unlabeled medium (for a final concentration of 5 μg/mL) and 100 μL of thapsigargin to an additional 100 mL of unlabeled medium (for a final concentration of 1 μM).
2. To six flasks of unlabeled cells, add 12 mL of the unlabeled medium containing tunicamycin. To the remaining six flasks of unlabeled cells, add 12 mL of the unlabeled medium containing thapsigargin.
3. To the labeled cells (12 culture flasks), add 12 mL of labeled medium.
4. Incubate two dishes of the labeled cells, one dish of the tunicamycin treated cells and one dish of the thapsigargin treated cells for each of the following amounts of time: 0 min, 1, 6, 12, and 24 h (see Note 5).

3.3. Cell Harvesting and Subcellular Fractionation

3.3.1. Cell Harvesting

1. Dissolve one protease inhibitor cocktail tablet in 10 mL of lysis buffer and keep on ice during use.
2. After each time point discard the conditioned medium from each paired control and treated culture flask and add 10–15 mL of PBS to each flask to wash the cells. Repeat the washing twice to remove any serum protein contaminants from the cell surface.

3. Add 2 mL of ice-cold lysis buffer to each flask and harvest the cells with a cell scraper while keeping the flask on ice. Transfer the cell suspensions to preweighed 10-mL polypropylene conical tubes and pellet the cells by gentle centrifugation for 5 min at 300 × *g* and 4°C.
4. Discard the supernatant and weigh the cell pellets using a precision balance. Mix equal amounts of labeled and unlabeled cells (w/w), add 1 mL of lysis buffer and process for subcellular fractionation (see Note 6).

3.3.2. ER Fractionation

1. Homogenize the cells by passing them 15 times through a 1-mL syringe with a 23 gauge needle and centrifuge for 10 min at 4,000 × *g* and 4°C.
2. Transfer the supernatant containing the microsomal (ER) fraction to a clean Eppendorf tube and further centrifuge at 13,000 × *g* for 20 min and 4°C to obtain the microsomal pellet.
3. Resuspend the pellet in a small volume of protein extraction buffer and vortex vigorously. Determine the protein concentration of each sample using Bio-Rad protein assay reagent (Bio-Rad, Hercules, CA) and store samples at – 80°C until analysis.

3.4. Prefractionation of Proteins by SDS–PAGE

1. Take aliquots containing 100 μg of total protein from each time point sample and reduce the volume to about 75 μL each by vacuum centrifugation. Desalt samples using Bio-spin 6 columns following the manufacturer's instructions.
2. Dry samples completely by vacuum centrifugation, then resuspend in 20 μL of Laemmli buffer with freshly added DTT (50 mM).
3. Boil samples for 5 min at 95°C.
4. Load samples and molecular weight marker into individual wells of a 10–20% Criterion Tris–HCl pre-cast gel. Run the gel with TGS buffer at 200 V (constant) until just after the dye front runs off the gel (45 min to 1 h).
5. Remove the gel from the cassette and cover with fixing solution. Incubate for 30 min at room temperature with gentle agitation.
6. Wash the gel three times for 5 min each in ddH_2O with gentle agitation.
7. Cover the gel with Bio-Safe Coomassie and stain for 1 h (this can be done at room temperature with gentle agitation, or overnight at 4°C).
8. Cover the gel with ddH_2O and destain for 1 h, replacing with clean water every 15 min.

9. With a razor blade, slice the gel on either side of the lane containing each sample. Then make horizontal slices to produce 30–40 gel bands per lane (see Note 7).

3.5. In-Gel Digestion

1. Wash gel slices twice by incubation in 50 μL of 50% ACN at room temperature, with vortexing, for 15 min each time.
2. Remove the 50% ACN and add 50 μL of 100% ACN. Wait for the gel pieces to shrink and turn white (this will happen almost immediately; some blue color may remain from the Coomassie).
3. Remove ACN and rehydrate gel pieces with 50 μL of 100 mM NH_4HCO_3.
4. Incubate at room temperature for 5 min.
5. Add 50 μL of 100% ACN (maintaining a 1:1 ratio with NH_4HCO_3).
6. Incubate at room temperature for 15 min, with vortexing.
7. Remove any liquids that did not absorb into the gel.
8. Add 50 μL of 100% ACN and wait for the gel pieces to shrink and turn white.
9. Remove all ACN.
10. Rehydrate gel pieces with 10–20 μL of digestion buffer and incubate on ice for 45 min.
11. Remove any excess digestion buffer.
12. Add 5 μL of 50 mM NH_4HCO_3.
13. Incubate overnight at 37°C (an incubator is preferable to a water bath).

3.6. Peptide Extraction (see Note 8)

1. Spin down the tubes to collect any condensation.
2. Add 25 μL of 25 mM NH_4HCO_3 and incubate at room temperature for 15 min.
3. Add 25 μL of 100% ACN and incubate for 15 min at room temperature, with vortexing.
4. Recover and save the supernatant containing extracted peptides.
5. Extract additional peptides from the gel piece by adding 30 μL of buffer comprising 50% ACN, 5% FA.
6. Incubate 10 min at room temperature, with vortexing.
7. Pool supernatant with that from the same gel piece in step 4.
8. Repeat steps 5–7.
9. Dry supernatants by vacuum centrifugation.
10. Resuspend peptides in 6 μL of 0.1% TFA in an autosampler vial, store at –80°C.

3.7. Mass Spectrometry Analysis

1. Externally calibrate and tune the LTQ mass spectrometer using the manufacturer's tune mixture and protocol.
2. Load the sample vials onto the autosampler and inject 6 μL into the LC-MS system using the Dionex-LC-Packings autosampler and loading pump.
3. Load peptide samples first onto a C18 trap connected in series with the C18 column and wash for 6 min using 0.1% TFA (A) before introducing them onto the C18 column.
4. Desalted peptides are turned in-line to the gradient column and eluted using a 100 min linear gradient from 5 to 60% B.
5. Introduce peptides to the mass spectrometer through a 10-μm silica tip at 1.7 kV and the heated capillary set to 160°C.
6. Operate the LTQ mass spectrometer continuously during the chromatographic elution.
7. Acquire a survey MS scan to determine the mass and intensity of eluting peptides.
8. Acquire data dependent MS/MS scans for the top five most intense peptides in the survey scan, which will be used for protein identification searches.
9. Acquire zoom scans (14 Da window) for each precursor mass to provide higher resolution data of the unlabeled and labeled peptide pairs for quantitation (non-zoom data on the LTQ is low resolution and not ideal for quantitation).

3.8. Protein Identification and Quantification

Unfortunately there is no universal software that can perform both identification and quantification of proteins generated by the different mass spectrometry instruments and the different proteomics strategies currently in use. In our study we used the SILAC strategy together with LC-MS/MS to generate raw data followed by analysis using a combination of Bioworks and ZoomQuant software for protein identification and quantification (see Fig. 2).

1. To streamline search time, use Bioworks software to index the Swiss-Prot database for proteins rising from the human species with fully enzymatic tryptic digestion and allowing up to two missed cleavages.
2. Search the raw mass spectral data using the Sequest algorithm within Bioworks and the indexed database. Set the search parameters as follows: signal threshold $\geq$ 1,000, peptide mass tolerance $\pm$ 1.5 Da, fragment ion tolerance $\pm$ 0.35 Da and differential modifications of 15.99492 Da for Met oxidation, 8.01420 Da for the $^{13}C_6$, $^{15}N_2$-Lys isotope, 6.02040 Da for the $^{13}C_6$-Arg isotope.

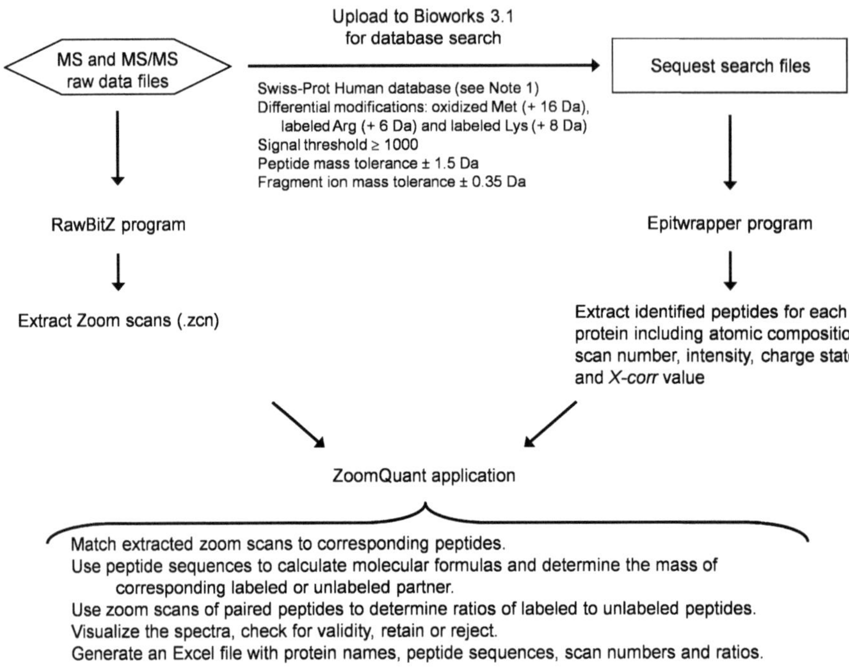

Fig. 2. Overall workflow used to identify and quantify protein ratios in this study.

3. Download and install the ZoomQuant software (see Note 9). The ZoomQuant package has three components: RawBitZ, Epitwrapper, and the ZoomQuant application.
4. Extract zoom scans by uploading the raw mass spectrometry data files to RawBitZ and saving the generated .zcn files in a new folder.
5. Upload the corresponding Sequest search files to Epitwrapper to filter identified peptides based on X-corr (1.9 for $z=1$, 2.5 for $z=2$, 3.5 for $z=3$), initial rank (50), ion match (0.2) and TIC or signal to noise ≥ 1,000. Save the new .colon files in the same folder as the zoom scan files from step 4.
6. Upload corresponding .zcn and .colon files to the ZoomQuant application as well as the label shift profile configured for differential SILAC labeling (this .lsp file will be found in the ZoomQuant program folder). The ZoomQuant application will generate a list of identified proteins with their corresponding peptides and ratios of labeled to unlabeled peptide pairs. One can select a peptide to view the quality of the corresponding zoom scan and choose valid labeled and unlabeled pairs to include in the analysis. After viewing and selecting data, save the report as HTML and Excel files.
7. Combine all Excel data files rising from one sample set (i.e., all the bands cut from one gel lane). Create a file that contains protein identifiers (e.g., accession number and/or name) in

the first column and the corresponding peptide ratios in the second column. Use the sample name (e.g., the time point and ER stressor used) as the first cell in the second column. Discard any proteins that are not represented by two or more unique peptides. Save this as a .txt file for GeneSpring analysis.

3.9. Data Normalization and Visualization (see Note 10)

1. Upload the generated .txt peptide ratio list from each time point to GeneSpring using accession numbers as identifiers and peptide ratios as signal intensities.
2. Use GeneSpring to normalize expression values for each time point by dividing individual peptide ratios by the median value of all ratios. This corrects for any unequal mixing of labeled and unlabeled cells that may have occurred before sample fractionation (see Note 11).
3. The GeneSpring algorithm recognizes the number of peptides per protein as it would array probe sets mapping to one gene and will determine an average ratio for each protein using the peptide count and generate *p*- or *z*-score values that can be used to filter significant data from nonsignificant data.
4. GeneSpring has several visualization options to facilitate data set comparison in a time series experiment. Set up the view depending on the type of display needed, filter for up and down-regulated proteins or show expression patterns for each single protein across different time points.

4. Notes

1. It is recommended to download the most updated FASTA format protein database. The UniProt knowledgebase consists of two sections: a section containing manually curated FASTA format protein sequences referred to as the "Swiss-Prot database," and a section with computationally analyzed records that await full manual annotation referred to as the "TrEMBL database." Most proteomics users prefer the Swiss-Prot database as opposed to the TrEMBL database since it contains a less redundant protein list.
2. In our experience the use of 10% serum in the culture medium does not interfere with the incorporation of the exogenous stable isotope labeled amino acids and thus the conventional 3 kDa cut off dialyzed serum is omitted. We found that the use of dialyzed serum is not adequate for primary cell cultures because it lacks several necessary low mass growth factors.
3. For most cell lines, approximately 97% incorporation of the stable isotope labeled amino acids is achieved after five cell

doublings. However, each cell line may behave differently and full labeling should be verified by mass spectrometry before the experiment proceeds. Usually, soluble proteins are extracted from an aliquot of cells by whole cell lysis and prepared as described in Subheadings 3.4–3.7. A Sequest search for labeled and unlabeled peptides is performed to determine the level of isotope incorporation.

4. Though twelve T-75 flasks of cells are needed for the final stage, it is wise to maintain additional flasks of labeled cells to be frozen down and stored for future use. Once the cells are fully labeled, they will remain so as long as they are always cultured in labeled medium. Therefore, cells seeded from frozen stocks are "ready to use" and do not have to go through the extensive passaging required for the initial labeling process.

5. Depending on the experiment and the subcellular organelle to be studied one can use alternate time points or drug doses. A pilot study should be performed to determine the optimal dose of ER stress agent to be used. In our experiments, the concentration of thapsigargin and tunicamycin were determined by using cell viability or cytotoxicity assays, such as the MTT assay (Promega) and/or LDH releasing assay (Sigma).

6. Equal amounts of wet cell material are established in each tube by removing cells from the tube containing the higher amount with a fine spatula.

7. Before cutting the gel, it is a good idea to scan an image of the gel. Then, as you cut the gel, number each band and mark their location on a print out of the image. This will allow you to match identified proteins with their approximate molecular weight on the gel. Also, dicing each band into smaller pieces before placing it into its corresponding numbered tube will increase the efficiency of tryptic digestion and peptide extraction in later steps.

8. In-gel digestion and peptide extraction can be stopped at any time and samples kept at –20°C for up to a few days.

9. There are only a few specialized software packages equipped to determine peptide ratios from a SILAC experiment. Unfortunately each instrument requires specific software. While software such as MaxQuant (7) and Census (8) are aimed at high resolution mass spectrometers (Sciex Qstar, Thermo LTQ-Orbitrap and Thermo LTQ-FTICR) fewer programs have been developed for low resolution mass spectrometers owing to the challenges in processing low resolution mass spectral data. For laboratories equipped with a low resolution LTQ there is an option to generate high resolution mass spectral data using the zoom scan capability and to analyze the data using ZoomQuant software (9). The zoom scan

events allow complete resolution of labeled and unlabeled peptide pairs, especially for triply and quadruply charged ions and thus facilitate intensity ratio measurements. All these software packages are publicly available and can be installed on any standard desktop PC. (Download MSQuant at *http://msquant.sourceforge.net/*, Census at *http://fields.scripps.edu/census/download.php?menu=6* and ZoomQuant at *http://proteomics.mcw.edu/zoomquant*). Detailed instructions are provided on how to install and run each of these programs.

10. Software platforms for proteome profiling and data visualization are still emerging. In the meantime, we used the mature GeneSpring analysis platform that was originally designed to process Affymetrix microarray data to help filter and visualize our time series proteomics data.

11. Usually, mixing labeled and unlabeled cells 1:1 using a high precision balance is very accurate, but sometimes slight errors may occur and will result in the overall peptide ratios being skewed too high or too low. This can be corrected for using internal normalization by dividing the ratio of labeled and unlabeled peptide pairs in each experiment by the median value of all ratios generated in the experiment.

References

1. Blagoev, B., Ong, S. E., Kratchmarova, I. and Mann, M. (2004) Temporal analysis of phosphotyrosine-dependent signaling networks by quantitative proteomics. *Nat Biotechnol* **22**, 1139–1145.
2. Fenselau, C. and Yao, X. (2009) 18O2-labeling in quantitative proteomic strategies: a status report. *J Proteome Res* **8**, 2140–2143.
3. Gygi, S. P., Rist, B., Gerber, S. A., Turecek, F., Gelb, M. H. and Aebersold, R. (1999) Quantitative analysis of complex protein mixtures using isotope-coded affinity tags. *Nat Biotechnol* **17**, 994–999.
4. Amanchy, R., Kalume, D. E. and Pandey, A. (2005) Stable isotope labeling with amino acids in cell culture (SILAC) for studying dynamics of protein abundance and posttranslational modifications. *Sci STKE* **2005**, pl2.
5. Mintz, M., Vanderver, A., Brown, K. J., Lin, J., Wang, Z., Kaneski, C., Schiffmann, R., Nagaraju, K., Hoffman, E. P. and Hathout, Y. (2008) Time series proteome profiling to study endoplasmic reticulum stress response. *J Proteome Res* **7**, 2435–2444.
6. Reeves, E. K., Gordish-Dressman, H., Hoffman, E. P. and Hathout, Y. (2009) Proteomic profiling of glucocorticoid-exposed myogenic cells: Time series assessment of protein translocation and transcription of inactive mRNAs. *Proteome Sci* **7**, 26.
7. Cox, J. and Mann, M. (2008) MaxQuant enables high peptide identification rates, individualized p.p.b.-range mass accuracies and proteome-wide protein quantification. *Nat Biotechnol* **26**, 1367–1372.
8. Park, S. K., Venable, J. D., Xu, T. and Yates, J. R., 3rd (2008) A quantitative analysis software tool for mass spectrometry-based proteomics. *Nat Methods* **5**, 319–322.
9. Halligan, B. D., Slyper, R. Y., Twigger, S. N., Hicks, W., Olivier, M. and Greene, A. S. (2005) ZoomQuant: an application for the quantitation of stable isotope labeled peptides. *J Am Soc Mass Spectrom* **16**, 302–306.

INDEX

A

B

C

Cathy H. Wu and Chuming Chen (eds.), *Bioinformatics for Comparative Proteomics*, Methods in Molecular Biology, vol. 694, DOI 10.1007/978-1-60761-977-2, © Springer Science+Business Media, LLC 2011

D

E

I

J

K

L

M

N

O

Q

R

S

T

MIX
Papier aus verantwortungsvollen Quellen
Paper from responsible sources
FSC® C105338

If you have any concerns about our products,
you can contact us on
ProductSafety@springernature.com

In case Publisher is established outside the EU,
the EU authorized representative is:
Springer Nature Customer Service Center GmbH
Europaplatz 3, 69115 Heidelberg, Germany

Printed by Libri Plureos GmbH
in Hamburg, Germany